VOLUME ONE HUNDRED AND FORTY

INTERNATIONAL REVIEW OF NEUROBIOLOGY

Animal Models for Examining Social Influences on Drug Addiction

INTERNATIONAL REVIEW OF NEUROBIOLOGY

VOLUME 140

SERIES EDITOR

PATRICIA JANAK

Janak Lab, Dunning Hall
Baltimore, MD, USA

PETER JENNER

Division of Pharmacology and Therapeutics
GKT School of Biomedical Sciences
King's College, London, UK

VOLUME ONE HUNDRED AND FORTY

INTERNATIONAL REVIEW OF NEUROBIOLOGY

Animal Models for Examining Social Influences on Drug Addiction

Edited by

M. FOSTER OLIVE
Behavioral Neuroscience Program,
Department of Psychology,
Arizona State University,
Tempe, AZ, United States

SEVEN E. TOMEK
Behavioral Neuroscience Program,
Department of Psychology,
Arizona State University,
Tempe, AZ, United States

ELSEVIER

ACADEMIC PRESS
An imprint of Elsevier

Academic Press is an imprint of Elsevier
50 Hampshire Street, 5th Floor, Cambridge, MA 02139, United States
525 B Street, Suite 1650, San Diego, CA 92101, United States
The Boulevard, Langford Lane, Kidlington, Oxford OX5 1GB, United Kingdom
125 London Wall, London, EC2Y 5AS, United Kingdom

First edition 2018

ISBN: 978-0-12-815469-4
ISSN: 0074-7742

Publisher: Zoe Kruze
Acquisition Editor: Sam Mahfoudh
Editorial Project Manager: Teresa Pons-Ferrer
Production Project Manager: Abdulla Sait
Cover Designer: Greg Harris

Typeset by SPi Global, India

CONTENTS

CONTRIBUTORS

Howard Becker
Department of Neurosciences; Department of Psychiatry, Medical University of South Carolina, Charleston, SC, United States

Lauren N. Beloate
Department of Neurobiology and Anatomical Sciences, University of Mississippi Medical Center, Jackson, MS, United States

Gabriella Cabrera-Brown
Department of Psychology, Arizona State University, Tempe, AZ, United States

Rosana Camarini
Department of Pharmacology, Institute of Biomedical Sciences, University of São Paulo, SP, Brazil

Lique M. Coolen
Department of Neurobiology and Anatomical Sciences; Department of Physiology and Biophysics, University of Mississippi Medical Center, Jackson, MS, United States

Stewart Cox
Department of Neurosciences, Medical University of South Carolina, Charleston, SC, United States

M. Foster Olive
Department of Psychology, Arizona State University, Tempe, AZ, United States

Cassandra D. Gipson
Department of Psychology, Arizona State University, Tempe, AZ, United States

Lauren Hood
University of Washington, Neuroscience Graduate Student, Seattle, WA, United States

Courtney King
Department of Neurosciences; Department of Psychiatry, Medical University of South Carolina, Charleston, SC, United States

Kah-Chung Leong
Department of Neurosciences, Medical University of South Carolina, Charleston, SC, United States

Jonna M. Leyrer-Jackson
Department of Psychology, Arizona State University, Tempe, AZ, United States

Priscila Marianno
Department of Pharmacology, Institute of Biomedical Sciences, University of São Paulo, SP, Brazil

Mark D. Namba
Department of Psychology, Arizona State University, Tempe, AZ, United States

Janet L. Neisewander
School of Life Sciences, Arizona State University, Tempe, AZ, United States

Broc A. Pagni
School of Life Sciences, Arizona State University, Tempe, AZ, United States

Jose A. Piña
Department of Molecular Sciences, Arizona State University, Tempe, AZ, United States

Gregory L. Powell
School of Life Sciences, Arizona State University, Tempe, AZ, United States

Mariana Rae
Department of Pharmacology, Institute of Biomedical Sciences, University of São Paulo, SP, Brazil

Carmela M. Reichel
Department of Neurosciences, Medical University of South Carolina, Charleston, SC, United States

Andrey E. Ryabinin
Department of Behavioral Neuroscience, Oregon Health & Science University, Portland, OR, United States

Samantha N. Scott
School of Life Sciences, Arizona State University, Tempe, AZ, United States

Akiko Shimamoto
Department of Biochemistry, Cancer Biology, Neuroscience, and Pharmacology, Meharry Medical College, Nashville, TN, United States

Seven E. Tomek
Department of Psychology, Arizona State University, Tempe, AZ, United States

Annika Vannan
School of Life Sciences, Arizona State University, Tempe, AZ, United States

Andre T. Walcott
Department of Behavioral Neuroscience, Oregon Health & Science University, Portland, OR, United States

Shigeru Watanabe
Keio University, Tokyo, Japan

PREFACE

Substance use disorders (SUDs), more colloquially referred to as drug addiction or chemical dependency, are multifaceted disorders that result from a highly complex interplay between cultural, biological, genetic, and social factors. With regards to the latter, social factors include characteristics and demographics of peer groups and families that can be a source of peer pressure, which influences the initiation and maintenance of drug-taking behaviors (Guise, Horyniak, Melo, McNeil, & Werb, 2017; Sudhinaraset, Wigglesworth, & Takeuchi, 2016). On the other hand, social groups such as support networks can also promote resilience against relapse and recidivism and facilitate the recovery process (Tracy & Wallace, 2016). However, largely as a result of their extremely complex and dynamic nature, social factors have been historically difficult to integrate into animal (e.g., rodent) models of addiction.

This book presents recent behavioral and neurobiological advances in animal models as they relate to the study of social factors and addictive drugs, as well as how they inform our understanding of the related human behaviors. Further, this volume aims to expand on well-established proposed conceptual frameworks and theories of social influences, and how their implication the development and maintenance of addiction. The preclinical research models thus far have made incredible technological strides, over the last 40 years, in their attempts to provide insight into the behavioral, physiological, and environmental basis of SUDs. These paradigms are for the most part reliable, replicable, and accurately capture various aspects of addiction in humans. Animal models such as self-administration, conditioned place preference (CPP), social defeat, environmental enrichment, and variable stress, to name a few, are reviewed in depth and their influence in relation to drugs such as alcohol, psychostimulants, opiates, nicotine, and cannabinoids.

Tobacco use remains one of the leading preventable causes of death worldwide, and in recent years, there has been a dramatic surge in use of its primary psychoactive and addictive component nicotine in the form of electronic cigarettes, particularly in adolescents and young adults. Social factors weigh heavily on the initiation of this and other forms of nicotine use. In the first chapter of this volume, Piña, Namba, Leyrer-Jackson, Cabrera-Brown, and Gipson discuss the role of family and peers, adolescent impulsivity and decision-making, socioeconomic status, and psychosocial stress as they contribute to nicotine use. They then review two primary paradigms that have been used to investigate social influences on nicotine intake—environmental enrichment and social facilitation of intravenous nicotine self-administration. They also discuss the neurobiological mechanisms underlying the anxiolytic and anxiogenic effects of nicotine as it relates to social interactions. Piña et al. conclude that the construct validity of rodent models of nicotine addiction could be vastly improved via the incorporation of social conspecifics, thus allowing researchers to better understand the vulnerabilities that promote the initiation and maintenance of nicotine use.

Alcohol use also significantly contributes to premature death, and like nicotine, its use is primarily initiated in adolescence and is facilitated by social pressures. In Chapter 2, Ryabinin and Walcott address the issue of the lack of significant incorporation of social influences in animal models of alcohol intake and alcohol abuse in the extant literature. Focusing primarily on methodological issues, the authors evaluate different methods for assessing alcohol intake in group vs. individually housed rodents, including overall and individual alcohol intake, as well as methods for tracking the behavior of individual animals in group settings such as video recording, radio frequency tracking, and use of the Intellicage system. The authors also review studies that have utilized social dyads housed together but separated by wire mesh which permits social contact but allows for assessment of individual drinking patterns. Finally, Ryabinin and Walcott highlight their recent unique work involving prairie voles, which are highly social creatures that are sensitive to the stressful effects of social isolation, which provide a unique opportunity to examine social factors regulating alcohol intake in a novel animal species.

In Chapter 3, Camarini, Marianno, and Rae continue on the subject of alcohol, but with a focus on behavioral sensitization, a phenomenon characterized by increasing behavioral sensitivity and responsiveness to repeated exposure to a single dose of a particular drug. The focus of this chapter is the influence of psychosocial stressors on behavioral sensitization to alcohol, including maternal (early life) separation, social defeat, housing conditions, and environmental enrichment. The studies reviewed here indicate important roles for neurochemical and neuroendocrine systems in stress effects on ethanol-induced behavioral sensitization, including monoaminergic transmission, neurotrophins, and various components of the hypothalamic–pituitary–adrenal (HPA) axis.

Next, in Chapter 4, Tomek and Olive address the opiate epidemic and often neglected social factors that contribute to opiate use. Following a review of common rodent paradigms to studying opiate addiction, the authors highlight the need for increased incorporation of social variables and procedures, such as environmental enrichment and introduction of conspecifics, similar to the issues raised in Chapter 1. Tomek and Olive also describe their recently developed prosocial "rescuing" paradigm, in which one rat in pair-housed conditions is allowed the opportunity to "free" its cagemate (or not) trapped in a plastic container, both before and after self-administration of heroin.

In Chapter 5, Watanabe describes social variables as they influence the rewarding effects of amphetamine (primarily methamphetamine), primarily as assessed by the CPP paradigm in mice. Recent studies show that the presence of a conspecific facilitates the rewarding effects of amphetamines. However, other factors, both pharmacological and non-pharmacological, also appear to be at play. For instance, Watanabe describes how sharing a common drug experience (i.e., exposure to methamphetamine) is crucial for social enhancement of methamphetamine reward. Watanabe also describes how amphetamine reward can also be influenced by the dominant/subordinate status of the animal, prior social defeat and isolation, as well as parental drug exposure.

In Chapter 6, Vannan, Powell, Scott, Pagni & Neisewander overview the effects of social stress on cocaine addiction-like behaviors in rodent models. The first form of social stress is early life stress (maternal deprivation), and gaps in the literature regarding the effects of cross-fostering, paternal deprivation, and cocaine effects on maternal care are identified. Next, the social defeat stress model is reviewed (also discussed in great detail in Chapter 10), with particular regard to the neurochemical and neuroendocrine substrates of its effects on cocaine intake. The authors also introduce the concept of social hierarchy models. Finally, as in other chapters, the existing literature on effects of environmental enrichment and social isolation are reviewed.

Turning to the topic of cannabis and other cannabinoids, in Chapter 7, Hood overviews the endogenous cannabinoid system and its role in natural social behavior. Social influences on cannabis use in humans are then discussed, followed by standard existing animal models of cannabinoid intake. Hood concludes that there is a virtual absence of published studies which specifically investigate social influences on cannabinoid intake in animal models. Given the general low levels of cannabinoid intake in standard models, Hood highlights the clear need for studies involving social variables such as the presence of conspecifics or introduction of psychosocial stress.

In Chapter 8, Leong, Stewart, King, Becker, and Reichel review studies examining the involvement of oxytocin, a neuroactive peptide often colloquially referred to as the "love molecule," in addictive behaviors. An overview of the physiology of the oxytocinergic system and the neural circuitry of social behavioral are provided, followed by a summary of published studies that have examined either genetic or pharmacological manipulation of the oxytocin system on either the intake or drug seeking behavior for methamphetamine, cocaine, heroin, and alcohol. The authors conclude with a proposal for oxytocinergic medications as novel therapeutic adjuncts for the treatment of SUDs.

In the penultimate chapter of this volume, Beloate and Coolen discuss in Chapter 9 their work examining the effects of sociosexual contact on behaviors induced by psychostimulants and opiates. The underlying, and often intersecting, neural mechanisms by which these effects occur is also reviewed, Finally, in Chapter 10, Shimamoto discusses in great detail the social defeat stress model as it has been applied to animal models of addiction. Attention is given to methodological details such as the frequency of social defeat (acute, repeated, chronic, and intermittent), as well as important individual and sex differences in the effects of this social stress paradigm.

We hope that the expert summaries and discussions provided in this volume will inform investigators in the addiction field as to how to implement critical refinements to existing animal models of drug addiction to include social variables. We underscore the importance of improving the face, construct, and predictive validity of existing models in order to generate novel areas of research that ultimately lead to effective socially-based approaches to treating SUDs, which will reduce the enormity of the global health burden these disorders impose on society.

SEVEN E. TOMEK
M. FOSTER OLIVE
Behavioral Neuroscience Program, Department of Psychology, Arizona State University, Tempe, AZ, United States

REFERENCES

Guise, A., Horyniak, D., Melo, J., McNeil, R., & Werb, D. (2017). The experience of initiating injection drug use and its social context: A qualitative systematic review and thematic synthesis. *Addiction*, *112*, 2098–2111. https://doi.org/10.1111/add.13957.

Sudhinaraset, M., Wigglesworth, C., & Takeuchi, D. T. (2016). Social and cultural contexts of alcohol use: Influences in a social-ecological framework. *Alcohol Research: Current Reviews*, *38*, 35–45.

Tracy, K., & Wallace, S. P. (2016). Benefits of peer support groups in the treatment of addiction. *Substance Abuse and Rehabilitation*, 7, 143–154. https://doi.org/10.2147/SAR.S81535.

FURTHER READING

Beloate, L. N., & Coolen, L. M. (2018). Effects of sexual experience on psychostimulant- and opiate- induced behavior and neural plasticity in the mesocorticolimbic pathway. In M. F. Olive & S. E. Tomek (Eds.), *Animal models for examining social influences on drug addiction: Vol. 140* (pp. 249–270). New York: Academic Press.

Camarini, R., Marianno, P., & Rae, M. (2018). Social factors in ethanol sensitization. In M. F. Olive & S. E. Tomek (Eds.), *Animal models for examining social influences on drug addiction: Vol. 140* (pp. 53–80). New York: Academic Press.

Hood, L. E. (2018). Social factors and animal models of cannabis use. In M. F. Olive & S. E. Tomek (Eds.), *Animal models for examining social influences on drug addiction: Vol. 140* (pp. 171–200). New York: Academic Press.

Leong, K.-C., Cox, S., King, C., Becker, H., & Reichel, C. M. (2018). Oxytocin and rodent models of addiction. In M. F. Olive & S. E. Tomek (Eds.), *Animal models for examining social influences on drug addiction: Vol. 140* (pp. 201–247). New York: Academic Press.

Piña, J. A., Namba, M. D., Leyrer-Jackson, J. M., Cabrera-Brown, G., & Gipson, C. D. (2018). Social influences on nicotine-related behaviors. In M. F. Olive & S. E. Tomek (Eds.), *Animal models for examining social influences on drug addiction: Vol. 140* (pp. 1–32). New York: Academic Press.

Ryabinin, A. E., & Walcott, A. T. (2018). Assessing social alcohol drinking in rodent models: Are we there yet? In M. F. Olive & S. E. Tomek (Eds.), *Animal models for examining social influences on drug addiction: Vol. 140* (pp. 33–51). New York: Academic Press.

Shimamoto, A. (2018). Social defeat stress, sex, and addiction-like behaviors. In M. F. Olive & S. E. Tomek (Eds.), *Animal models for examining social influences on drug addiction: Vol. 140* (pp. 271–313). New York: Academic Press.

Tomek, S. E., & Olive, M. F. (2018). Social influences in animal models of opiate addiction. In M. F. Olive & S. E. Tomek (Eds.), *Animal models for examining social influences on drug addiction: Vol. 140* (pp. 81–107). New York: Academic Press.

Vannan, A., Powell, G. L., Scott, S. N., Pagni, B. A., & Neisewander, J. L. (2018). Animal models of the impact of social stress on cocaine use disorders. In M. F. Olive & S. E. Tomek (Eds.), *Animal models for examining social influences on drug addiction: Vol. 140* (pp. 131–169). New York: Academic Press.

Watanabe, S. (2018). Social modification of amphetamine reward. In M. F. Olive & S. E. Tomek (Eds.), *Animal models for examining social influences on drug addiction: Vol. 140* (pp. 109–129). New York: Academic Press.

CHAPTER ONE

Social Influences on Nicotine-Related Behaviors

Jose A. Piña*, Mark D. Namba†, Jonna M. Leyrer-Jackson†, Gabriella Cabrera-Brown†, Cassandra D. Gipson†,1

*Department of Molecular Sciences, Arizona State University, Tempe, AZ, United States
†Department of Psychology, Arizona State University, Tempe, AZ, United States
[1]Corresponding author: e-mail address: cgipsonr@asu.edu

Contents

Abstract

Cigarette smoking is the leading preventable cause of death in the United States. The number of new smokers, specifically among adolescents, has risen rapidly in recent years. Thus, understanding the role of social influences on patterns of nicotine and tobacco use is important. Clinical studies have addressed the impact social relationships

International Review of Neurobiology, Volume 140
ISSN 0074-7742
https://doi.org/10.1016/bs.irn.2018.07.001

such as family members and peers have on smoking acquisition and susceptibility. As well, preclinical animal models have examined the impact of social factors on drug intake, acquisition, maintenance, and relapse. For example, environmental enrichment (EE) is a multi-faceted model that includes social factors, exercise, and novelty, among others. This model has elucidated addiction-related neurobehavioral effects of these different factors. However, there is a dearth of literature examining the impact of social partners on nicotine addiction and underlying neural mechanisms. Here we discuss the importance of social factors on nicotine addiction vulnerability, and propose new directions for addiction research that integrate social aspects of nicotine use.

1. INTRODUCTION

Cigarette smoking and other variations of nicotine use (e.g., electronic cigarettes (e-cigarettes) and chewing tobacco) is a major public health issue. Cigarette smoke contains harmful aldehydes such as acetaldehyde, acrolein, and formaldehyde, which may expose users to an increased risk of cardiovascular disease (Ogunwale et al., 2017). Notably, social influences can be a possible determinant of nicotine use, especially among adolescents. Several studies have addressed connections between social interactions and acquisition of nicotine use (Gecková et al., 2005; Leatherdale, Mcdonald, Cameron, & Brown, 2005; West, Sweeting, & Ecob, 1999), which place an emphasis on smoking behaviors that may be mediated by various types of social interactions among family members and peers. Given the recent surge in e-cigarette use among adolescents, it is essential to further our and understanding of how social influences can impact nicotine addiction susceptibility, especially during this critical developmental period.

Preclinical animal models have addressed the relationship between proximal versus distal social contact in relation to drug addiction (described in detail below), where distal social contact refers to no social influences present when drug use occurs. Conversely, proximal social contact refers to existent social influences present when drug use occurs. In the context of distal social contact, environmentally enriched housing conditions during development (discussed in further detail below) serve as a facilitator of social interaction and decrease drug intake. For instance, in rodents, EE suppresses nicotine's stimulant effects (Green, Cain, Thompson, & Bardo, 2003), whereas social isolation conditions induce a higher consumption of morphine (Hadaway, Alexander, Coambs, & Beyerstein, 1979). Although animals are placed in enriched housing conditions in a home cage, drug use typically occurs in isolation within an operant chamber. Thus, it is unclear what impact enrichment (such as social contact) during drug use would have on intake due to difficulty of conducting self-administration studies with multiple animals present. Thus, there is a large gap in our understanding of the neurobehavioral impact of social enrichment (SE) and EE on drug-seeking behavior. To date, few studies have shown the distinct role of social interaction regarding neurobehavioral and neurobiological mechanisms of nicotine addiction. In this chapter, we will explore the relationship between nicotine addiction and social influences by evaluating clinical and preclinical literature. As well, we will describe possible underlying neurobiological mechanisms. In addition to social interaction studies involving nicotine, we will emphasize nicotine's anxiolytic and anxiogenic effects, as anxiety/stress serve as a mediating factors of social interaction. Specifically, we will discuss regions of the brain such as the nucleus accumbens (NA), prefrontal cortex (PFC), hippocampus (HPA), dorsal raphe nuclei (DRN), and the medial habenula (MHB) and their distinct roles in modulating anxiety/stress when considering social interaction, addiction, and anxiety. Lastly, we aim to discuss both potential gaps in the literature where warranted that may further the understanding of this complex interaction, as well as propose future directions that better implement social factors in drug addiction models.

2. CLINICAL EXAMINATION OF THE RELATIONSHIP BETWEEN SOCIAL INFLUENCE AND NICOTINE USE

2.1 Social Smoking, Detection, and Intervention

Individual differences exist in patterns of smoking. Individuals who smoke socially are still defined as "smokers," determining if an individual meets criteria for a tobacco use disorder is difficult, as this subset of individuals typically do not self-identify as "smokers" (Waters, Harris, Hall, Nazir, & Waigandt, 2006). Thus, it has been recommended that studies using self-report surveys avoid smoking and addiction self-identification terms (e.g., "Are you a smoker?"), and instead use numerical frequency options to characterize behavior (e.g., "Have you smoked in the last 30 days?") (Schane, Glantz, & Ling, 2009). Specifically evaluating and targeting characteristics of self-perception to personalize cessation techniques is likely to be more effective and practical for social smokers. For example, partitioning social smokers into categories of either "behavioral" social smokers or "self-identified" social smokers could predict cessation success.

While it has been reported that "self-identified" smokers are less likely to be successful during cessation attempts and "behavioral" smokers are more likely to be successful, the effects of the treatment that lead to participants' success may still be variable within the "self-identified" and "behavioral" subcategories (Song & Ling, 2011).

The inability to categorize differences between daily and non-daily smokers could cause problems for social smokers who remain largely unaffected by treatments targeted toward daily smokers (Schane et al., 2009). This disparity between treatment options available for daily and non-daily/social smokers highlights the need for devising relevant therapeutic interventions. Regimented smoking cessation programs are not successful in a majority of participants, though 25% of all participating individuals show reductions in smoking (Anthonisen et al., 2005). Program efficacy for less than a quarter of participants suggests that certain smoking motivations may not be accurately targeted by typical intervention programs. Specifically, regular weekly check-ups might not be sufficient for individuals who smoke socially, as smoking patterns within this subpopulation of individuals are more difficult to predict (Schane et al., 2009). Thus, smoking cessation programs should be designed to better target social smokers.

2.2 Social Influences of Family and Peers on Smoking Behavior in Adolescents

Clinical data suggest that social relationships during adolescence can impact choices to use drugs, which in turn can impact health. For instance, close family members smoking behaviors can be associated with increased cigarette smoking behavior in adolescents (Leatherdale et al., 2005). Similarly, adolescent smoking behavior is more prevalent in individuals with close friends who smoke cigarettes (West et al., 1999). A unique and critical period for vulnerability to adolescent smoking prevalence exists between the ages of 15 and 18. Interestingly, if an individual surrounds themselves with friends who are also non-smokers within this age range, their likelihood of initiating smoking behavior is much lower than in those with friends that smoke regularly (West et al., 1999). In another recent study, adolescents were more likely to develop smoking-related behaviors if they live in a pro-smoking environment, where the beliefs of peers and parents were pro-smoking, when compared to adolescents who lived in a non-smoking social environment (Scalici & Schulz, 2017). Taken together, these studies suggest that the smoking patterns of peers and family members can have a significant influence on an adolescent's smoking susceptibility.

2.3 Impulsivity, Adolescence, and Smoking

Adolescence a critical developmental phase in which peer influence can profoundly impact impulsive risk taking behavior (Blakemore, 2008; Spear, 2000), where risk taking is much more likely to occur in the presence of peers (Chassin, Hussong, & Beltran, 2009; Simons-Morton, Lerner, & Singer, 2005). Impulsive risk taking likely has its roots in evolution. For instance, adolescence-limited increases in emotionality and risk taking have been found to occur across species (Spear, 2000). This may be adaptive for helping adolescents to engage in the next steps in development, which includes leaving the natal home. As well, adolescent spikes in facets of impulsivity, such as sensation seeking, may be adaptive for procreation, as these spikes may enable adolescents to take risks for the survival of the species. A "normal" impulsivity phenotype may be adaptive for finding new food sources when food is scarce, as it entails taking risks in eating new foods that may be harmful. According to Dickman (1990), impulsivity can be broken down into two types, both functional and dysfunctional. These both lead to fast and error-prone action, but are quite different in meaning. Functional impulsivity refers to the willingness and ability to take risks in appropriate and necessary situations in which this behavior results in a benefit for survival. Dysfunctional impulsivity, however, refers to the tendency to engage in thoughtlessness and the inability to plan ahead and anticipate negative consequences.

There is a growing argument that adolescent impulsivity and substance use is perhaps rashly dismissed as completely negative (Gullo & Dawe, 2008). Although it is widely accepted that risk propensity is a component of impulsivity, and that impulsivity contributes to substance misuse, there is evidence that high reward-seeking drive in adolescence is associated with positive results in adulthood, such as entrepreneurialism (Stewart & Roth, 2001). According to Daruna and Barnes (1993), "impulsivity encompasses actions that appear poorly conceived, prematurely expressed, unduly risky, or inappropriate to the situation and that often result in undesirable consequences. When such actions have positive outcomes, they tend not to be seen as signs of impulsivity, but as indicators of boldness, quickness, spontaneity, courageousness, or unconventionality." This definition of impulsivity, according to Evenden (1999), is accepted widely, encompasses many forms of impulsivity, and emphasizes that impulsivity can be good and bad, but only impulsivity that results in negative outcomes is considered problematic. It is important to consider impulsivity as having an important evolutionary basis and may not always be maladaptive. According to Evenden (1999), impulsivity is an aspect of normal behavior, and normal individuals can exhibit some level of impulsivity. Thus, it is important to distinguish when these levels become problematic, and are not simply part of an individual's normal behavioral repertoire. Understanding the different facets of impulsivity, as well as individual differences in level of risk, is important not only for understanding psychopathology, but also to better identify when risk propensity is within a normal range, especially in adolescence. These fundamental characteristics underlying impulsivity thus provide a useful framework for dissecting the role of risky decision making among adolescents and how smoking behaviors may be transmitted between peers.

During the developmental period of adolescence, personality traits are not yet consolidated into a more permanent and holistic self-identity (Hampson & Goldberg, 2006). As mentioned above, adolescent behavior can be impacted by personality traits such as impulsivity (Elkins, King, McGue, & Iacono, 2006). The risk for smoking cigarettes is positively correlated with measures related to heightened impulsivity, and the use e-cigarettes is associated with a significantly higher risk for smoking later in life (Bold et al., 2017; Burris, Riley, Puleo, & Smith, 2017). Given the alarming rise in e-cigarette use among adolescents (Bold et al., 2017), it is critical to understand the effects of impulsive choice behavior and social transmission of these behaviors among adolescent peers. Notably, adolescents are more likely than adults to engage in risky behavior, such as substance abuse (Chassin et al., 2009), when in the presence of peers. Moreover, it has been demonstrated that the presence of peers is sufficient to influence adolescent decision making (Gardner & Steinberg, 2005; O'Brien, Albert, Chein, & Steinberg, 2011).

Decision-making and reward/risk-assessment processes involve interconnected brain structures such as the orbitofrontal cortex (OFC) and subcortical limbic structures such as the NA (Schoenbaum, Roesch, & Stalnaker, 2006). Interestingly, it has been observed that adolescents (relative to adults) exhibit increased activation of the NA and OFC when assessing risk in a computer-based task, but only when being observed by peers (Chein, Albert, O'Brien, Uckert, & Steinberg, 2011). These results highlight the neurobehavioral interconnectivity between social influence and risky decision making, which could have a profound impact on the acquisition of smoking among adolescents. Beyond acquisition of habitual smoking among adolescents, impulsivity and risky behaviors among peers may impede cessation attempts if these behaviors are not considered when designing adolescent prevention and intervention programs. For example, the progression of peer smoking (from non-smoker to frequent smoker) has been demonstrated to have a direct effect on the progression of adolescent smoking behavior (Audrain-McGovern, Rodriguez, Tercyak, Neuner, & Moss, 2005). In addition, self-report measures of impulsivity (e.g., Barratt impulsiveness scale (BIS-II)) have been shown to correlate with the ability to maintain abstinence among adolescent smokers, where more impulsive individuals may be at higher risk for relapse (Krishnan-Sarin et al., 2007). This effect has also been shown in adult smokers (Doran, Spring, McChargue, Pergadia, & Richmond, 2004). Taken together, it is clear that impulsivity on its own is a significant factor contributing to the onset and maintenance of smoking and that peer influence among adolescents is an important factor related to the social transmission of impulsive choice behavior and risky decision making. Further research is needed to more clearly define this relationship and how impulsivity among peers influences substance abuse behavior, particularly among adolescents.

2.4 Socioeconomic Status and Impulsivity

Socioeconomic status (SES) itself does not typically have a direct causal influence on smoking behavior among adolescents, however, it is mediated by other factors such as education level of parents and smoking behavior of parents/peers. (Gecková et al., 2005). These data suggest that SES status is not the direct cause of smoking, but rather integration into low SES social networks may increase risk of smoking (Gecková et al., 2005). Additionally, one study examining social networks of individuals and smoking habits found that individuals with at least 1 year of college were more likely to start smoking as a result of peer influence, and more likely to influence others with similar levels of education in their peer group (Christakis & Fowler, 2008). Interestingly, individuals with a high school diploma or less were less prone to peer influence of smoking than their more educated counterparts (Christakis & Fowler, 2008). This pattern was postulated to be a result of trends forming within a class of individuals that are relatively cut off from other SES classes. Thus, education is suggested to be a bidirectional mediator between smoking initiation, cessation, and SES—where groups of young college students are more likely to initiate smoking if they are engaged in peer interactions with others who smoke (Christakis & Fowler, 2008).

Education alone is not the only social factor in SES that influences smoking behavior. Impulsivity is negatively correlated with SES and education, and is closely linked to drug use and abuse and is both a determinant and a consequence of drug use (de Wit, 2008; Lynam et al., 2000). According to de Wit (2008), impulsivity is a risk factor for initial drug experimentation and an inability to abstain from drug use. Alternatively, drug use may increase maladaptive behaviors (increasing impulsivity) through acute effects or through long-term use. Impulsivity is correlated with a higher likelihood of becoming addicted to nicotine, as well as a lower rate of long-term abstinence (Blum et al., 2000; Kreek, Nielsen, Butelman, & LaForge, 2005). In adolescents, the presence of peers can increase risky behavior through activation of the brain's reward pathways, and such increased impulsivity and reward center activation can increase smoking behavior (Chein et al., 2011; Reynolds et al., 2007).

Pre-existing personality traits like impulsivity could help to explain why higher SES individuals still smoke less than those of low SES despite the fact that their higher education leads to an increased likelihood to smoke based on smoking peers. Individuals who have a natural tendency to be impulsive have a higher likelihood of becoming addicted to nicotine, as well as a lower rate of long-term abstinence (Blum et al., 2000; Kreek et al., 2005). In adolescents, the presence of peers can increase risky behavior through activation of the brain's reward pathways, and such increased impulsivity and reward center activation can increase smoking behavior (Chein et al., 2011; Reynolds et al., 2007). Low SES individuals are likely to be more impulsive and less future-oriented, and are therefore far more susceptible to becoming addicted independent of peer influence (Peretti-Watel, L'Haridon, & Seror, 2013). High SES individuals have better long-term outcomes when they are in stable environments with few impulsive friends, however, they are not resilient as they are increasingly at risk when exposed to impulse-driven peers. Impulsive drug-seeking behavior therefore seems to be derived intrinsically for low SES individuals, and extrinsically for high SES individuals.

These trends in SES and impulsivity may help to explain why highly educated and high SES individuals have lower overall rates of smoking compared to less educated and low SES individuals, even though being highly educated puts people at a higher risk for being influenced by peers smoking (Christakis & Fowler, 2008; Hiscock, Bauld, Amos, Fidler, & Munafò, 2012). When high SES, less impulsive individuals self-select into universities and more educated social circles, and are less likely to encounter highly impulsive individuals who have a higher propensity to be addicted to nicotine and other substances (Spinella, 2004).

2.5 The Relationship Between Smoking Behavior and Psychosocial Stress

Some studies have indicated that certain mental illnesses are highly influenced by social factors. For example, increased depression and symptoms of social anxiety have been linked to lower perceived social support, and engagement in fewer social activities (Dumont & Provost, 1999). Decreased social support has long been considered a factor in life stress, where decreased social support is correlated with depression and increasing perceived social support has been a tactic used in depression therapy (Cobb, 1976; George, Blazer, Hughes, & Fowler, 1989). Moreover, social support and social activities may decrease stress and depression in adolescents (Dumont & Provost, 1999). The comorbidity of depression and nicotine addiction has been well established in both adults and adolescents (Breslau, Kilbey, & Andreski, 1993). Though depression and low perceived social support negatively impact addiction-related behaviors, research has shown that depression does not impede cessation (John, Meyer, Rumpf, & Hapke, 2004). Thus, it is possible that social support is a justifiable and feasible target for smoking cessation therapy through its influence on depression (Murray, Johnston, Dolce, Lee, & O'Hara, 1995).

Many studies have highlighted the social component of addiction for non-daily or social smokers; individuals who do not smoke on a regular schedule have smoking habits which can typically be predicted by social interaction. (Levinson et al., 2007), where some individuals only smoke solely in the context of social interactions. In one study, researchers found that fear and anxiety in social settings are likely to exacerbate smoking and addiction to nicotine in adolescents and young adults, even when controlling for frequently comorbid depressive disorders (Sonntag, Wittchen, Höfler, Kessler, & Stein, 2000). Identifying social smokers by the affective state (i.e., social fear or anxiety) leading to non-daily and social smoking helps to characterize previously unexplained variance in smoking motivation in the smoking population, and can therefore better inform smoking cessation intervention strategies targeting this unique group of social smokers. A uniform cessation strategy might not be as efficacious for non-daily smokers because there are specific periods of elevated risk for smoking in non-daily smokers in social contexts associated with smoking. That is, it might not be necessary to use nicotine patches or other daily forms of pharmacological nicotine replacement therapies used for daily smokers when the risk is context-dependent.

As described above, social smoking habits are generally context-dependent, and these habits are often accompanied by use of other substances, particularly alcohol (Jiang, Lee, & Ling, 2014). This co-abuse of substances may represent one unique feature of social smoking that distinguishes it from other forms of smoking that are not dependent on social context. Feelings of social anxiety or a motivation to conform could be factors underlying negative urgency (i.e., the inclination to act impulsively when distressed) that lead to smoking because it may ameliorate psychosocial stress associated with successfully integrating into a social group (Guillot, Pang, & Leventhal, 2014; Settles et al., 2012). However, positive urgency from rewarding social interactions could motivate individuals to seek nicotine to augment or facilitate the reinforcing effects of positive social interaction (Spillane, Smith, & Kahler, 2010; Zapolski, Cyders, & Smith, 2009). Regardless, it is clear that among social smokers, psychosocial therapeutic interventions, as opposed to pharmacological treatment interventions, may represent a more sensible approach toward achieving abstinence in non-daily social smokers. Beyond the known negative health consequences associated with nicotine addiction, there is sufficient evidence that non-daily smoking can progress to daily smoking, particularly among some racial ethnic minorities, and non-daily smoking is associated with alcohol use disorders and dangerous drinking behavior (Harrison & McKee, 2011; Kandel, Kiros, Schaffran, & Hu, 2004). Therefore, specialized treatment for non-daily smokers may be needed to provide protection against escalation to daily smoking.

2.6 Emergence of E-Cigarettes in Relation to Social Interaction

An apparent rise in e-cigarette use among adolescents has increased dramatically from 4.5% in 2011 to 27.4% in 2014 (The Centers for Disease Control and Prevention, 2015). E-cigarettes are distinct from conventional cigarettes as they are battery powered and function by heating a liquid substance to generate a vapor that is inhaled by the user. This recent surge of e-cigarette use in adolescents may be partly promoted by the perception that e-cigarettes are associated with fewer health risks (Barrington-Trimis et al., 2016; Persoskie, O'Brien, Nguyen, & Tworek, 2017). This perception is possibly due to e-cigarettes being marketed as a safer alternative as they do not contain many of the harmful constituents found in conventional cigarettes (Henningfield & Zaatari, 2010), Additionally, e-cigarette use among adolescents has become socially acceptable in public places compared to combustible cigarettes, thus promoting their use. The inclusion of flavorings, such as cherry and bubblegum among many others, also targets younger generations. Thus, although there has been a decrease in use of combustible cigarettes, there has been a concomitant increase in use of e-cigarettes (Lauterstein et al., 2014). This increase may exacerbate nicotine intake and addiction susceptibility in adolescents, as more socially acceptable forms of nicotine intake become available. Interestingly, susceptibility to smoking may increase in adolescents if three or more of their friends also use e-cigarettes, as compared to users who do not engage with peers who use e-cigarettes (Barrington-Trimis et al., 2016). Due to their recent development and popularity, the detrimental effects of e-cigarette use are largely unknown.

3. PRECLINICAL MODELS OF SOCIAL INFLUENCE ON NICOTINE ADDICTION

Animal models of drug abuse have provided unique insights into the behavioral and neurobiological etiology and maintenance of substance use disorders. As well, these models have facilitated the development of pharmacotherapeutics that have proven clinically efficacious (Burke, Hays, & Ebbert, 2016). For example, early preclinical work investigating the role of β_2-containing nicotinic acetylcholine receptors (nAChRs) in the rewarding and reinforcing properties of nicotine served as the foundation for the development of the partial $\alpha_4\beta_2$ agonist varenicline (i.e., Chantix®) (Picciotto et al., 1998). Social influence is a powerful and reliable predictor of drug use among adolescents (Bahr, Hoffmann, & Yang, 2005; Simons-Morton & Chen, 2006; Walden, McGue, lacono, Burt, & Elkins, 2004) and preclinical animal models examining the effects of social contact on drug-seeking behavior support these findings. However, the distinction between proximal versus distal social contact may be an important factor to consider when studying the role of social influence on addiction processes. As mentioned above, proximal social influence refers to factors that are immediately present at the time of drug-taking, such as a conspecific social partner within an adjacent operant self-administration chamber (Fig. 1). Conversely, distal social influence refers to factors that are present outside of the drug-taking context, such as with cage mates within social housing conditions. EE (i.e., distal social influence), which provides social interaction

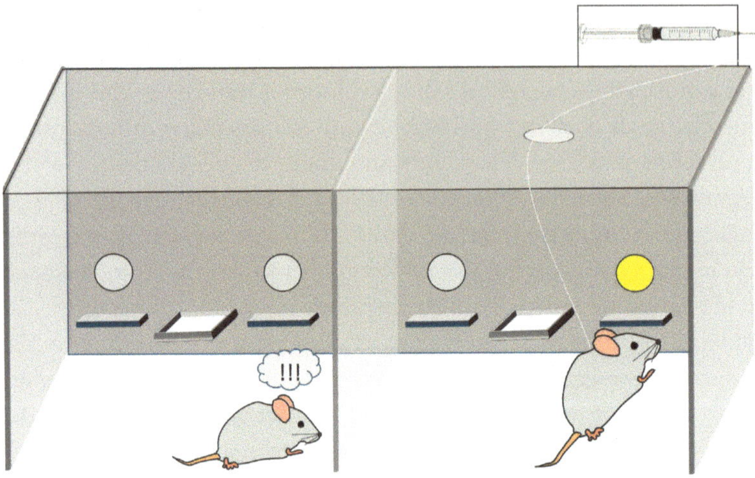

Fig. 1 Illustration of self-administration operant chambers used to examine social influence. Two rats are paired in different chambers separated by a plexiglas wall. Each operant chamber includes a cue light paired with each lever and a food receptacle per chamber. Amphetamine was administered via IV as the number of each infusion per session varied dependent upon on the number of lever presses.

and novel objects within the rats' home cage, inhibits the stimulant effects of nicotine as compared to isolated rats (Green et al., 2003). In contrast, a recent study utilizing conditioned place preference (CPP; where animals learn to associate non-contingent drug injections with a specific context) has demonstrated that environmentally enriched rats acquired CPP of nicotine at 0.4, 0.6, and 0.8 mg/kg and reinstated the CPP response at the two highest doses, as opposed to isolated rats that did not acquire nicotine CPP at these doses (Ewin, Kangiser, & Stairs, 2015). It has been argued that CPP as compared to self-administration measures different aspects of the reinforcing properties of stimulant drugs, where EE may exert differential effects over the primary and secondary effects of stimulants (Bardo, Neisewander, & Kelly, 2013). As well, it has been suggested that nicotine and social rewards may interact synergistically, where the combination of social interaction and non-contingent nicotine administration produces more robust CPP than either treatments alone (Thiel, Sanabria, & Neisewander, 2009). Indeed, it is known that social interaction is inherently rewarding in rodents (Zernig, Kummer, & Prast, 2013), particularly in female rats (Wang, Han, & Chen, 2016). Collectively, these data support the conclusion that distal social interaction, such as in the form of EE could have a protective effect on nicotine abuse, whereas proximal social influence (described in more detail below) could potentially enhance nicotine consumption, particularly during the early phases of nicotine acquisition.

Studies examining proximal social contact during self-administration have demonstrated that this form of social influence can exert overlapping effects on drug-seeking behavior depending on the drug of abuse and whether animals were housed in isolated or social conditions. For example, one of the earliest studies to examine this phenomenon showed that socially isolated rats consumed higher amounts of a morphine-sucrose solution as compared to socially housed rats (Hadaway et al., 1979). In another study examining amphetamine self-administration in isolated rats, the presentation of a novel same-sex conspecific social partner (separated by a plexiglas partition) facilitated responding for amphetamine at a high dose (Gipson et al., 2011). Similarly, social contact during nicotine self-administration facilitates acquisition of operant responding and nicotine intake; however, these animals were socially housed (Chen, Sharp, Matta, & Wu, 2011). A more recent study examining social context in the acquisition and maintenance of nicotine self-administration found that social interaction among pair-housed males transitioning into adulthood enhances the initial reinforcing effects of nicotine during the early phases of acquisition. However, this social interaction appears to protect against nicotine intake during the later sessions of nicotine self-administration, which is more pronounced in females (Peartree et al., 2017). Thus, age and sex are likely important factors that shape the modulatory role of social influence over nicotine-seeking behavior.

As highlighted above, it is possible that social factors could facilitate nicotine-seeking behavior, which may be due to the inherently rewarding and anxiolytic aspects of social interaction. Social interaction has been demonstrated to trigger the release of oxytocin (OXT) from neurons in the paraventricular nucleus of the hypothalamus projecting to the ventral tegmental area (VTA), which gates the rewarding properties of social interaction (Hung et al., 2017). OXT has also been shown to suppress nicotine aversion during the early stages of nicotine use and may also inhibit anxiety-like behavior (Lee, Jang, & Noh, 2017). In line with these findings, an earlier study demonstrated that OXT reduces somatic signs of precipitated nicotine withdrawal in nicotine-dependent rats. However, this study also showed that OXT failed to attenuate withdrawal-induced elevations in intracranial self-stimulation thresholds (Manbeck, Shelley, Schmidt, & Harris, 2014). Provided these findings, it seems likely that OXT may contribute to the positive reinforcing effects of social interaction, which could inhibit the negative aversive effects of nicotine during the early phases of acquisition. There is a distinct relationship between OXT and social bonding, where OXT is thought to promote sociability (Kosfeld, Heinrichs, Zak, Fischbacher, & Fehr, 2005; Lim & Young, 2006). It is thought that social bonding and the ability to maintain stable social relationships may prevent and even counteract drug addiction (Blackson & Tarter, 1994; Johnson & Pandina, 1991). Interestingly, intranasal OXT decreases cigarette cravings in daily smokers during exposure to smoking-related cues (Miller, Bershad, King, Lee, & Wit, 2016). This suggests that pro-social interaction (akin to social enrichment in rodents) that augments OXT signaling may reduce nicotine seeking and relapse. Nevertheless, further research is needed to better understand the neurobiological mechanisms underlying OXT and its role in mediating social behaviors and reward seeking.

In addition to the effects of positive social influence on nicotine-seeking behavior, nicotine appears to protect against cognitive deficits in learning and memory associated with social stress. For example, social stress produced by interrupting social hierarchies among male rats induces deficits in spatial memory and long-term potentiation (LTP) within the CA1 region of the HPA, which is inhibited by treatment with nicotine (Aleisa, Alzoubi, Gerges, & Alkadhi, 2005). These results parallel clinical findings demonstrating a positive correlation between subjective social status and duration of smoking abstinence (Reitzel et al., 2011) and that social stress and anxiety are positively correlated with urge to smoke (Niaura, Shadel, Britt, & Abrams, 2002). Taken together, it is possible that social stress could produce impairments in corticolimbic structure and function that are ameliorated by nicotine. Indeed, nicotine's effects on cholinergic signaling and its cognition-enhancing properties could underlie a "self-medication" mechanism through which individuals seek relief from psychosocial stress and anxiety. Relatedly, it has been hypothesized that the particularly high rate of smoking among schizophrenia patients, for example (see de Leon, 1996), could be the result of self-medication to alleviate (albeit transiently) the negative symptoms associated with schizophrenia (Kumari & Postma, 2005; Smith, 2002). Intriguingly, social anxiety has been demonstrated to be highly prevalent among schizophrenia patients and is also associated with greater past substance/alcohol abuse (Pallanti, Quercioli, & Hollander, 2004). Nevertheless, recent evidence has raised doubts about this proposed self-medication hypothesis (Hahn et al., 2013; Manzella, Maloney, & Taylor, 2015). Likely, chronic stress exerted through negative social influence may represent a general addiction vulnerability that predisposes individuals to chronic, compulsive drug use (Chambers, 2009). In any case, it is clear that social influence plays a critical role in modulating nicotine addiction vulnerability and further preclinical studies that dissect the neurobehavioral mechanisms of this phenomenon are warranted.

4. NEUROBIOLOGICAL MECHANISMS UNDERLYING ANXIOLYTIC AND ANXIOGENIC EFFECTS OF NICOTINE AND ITS ROLE IN SOCIAL INTERACTION

Anxiety disorders affect one in every 13 people globally (ADAA, 2018). Individuals with anxiety disorders often suffer symptoms including panic, fear and most importantly, social withdrawal. These individuals also experience heightened startle and fear responses triggered by stressful situations. Interestingly, individuals suffering from anxiety-like disorders often use tobacco or other drugs of abuse due to their stress-alleviating properties. Nicotine is known to exhibit anxiolytic effects in individuals suffering from anxiety disorders (Benaroya-Milshtein et al., 2004). However, it is interesting to note that at high doses, nicotine has anxiogenic affects that lead to tobacco aversion.

Nicotine has been shown to attenuate stress and decreases anxiety-like symptoms (Benaroya-Milshtein et al., 2004). Because anxiety decreases social interactions, it may be possible that nicotine can increase social interactions by alleviating anxiety. To fully understand the effects of nicotine on social interactions, it is essential to understand the effects of nicotine on anxiety, a mediator of social interactions. Links between nicotine use, anxiety, and social interactions have been separately characterized using a variety of measurements, yet studies examining the complex relationship between all three are limited (File, Kenny, & Ouagazzal, 1998; Irvine, Cheeta, & File, 1999; Irvine, Cheeta, Marshall, & File, 2001). Primarily, these studies have examined the effects of nicotine on anxiety using social interaction tests. Here we aim to clarify the role of nicotine in alleviating anxiety, attenuating stress, and increasing social interaction.

4.1 Brain Regions and Pathways Involved in Mediating Anxiolysis

The effects of addictive substances, specifically nicotine, on anxiety, stress, and social interaction are mediated by various brain regions with a high degree of overlap. In order to understand these interactions, brain regions regulating these functions must be characterized in addition to how they are altered with drug consumption. Although the literature identifying the brain regions involved in these functions remains incomplete, we highlight known information regarding functionality and interconnectivity underlying these complex behaviors. The NA, PFC, HPA, DRN, and MHB are key brain regions implicated in addiction, anxiety-/stress-induced behaviors, and social interactions. Below, we will discuss these brain areas, including the behavioral effects and subcellular mechanisms that may regulate these phenotypes and behaviors.

4.2 The Prefrontal Cortex

The prefrontal cortex (PFC) plays an essential role in mediating executive functioning and working memory. It has also been established that this area plays a role in mediating addictive behaviors as well as anxiety disorders. Notably, the prelimbic (PL) area of the PFC promotes drug-seeking behavior, whereas the infralimbic (IL) region inhibits this behavior (Peters, Kalivas, & Quirk, 2009; Peters, LaLumiere, & Kalivas, 2008). Social interactions are also mediated by the IL, where animals that had recently participated in a social interaction task showed increased levels of the neuronal activity marker c-fos within this area (Avale et al., 2011). Further, lesions of the IL have been shown to decrease exploratory behavior (Avale et al., 2011), a phenotype often observed in animals experiencing anxiety accompanied by a decrease in social behaviors.

It has been suggested that predispositions for anxiety disorders and addiction may be mediated by pathological abnormalities in ventromedial PFC outputs targeting downstream limbic structures, limiting their ability to effectively communicate with these targets. Thus, limbic areas such as the NA may have the inability to recognize synaptic activity from the PFC (Robinson & Kolb, 1999), leading to behavioral consequences induced by repeated drug exposure. Human functional magnetic resonance imaging studies also show hypofrontality in individuals with substance use disorders compared with non-addicted individuals (for review see Goldstein & Volkow, 2011) and modafinil, a drug used for blunting cocaine addiction, and known to agonize α-receptors (Ballon & Feifel, 2006), alleviates hypofrontality of the PFC in cocaine abusers (Dackis, Kampman, Lynch, Pettinati, & O'Brien, 2005; Dackis et al., 2003; Dackis & O'Brien, 2003). While the complete mechanism of modafinil is far from understood, it is thought that modafinil enhances glutamatergic activity, due to its effects on cysteine–glutamate exchange and voltage-dependent calcium channels

(Mahler et al., 2014), blunting the euphoric effects of cocaine (Dackis et al., 2003). Due to these characteristics, the PFC has become a potential therapeutic target for drug addiction, which may aide in alleviating social withdrawal and anxiety-like symptomology also observed with regional dysregulation.

4.3 The Dorsal Raphe Nucleus and Medial Habenula

Like the PFC, the DRN plays a role in regulating learning, memory, and affect, while the habenula plays a role in reproductive behavior, stress responses, and learning. Similar to other brain regions, these areas are impacted by nicotine which has robust effects on social interaction. Interestingly, low-dose infusions of nicotine into the DRN increases the time animals spend in social interaction (Cheeta, Tucci, & File, 2001), suggesting that nicotine exerts a modulatory role over social influence through interactions with the DRN. Further, coadministration of dihydro-β-erythroidine hydrobromide (DHβE), a nicotine receptor antagonist specific for $\alpha_4\beta_2$-containing nAChRs, prevents these effects. These results indicate that nicotine acts on this subset of nicotinic receptors to enhance social interactions. During stress, the anxiogenic neuropeptide corticotropin releasing hormone (also known as corticotropin releasing factor; CRF) is released. CRF has been shown to act on serotonin (5-HT) neurons located within the DRN, an effect that can be alleviated by nicotinic activation at $\alpha_4\beta_2$-containing receptors. Thus, these studies indicate that nicotine not only increases social interactions, but also has the ability to overcome the anxiogenic properties of CRF.

Nicotine withdrawal has also been shown to induce anxiety-like symptoms (Zhao-Shea et al., 2015), which are mediated in part by the MHB. Specifically, activation of this area increases activity of the interpeduncular intermediate of the interpeduncular nucleus, an area rich in $\alpha_3\beta_4$ nAChRs. Taken together, activation of $\alpha_3\beta_4$ and $\alpha_4\beta_2$ nAChRs may decrease anxiety elicited by the release of CRF under stressful situations.

4.4 The Hippocampus

As a mediator of memory function, the HPC is hypothesized to be involved in the consolidation of memories underlying addiction and stress that can trigger anxiety. Rich in nAChRs, it is not surprising that the HPC is thought to play a role in mediating pathways underlying both nicotine addiction and anxiety. Unlike other brain areas, intracranial nicotine administration into the dorsal HPC has anxiogenic properties, especially during social tasks (File et al., 1998) where a decrease of time spent in social interaction is observed. Some suggest that nicotine evokes acetylcholine release within the HPC, activating postsynaptic muscarinic receptors and inducing anxiety (Kenny, Cheeta, & File, 2000). While it is believed that endogenous acetylcholine release reduces anxiety and promotes socialization (File et al., 1998), administration of the muscarinic Ach receptor antagonist scopolamine increases anxiety (Kenny et al., 2000). Together, these findings suggest that nicotine has anxiogenic effects when downstream release of acetylcholine acts on muscarinic receptors within the HPC. However, concentrations of nicotine found in cigarettes have been shown to desensitize α_7 nAChRs in CA1 interneurons of the HPA, which may prevent the anxiogenic effects described above (Alkondon, Pereira, Almeida, Randall, & Albuquerque, 2000). Thus, nicotine within the HPC may promote or inhibit anxiety dependent on concentration and receptors activated (File et al., 1998; Fowler, Arends, & Kenny, 2008; Fowler, Lu, Johnson, Marks, & Kenny, 2011).

As addiction is heavily reliant on learning and memory, functions mediated by the HPC and PFC, it is essential to uncover the cellular mechanisms underlying these functions and interactions. Thus, it is important to understand how nicotinic activation of the HPC regulates and modifies downstream targets such as PFC and the NA. Under normal circumstances, hippocampal afferents targeting the NA regulate the glutamatergic tone of medium spiny neurons (MSNs; Goto & O'Donnell, 2001). However, nicotinic activation of nAChRs within the HPC disrupts HPC-NA homeostatic signaling, which persists after nicotine administration is halted (Mansvelder, Mertz, & Role, 2009). Yet, other studies have reported that activation of nAChRs within the HPC elicits synaptic facilitation within the NA (Zhong et al., 2008). In addition to downstream effects on the NA, postsynaptic activation of nicotine at CA1 pyramidal cells of the HPC can turn short-term plasticity into LTP within Schaffer collateral synapses (Jay & Witter, 1991). This LTP promotes activation of interneurons that synapse onto pyramidal neurons of the PFC, preventing modifications of glutamatergic synapses within the PFC (Jay & Witter, 1991) and therefore functions mediated by this region. Although the effects of nicotine on these pathways are far from understood and are likely complicated, these studies provide beneficial information for an overall understanding of nicotinic effects on downstream targets of brain regions dense in nAChRs.

4.5 The Nucleus Accumbens

The NA is a relay between prefrontal cortical control and motor output within the mesocorticolimbic pathway (Kalivas, 2009). Due to its ability to integrate input from multiple regions, the NA plays an important role in processing motivation, aversion, reward, and reinforcement learning. Thus, it has become one of the most widely studied brain regions within the field of addiction. The NA undergoes significant changes in synaptic plasticity during addiction, and is thought to play a role in relapse vulnerability. Nicotine and other addictive substances modify cellular morphology and electrophysiological characteristics within the NA. For example, increases in dendritic spine density on MSNs within the NA (Russo et al., 2010) and modification of glutamatergic currents (as measured by AMPA/NMDA ratios) have been observed with nicotine (Gipson et al., 2013), cocaine (Dobi, Seabold, Christensen, Bock, & Alvarez, 2011; Mameli et al., 2009), heroin (Shen, Moussawi, Zhou, Toda, & Kalivas, 2011) and amphetamine (Saal, Dong, Bonci, & Malenka, 2003), leading to alterations in glutamatergic signaling between cells within this region.

The NA plays an important role in mediating anxiety- and stress-like behaviors. Like addiction, spine density and dendritic length of neurons within the NA are altered by prenatal stress (Martínez Téllez, Hernández-Torres, Gamboa, & Flores, 2009), which is thought to predispose individuals to addictive traits. Further, neuropeptides like CRF mediate stress responses by acting on CRF receptors located within the NA. CRF has bi-directional actions within the NA, where stress-induced release of CRF increases dopamine release via co-activation of receptors 1 (CRFR1) and 2 (CRFR2) (Lemos et al., 2012). Conversely, under severe stress, CRF loses its capacity to regulate dopamine release in the NA, leading to aversion to social bonding and decreased motivation (Lemos et al., 2012). Taken together, these data support the conclusion that cellular changes within the NA in response to anxiety and stress lead to the behavioral phenotypes commonly observed with anxiety disorders. Thus, through its anxiolytic effects, nicotine may further alter cellular changes within the NA resulting in therapeutic alleviation of stress.

For an abbreviated representation of brain region connectivity and neurotransmitter release, see Fig. 2.

4.6 Nicotinic Receptor Localization in the Mammalian Brain

Nicotinic receptors are heavily located throughout the mammalian brain including heavy expression within the HPA, PFC, MHB, striatum, VTA,

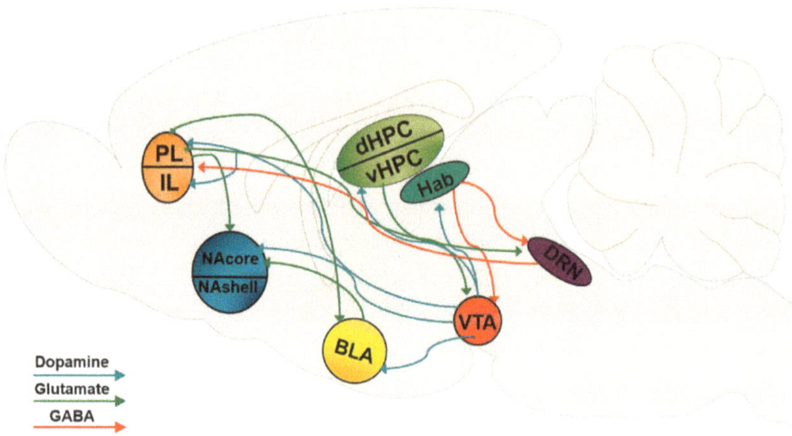

Fig. 2 Illustration of afferent and efferent connections of brain regions underlying nicotine addiction and social interaction. Prelimbic region of the prefrontal cortex (PL); infralimbic region of the prefrontal cortex (IL); nucleus accumbens core (NAcore); nucleus accumbens shell (NAshell); basolateral amygdala (BLA); ventral tegmental area (VTA); dorsal hippocampus (dHPC); ventral hippocampus (vHPC); habenula (Hab); dorsal raphe nuclei (DRN). Arrows indicate connectivity between regions, color coded by neurotransmitter.

and the hypothalamus. However, the composition and ligand binding sites of these receptors vary greatly depending on the brain region in which they are found. For example, α_3, α_4, β_2 and β_4 subunits are heavily expressed within the cortex, yet are unequally distributed among its different layers; whereas the HPA contains $\alpha_{3,4,5,7}$ and $\beta_{2,3,4}$ subunits. For a review of all brain regions the reader is referred to Clementi, Fornasari, and Gotti (2000). While subunit localization differs across brain regions as well as distribution within the brain region itself, nAChR subunit composition also varies across brain regions. Specifically, $\alpha_4\beta_2$ receptors are found within many brain regions including the cortex, HPA, striatum, hypothalamus, VTA, and the MHB, whereas α_7 receptors are only present in a subset of these regions including the PFC, HPA and VTA. Further, $\alpha_3\beta_4$ containing nAChRs are located primarily within the lateral habenula, HPA and VTA (see Gotti, Zoli, & Clementi, 2006 for a complete review on receptor composition variations across brain regions). Additionally, it is likely that receptor subtypes and combinations located within each brain region have both pre- and postsynaptic localization, and may likely be located on different neuronal subtypes. Thus, due to the complexity of each brain region, receptor subunit composition and varied localization, the effects of nicotine within each specific region is far from understood. Therefore, we aim to shed light on nAChR compositions within various brain regions and their roles in mediating anxiety-like behaviors and nicotine consumption.

4.7 Alterations in Subunits of Nicotinic Acetylcholine Receptors

As described above, nAChRs exist in many subunit compositions. Specific compositions and subunit subtypes can have vastly different effects on nicotine addiction, anxiety disorders, and social interactions. For example, nAChRs containing β_2 and β_7 subunits play a role in depression and anxiety disorders, which lead to detrimental effects on social interactions and motivation (Mineur et al., 2015). While mice lacking β_2 subunit-containing nAChRs display an increase in time spent engaging in social interactions as well as exaggerated approach movements toward novel conspecifics, these effects were only observed during conflicting motivation (i.e., the choice between food and social interactions) (Avale et al., 2011). At high doses, nicotine elicits aversive effects, mediated by nAChRs containing α_5 subunits (Fowler et al., 2011). Manipulation of these subunits can prevent this aversion, where a decrease in α_5 subunits prevents aversion at high concentrations and increases nicotine consumption (Fowler et al., 2011). Thus, while some studies show that nAChRs mediate aversive effects toward high concentrations of nicotine (through α_5 subunits) as well as play a role in depression and anxiety disorders (β_2 and β_7), others provide strong evidence for nicotinic alleviation of anxiety-like behaviors (i.e., $\alpha_3\beta_4$ and $\alpha_4\beta_2$ nAChRs with the dorsal raphe nucleus (see above)). Thus, it is likely that nicotine's role in mediating anxiety is dependent upon nAChR subunit composition, site of action, and may also be dose-dependent. While the actions of nicotine at specific receptor subunits have been reported (as discussed above), further research identifying the effects of nicotine on all nAChR subunits, receptor compositions, as well as regional specificity is necessary for a complete understanding on how nicotine may mediate anxiety-like behaviors.

4.8 Neurobiological Underpinnings of Environmental Enrichment and Its Effects on Anxiety and Nicotine Consumption

EE is defined as the stimulation of the brain by enhancing an individual or animals social or physical surroundings (Mellen & Sevenich MacPhee, 2001). EE often consists of social enhancement (pair housed) or cage enhancements, including toys, exercise wheels, and/or housing structures

Fig. 3 Environmental enrichment consists of introducing novel objects, social housing, and exercise opportunities to the home cage. For example, rats in the enriched condition are typically given objects arranged to maximize exploration and exercise. These objects can include plastic tubes, various children's toys, metal running wheels, and large plastic balls. The objects typically vary in color, size, shape, and some toys have holes for rats to explore inside the toys.

(e.g., PVC piping; see Fig. 3). Animals with EE show alterations in brain structure and function, including receptor sensitization, dendritic branching, and connectivity. EE is also known to have an effect on many of the same areas altered in drug addiction including the HPA (Kempermann, Kuhn, & Gage, 1997), the PFC (Segovia, Arco, & Mora, 2009) and the NA (Wood, Buse, Wellman, & Rebec, 2005). Additionally, EE has been shown to promote social behaviors, likely through modifications of these areas. Notably, EE induces stress resiliency to social defeat mediated by an IL-prefrontal cortical pathway (Lehmann & Herkenham, 2011). As discussed above, social interactions increase c-fos activity within the

IL, thus it is not surprising that EE may increase IL activation to overcome social defeat. EE also increases the time spent in socialization following stressors compared to non-enriched animals (Segovia et al., 2009). Interestingly, prenatal stress alters cellular plasticity, enhancing long-term depression and decreasing LTP in CA1 hippocampal neurons, thus leading to cognitive and working memory deficits which can be restored with EE (Yang et al., 2007). Moreover, these prenatally-stressed animals may be more prone to social abnormalities and addictive traits, which may be offset with EE. Furthermore, EE throughout development has been shown to protect against escalation of stimulant use with low but not high doses of cocaine (Gipson et al., 2011).

Overall, EE has been shown to have a protective effect against drug abuse (for review see Stairs & Bardo, 2009), where animals are less sensitive to the reinforcing properties of self-administered drugs (i.e., amphetamines). Additionally, rats reared in EE conditions are less sensitive to both acute and repeated stimulant effects of nicotine (Green et al., 2012) and are less sensitive to the discriminative stimulatory effects of nicotine (Bockman et al., 2018). EE-induced alterations in nAChRs are suggested to mediate sensitivity to the discriminative stimulatory effects of nicotine, where the noncompetitive nAChR antagonist mecamylamine prevents this loss in sensitivity (Bockman et al., 2018). EE has also been shown to block cocaine induced reinstatement, where 30 days of EE after 10 days of drug exposure completely abolishes behavioral sensitization and place preference for cocaine (Solinas, Chauvet, Thiriet, El Rawas, & Jaber, 2008) Thus, EE can have profound effects on drug addiction susceptibility, generally resulting in decreased drug seeking, dependence, and sensitivity.

5. CONCLUSIONS

Although only a subset of the population develops nicotine abuse or dependence, it is unclear what precise factors contribute to this susceptibility. Given the importance of social factors in the development and maintenance of tobacco use disorders, this factor should be regularly incorporated into preclinical animal models of nicotine addiction and relapse. Although some preclinical research has begun to unravel the impact of conspecifics on drug self-administration (see Strickland & Smith, 2015 for a more detailed review), the large majority of studies include animals raised and trained to self-administer drugs of abuse in isolation.

Here we propose that preclinical models of nicotine self-administration can increase construct validity (i.e., increasing the similarity of the behavior in the model with the behavior in the modeled condition) by including social conspecifics within the self-administration context. One limiting factor of successfully implementing social factors is that the current paradigm used for nicotine self-administration requires an intravenous jugular catheter mounted to a leash, as well as the acquisition of a non-dominant response to an operandum. These aspects of the intravenous self-administration model pose difficulties to including conspecifics within the same context. Importantly, social facilitation of drug use may be due to a drive state that increases the probability of the dominant response to self-administer the drug (Zajonc, 1965). Social learning theory would predict that when a rat is placed in an operant chamber, the learned lever press response for drug in a well-trained animal may be enhanced in the presence of a conspecific (Levine & Zentall, 1974). However, prior to lever pressing becoming the dominant behavior (i.e., prior to learning the operant response to self-administer nicotine), the presence of a conspecific may impair performance. Thus, animal models of nicotine self-administration may be enhanced by placing conspecifics within the self-administration context, but one limitation may be that the animal must have prior training of the lever press behavior in order to avoid impeding acquisition of nicotine self-administration. A potential future direction that may alleviate these issues is nicotine vapor self-administration, as sniffing is a respiratory action that commonly occurs during motivated and social behaviors (Clarke & Trowill, 1971; Grant, 1963). Thus, as opposed to learning a lever press or nose poke, inhalation of nicotine vapor may require less training, have fewer technical difficulties than intravenous models, and be less susceptible to impairment of acquisition from the presence of a conspecific.

An additional limitation to current intravenous self-administration models of nicotine addiction is the large amount of training required for rodents to acquire nicotine self-administration. This along with the surgical procedure to implant the jugular catheter makes it difficult to capture acquisition of nicotine intake during the short adolescent timeframe within a rodent, which is conservatively defined as the time between postnatal days 28–42. During this age range of less than 20 days, rodents exhibit neurobehavioral characteristics typical of adolescence (Spear, 2000). Self-administration procedures typically require at least 1 week of recovery from surgery, followed by multiple weeks of training prior to successful acquisition of the operant response to self-administer nicotine. Additionally, rats grow at faster rates during this developmental phase, decreasing the probability of maintaining catheter patency. Due to these limitations, capturing acquisition of intravenous nicotine self-administration during this short critical window is difficult to achieve. Nicotine vapor self-administration may allow for examination of nicotine acquisition during this short time as surgery would be unnecessary.

Clinically, there is limited evidence on the direct social transmission of impulse choice behavior among adolescents regarding substance abuse, and thus remains an important aspect to evaluate, especially pertaining to treatment. Further, while nicotine intake is highly variable between individuals, it is essential that treatment strategies for patients be carefully tailored to each individual. Specifically, we propose that it would be beneficial for treatments to account for whether the individual is a social, non-daily smoker, where smoking is highly context-dependent, versus a chronic daily user.

In conclusion, we propose that implementing social factors into models of various phases of the addiction cycle is necessary to better understand vulnerabilities that facilitate the development and maintenance of nicotine use. Given the impact of social interaction on decision-making processes, especially in adolescence, it is important to determine what environmental factors contribute to individual differences in nicotine abuse vulnerability. Improving preclinical models of nicotine addiction by capturing these critical aspects of the addiction cycle will be necessary to further our understanding of the neurobehavioral mechanisms that underlie the interactions between social factors and nicotine use.

REFERENCES

ADAA. (2018). *Facts & statistics*. Anxiety and Depression Association of America, ADAA. Retrieved from https://adaa.org/about-adaa/press-room/facts-statistics#.

Aleisa, A., Alzoubi, K., Gerges, N., & Alkadhi, K. (2005). Nicotine blocks stress-induced impairment of spatial memory and long-term potentiation of the hippocampal CA1 region. *The International Journal of Neuropsychopharmacology*, *9*(4), 417.

Alkondon, M., Pereira, E., Almeida, L., Randall, W., & Albuquerque, E. (2000). Nicotine at concentrations found in cigarette smokers activates and desensitizes nicotinic acetylcholine receptors in CA1 interneurons of rat hippocampus. *Neuropharmacology*, *39*(13), 2726–2739.

Anthonisen, N. R., Skeans, M. A., Wise, R. A., Manfreda, J., Kanner, R. E., & Connett, J. E. (2005). The effects of a smoking cessation intervention on 14.5-year mortality: A randomized clinical trial. *Annals of Internal Medicine*, *142*(4), 233–239.

Audrain-McGovern, J., Rodriguez, D., Tercyak, K., Neuner, G., & Moss, H. (2005). The impact of self-control indices on peer smoking and adolescent smoking progression. *Journal of Pediatric Psychology*, *31*(2), 139–151.

Avale, M., Chabout, J., Pons, S., Serreau, P., De Chaumont, F., Olivo-Marin, J., et al. (2011). Prefrontal nicotinic receptors control novel social interaction between mice. *The FASEB Journal*, *25*(7), 2145–2155.

Bahr, S., Hoffmann, J., & Yang, X. (2005). Parental and peer influences on the risk of adolescent drug use. *The Journal of Primary Prevention*, *26*(6), 529–551.
Ballon, J. S., & Feifel, D. (2006). A systematic review of Modafinil: Potential clinical uses and mechanisms of action. *Journal of Clinical Psychiatry*, *64*(4), 554–566.
Bardo, M., Neisewander, J., & Kelly, T. (2013). Individual differences and social influences on the neurobehavioral pharmacology of abused drugs. *Pharmacological Reviews*, *65*(1), 255–290.
Barrington-Trimis, J., Berhane, K., Unger, J., Cruz, T., Huh, J., Leventhal, A., et al. (2016). Psychosocial factors associated with adolescent electronic cigarette and cigarette use. *Pediatrics*, *136*(2), 308–317.
Barrington-Trimis, J., Berhane, K., Unger, J., Cruz, T., Urman, R., Chou, C., et al. (2016). The E-cigarette social environment, E-cigarette use, and susceptibility to cigarette smoking. *Journal of Adolescent Health*, *59*(1), 75–80.
Benaroya-Milshtein, N., Hollander, N., Apter, A., Kukulansky, T., Raz, N., Wilf, A., et al. (2004). Environmental enrichment in mice decreases anxiety, attenuates stress responses and enhances natural killer cell activity. *European Journal of Neuroscience*, *20*(5), 1341–1347.
Blackson, T., & Tarter, R. (1994). Individual, family, and peer affiliation factors predisposing to early-age onset of alcohol and drug use. *Alcoholism: Clinical and Experimental Research*, *18*(4), 813–821.
Blakemore, S. J. (2008). The social brain in adolescence. *Nature Reviews. Neuroscience*, *9*(4), 266–267.
Blum, K., Braverman, E., Holder, J., Lubar, J., Monastra, V., Miller, D., et al. (2000). The reward deficiency syndrome: A biogenetic model for the diagnosis and treatment of impulsive, addictive and compulsive behaviors. *Journal of Psychoactive Drugs*, *32*(1), 1–112.
Bockman, C., Zeng, W., Hall, J., Mittelstet, B., Schwarzkopf, L., & Stairs, D. (2018). Nicotine drug discrimination and nicotinic acetylcholine receptors in differentially reared rats. *Psychopharmacology*, *235*(5), 1415–1426.
Bold, K., Morean, M., Kong, G., Simon, P., Camenga, D., Cavallo, D., et al. (2017). Early age of e-cigarette use onset mediates the association between impulsivity and e-cigarette use frequency in youth. *Drug and Alcohol Dependence*, *181*, 146–151.
Breslau, N., Kilbey, M. M., & Andreski, P. (1993). Nicotine dependence and major depression: New evidence from a prospective investigation. *Archives of General Psychiatry*, *50*(1), 31–35.
Burke, M., Hays, J., & Ebbert, J. (2016). Varenicline for smoking cessation: A narrative review of efficacy, adverse effects, use in at-risk populations, and adherence. *Patient Preference and Adherence*, *10*, 435–441.
Burris, J., Riley, E., Puleo, G., & Smith, G. (2017). A longitudinal study of the reciprocal relationship between ever smoking and urgency in early adolescence. *Drug and Alcohol Dependence*, *178*, 519–526.
Centers for Disease Control and Prevention. (2015). *National youth tobacco survey (NYTS)*. [Web 29 March 2018].
Chambers, R. A. (2009). A nicotine challenge to the self-medication hypothesis in a neurodevelopmental animal model of schizophrenia. *Journal of Dual Diagnosis*, *5*(2), 139–148.
Chassin, L., Hussong, A., & Beltran, I. (2009). Adolescent substance use. In R. Lerner & L. Steinberg (Eds.), *Handbook of adolescent psychology: Vol. 1* (pp. 723–763). Hoboken, NJ: Wiley.
Cheeta, S., Tucci, S., & File, S. E. (2001). Antagonism of the anxiolytic effect of nicotine in the dorsal raphé nucleus by dihydro-β-erythroidine. *Pharmacology, Biochemistry, and Behavior*, *70*(4), 491–496.
Chein, J., Albert, D., O'Brien, L., Uckert, K., & Steinberg, L. (2011). Peers increase adolescent risk taking by enhancing activity in the brain's reward circuitry. *Developmental Science*, *14*(2), F1–F10.

Chen, H., Sharp, B., Matta, S., & Wu, Q. (2011). Social interaction promotes nicotine self-administration with olfactogustatory cues in adolescent rats. *Neuropsychopharmacology, 36*(13), 2629–2638.

Christakis, N. A., & Fowler, J. H. (2008). The collective dynamics of smoking in a large social network. *New England Journal of Medicine, 358*(21), 2249–2258.

Clarke, S., & Trowill, J. (1971). Sniffing and motivated behavior in the rat. *Physiology & Behavior, 6*(1), 49–52.

Clementi, F., Fornasari, D., & Gotti, C. (2000). Neuronal nicotinic receptors, important new players in brain function. *European Journal of Pharmacology, 393*(1–3), 3–10.

Cobb, S. (1976). Social support as a moderator of life stress. *Psychosomatic Medicine, 38*(5), 300–314.

Dackis, C., Kampman, K., Lynch, K., Pettinati, H., & O'Brien, C. (2005). A double-blind, placebo-controlled trial of modafinil for cocaine dependence. *Neuropsychopharmacology, 30*(1), 205–211.

Dackis, C., Lynch, K., Yu, E., Samaha, F., Kampman, K., Cornish, J., et al. (2003). Modafinil and cocaine: A double-blind, placebo-controlled drug interaction study. *Drug and Alcohol Dependence, 70*(1), 29–37.

Dackis, C., & O'Brien, C. (2003). Glutamatergic agents for cocaine dependence. *Annals of the New York Academy of Sciences, 1003*(1), 328–345.

Daruna, J. H., & Barnes, P. A. (1993). A neurodevelopmental view of impulsivity. In W. G. McCown, J. L. Johnson, & M. B. Shure (Eds.), *The impulsive client: Theory, research and treatment* (pp. 23–37). American Psychological Association.

de Leon, J. (1996). Smoking and vulnerability for schizophrenia. *Schizophrenia Bulletin, 22*(3), 405–409.

de Wit, H. (2008). Impulsivity as a determinant and consequence of drug use: A review of underlying processes. *Addiction Biology, 14*(1), 22–31.

Dickman, S. (1990). Functional and dysfunctional impulsivity: Personality and cognitive correlates. *Journal of Personality and Social Psychology, 58*(1), 95–102.

Dobi, A., Seabold, G., Christensen, C., Bock, R., & Alvarez, V. (2011). Cocaine-induced plasticity in the nucleus accumbens is cell specific and develops without prolonged withdrawal. *Journal of Neuroscience, 31*(5), 1895–1904.

Doran, N., Spring, B., McChargue, D., Pergadia, M., & Richmond, M. (2004). Impulsivity and smoking relapse. *Nicotine & Tobacco Research, 6*(4), 641–647.

Dumont, M., & Provost, M. A. (1999). Resilience in adolescents: Protective role of social support, coping strategies, self-esteem, and social activities on experience of stress and depression. *Journal of Youth and Adolescence, 28*(3), 343–363.

Elkins, I., King, S., McGue, M., & Iacono, W. (2006). Personality traits and the development of nicotine, alcohol, and illicit drug disorders: Prospective links from adolescence to young adulthood. *Journal of Abnormal Psychology, 115*(1), 26–39.

Evenden, J. (1999). Impulsivity: A discussion of clinical and experimental findings. *Journal of Psychopharmacology, 13*(2), 180–192.

Ewin, S., Kangiser, M., & Stairs, D. (2015). The effects of environmental enrichment on nicotine condition place preference in male rats. *Experimental and Clinical Psychopharmacology, 23*(5), 387–394.

File, S., Kenny, P., & Ouagazzal, A. (1998). Bimodal modulation by nicotine of anxiety in the social interaction test: Role of the dorsal hippocampus. *Behavioral Neuroscience, 112*(6), 1423–1429.

Fowler, C., Arends, M., & Kenny, P. (2008). Subtypes of nicotinic acetylcholine receptors in nicotine reward, dependence, and withdrawal: Evidence from genetically modified mice. *Behavioural Pharmacology, 19*(5–6), 461–484.

Fowler, C., Lu, Q., Johnson, P., Marks, M., & Kenny, P. (2011). Habenular α5 nicotinic receptor subunit signalling controls nicotine intake. *Nature, 471*(7340), 597–601.

Gardner, M., & Steinberg, L. (2005). Peer influence on risk taking, risk preference, and risky decision making in adolescence and adulthood: An experimental study. *Developmental Psychology*, *41*(4), 625–635.
Gecková, A. M., Stewart, R., Dijk, J. P., Orosová, O., Groothoff, J. W., & Post, D. (2005). Influence of socio-economic status, parents and peers on smoking behaviour of adolescents. *European Addiction Research*, *11*(4), 204–209.
George, L. K., Blazer, D. G., Hughes, D. C., & Fowler, N. (1989). Social support and the outcome of major depression. *The British Journal of Psychiatry*, *154*(4), 478–485.
Gipson, C., Kupchik, Y., Shen, H., Reissner, K., Thomas, C., & Kalivas, P. (2013). Relapse induced by cues predicting cocaine depends on rapid, transient synaptic potentiation. *Neuron*, *77*(5), 867–872.
Gipson, C., Yates, J., Beckmann, J., Marusich, J., Zentall, T., & Bardo, M. (2011). Social facilitation of D-amphetamine self-administration in rats. *Experimental and Clinical Psychopharmacology*, *19*(6), 409–419.
Goldstein, R., & Volkow, N. (2011). Oral methylphenidate normalizes cingulate activity and decreases impulsivity in cocaine addiction during an emotionally salient cognitive task. *Neuropsychopharmacology*, *36*(1), 366–367.
Goto, Y., & O'Donnell, P. (2001). Synchronous activity in the hippocampus and nucleus accumbens in vivo. *The Journal of Neuroscience*, *21*(4).
Gotti, G., Zoli, M., & Clementi, F. (2006). Brain nicotinic acetylcholine receptors: Native subtypes and their relevance. *Trends in Pharmacological Sciences*, *27*(9), 482–491.
Grant, E. (1963). An analysis of the social behaviour of the male laboratory rat. *Behaviour*, *21*(3), 260–281.
Green, T., Cain, M., Thompson, M., & Bardo, M. (2003). Environmental enrichment decreases nicotine-induced hyperactivity in rats. *Psychopharmacology*, *170*(3), 235–241.
Green, H., Horta, M., Haye, K., Tucker, J., Kennedy, D., & Pollard, M. (2012). Peer influence and selection processes in adolescent smoking behavior: A comparative study. *Nicotine & Tobacco Research*, *15*(2), 534–541.
Guillot, C. R., Pang, R. D., & Leventhal, A. M. (2014). Anxiety sensitivity and negative urgency: A pathway to negative reinforcement-related smoking expectancies. *Journal of Addiction Medicine*, *8*(3), 189.
Gullo, M., & Dawe, S. (2008). Impulsivity and adolescent substance use: Rashly dismissed as "all-bad"? *Neuroscience & Biobehavioral Reviews*, *32*(8), 1507–1518.
Hadaway, P., Alexander, B., Coambs, R., & Beyerstein, B. (1979). The effect of housing and gender on preference for morphine-sucrose solutions in rats. *Psychopharmacology*, *66*(1), 87–91.
Hahn, B., Harvey, A., Concheiro-Guisan, M., Huestis, M., Holcomb, H., & Gold, J. (2013). A test of the cognitive self-medication hypothesis of tobacco smoking in schizophrenia. *Biological Psychiatry*, *74*(6), 436–443.
Hampson, S., & Goldberg, L. (2006). A first large cohort study of personality trait stability over the 40 years between elementary school and midlife. *Journal of Personality and Social Psychology*, *91*(4), 763–779.
Harrison, E. L., & McKee, S. A. (2011). Non-daily smoking predicts hazardous drinking and alcohol use disorders in young adults in a longitudinal US sample. *Drug and Alcohol Dependence*, *118*(1), 78–82.
Henningfield, J., & Zaatari, G. (2010). Electronic nicotine delivery systems: Emerging science foundation for policy. *Tobacco Control*, *19*(2), 89–90.
Hiscock, R., Bauld, L., Amos, A., Fidler, J. A., & Munafò, M. (2012). Socioeconomic status and smoking: A review. *Annals of the New York Academy of Sciences*, *1248*(1), 107–123.
Hung, L., Neuner, S., Polepalli, J., Beier, K., Wright, M., Walsh, J., et al. (2017). Gating of social reward by oxytocin in the ventral tegmental area. *Science*, *357*(6358), 1406–1411.

Irvine, E., Cheeta, S., & File, S. (1999). Time-course of changes in the social interaction test of anxiety following acute and chronic administration of nicotine. *Behavioural Pharmacology, 10*(6), 691–697.

Irvine, E., Cheeta, S., Marshall, M., & File, S. (2001). Different treatment regimens and the development of tolerance to nicotine's anxiogenic effects. *Pharmacology, Biochemistry, and Behavior, 68*(4), 769–776.

Jay, T., & Witter, M. (1991). Distribution of hippocampal CA1 and subicular efferents in the prefrontal cortex of the rat studied by means of anterograde transport of Phaseolus vulgaris-leucoagglutinin. *The Journal of Comparative Neurology, 313*(4), 574–586.

Jiang, N., Lee, Y., & Ling, P. (2014). Young adult social smokers: Their co-use of tobacco and alcohol, tobacco-related attitudes, and quitting efforts. *Preventive Medicine, 69*, 166–171.

John, U., Meyer, C., Rumpf, H. J., & Hapke, U. (2004). Depressive disorders are related to nicotine dependence in the population but do not necessarily hamper smoking cessation. *The Journal of Clinical Psychiatry, 65*(2), 169–176.

Johnson, V., & Pandina, R. (1991). Effects of the family environment on adolescent substance use, delinquency, and coping styles. *The American Journal of Drug and Alcohol Abuse, 17*(1), 71–88.

Kalivas, P. (2009). The glutamate homeostasis hypothesis of addiction. *Nature Reviews. Neuroscience, 10*(8), 561–572.

Kandel, D. B., Kiros, G. E., Schaffran, C., & Hu, M. C. (2004). Racial/ethnic differences in cigarette smoking initiation and progression to daily smoking: A multilevel analysis. *American Journal of Public Health, 94*(1), 128–135.

Kempermann, G., Kuhn, H., & Gage, F. (1997). More hippocampal neurons in adult mice living in an enriched environment. *Nature, 386*(6624), 493–495.

Kenny, P., Cheeta, S., & File, E. (2000). Anxiogenic effects of nicotine in the dorsal hippocampus are mediated by 5-HT1A and not by muscarinic M1 receptors. *Neuropharmacology, 39*(2), 300–307.

Kosfeld, M., Heinrichs, M., Zak, P., Fischbacher, U., & Fehr, E. (2005). Oxytocin increases trust in humans. *Nature, 435*(7042), 673–676.

Kreek, M. J., Nielsen, D. A., Butelman, E. R., & LaForge, K. S. (2005). Genetic influences on impulsivity, risk taking, stress responsivity and vulnerability to drug abuse and addiction. *Nature Neuroscience, 8*(11), 1450.

Krishnan-Sarin, S., Reynolds, B., Duhig, A., Smith, A., Liss, T., McFetridge, A., et al. (2007). Behavioral impulsivity predicts treatment outcome in a smoking cessation program for adolescent smokers. *Drug and Alcohol Dependence, 88*(1), 79–82.

Kumari, V., & Postma, P. (2005). Nicotine use in schizophrenia: The self-medication hypotheses. *Neuroscience & Biobehavioral Reviews, 29*(6), 1021–1034.

Lauterstein, D., Hoshino, R., Gordon, T., Watkins, B., Weitzman, M., & Zelikoff, J. (2014). The changing face of tobacco use among United States youth. *Current Drug Abuse Reviews*, 7(1), 29–43.

Leatherdale, S. T., Mcdonald, P. W., Cameron, R., & Brown, K. S. (2005). A multilevel analysis examining the relationship between social influences for smoking and smoking onset. *American Journal of Health Behavior, 29*(6), 520–530.

Lee, H., Jang, M., & Noh, J. (2017). Oxytocin attenuates aversive response to nicotine and anxiety-like behavior in adolescent rats. *Neuroscience Research, 115*, 29–36.

Lehmann, M., & Herkenham, M. (2011). Environmental enrichment confers stress resiliency to social defeat through an infralimbic cortex-dependent neuroanatomical pathway. *Journal of Neuroscience, 31*(16), 6159–6173.

Lemos, J., Wanat, M., Smith, J., Reyes, B., Hollon, N., Van Bockstaele, E., et al. (2012). Severe stress switches CRF action in the nucleus accumbens from appetitive to aversive. *Nature, 490*(7420), 402–406.

Levine, J. M., & Zentall, T. R. (1974). Effect of conspecific's presence on deprived rats' performance: Social facilitation vs. distraction/imitation. *Animal Learning and Behavior, 2*, 119–122.
Levinson, A. H., Campo, S., Gascoigne, J., Jolly, O., Zakharyan, A., & Vu Tran, Z. (2007). Smoking, but not smokers: Identity among college students who smoke cigarettes. *Nicotine & Tobacco Research, 9*(8), 845–852.
Lim, M., & Young, L. (2006). Neuropeptidergic regulation of affiliative behavior and social bonding in animals. *Hormones and Behavior, 50*, 506–517.
Lynam, D., Caspi, A., Moffit, T., Wikström, P., Loeber, R., & Novak, S. (2000). The interaction between impulsivity and neighborhood context on offending: The effects of impulsivity are stronger in poorer neighborhoods. *Journal of Abnormal Psychology, 109*(4), 563–574.
Mahler, S. V., Hensley-Simon, M., Tahsili-Fahadan, P., LaLumier, R. T., Thomas, C., Fallon, R. V., et al. (2014). Modafinil attenuates reinstatement of cocaine seeking: Role for cysteine-glutamate exchange and metabotrophic glutamate receptors. *Addiction Biology, 19*(1), 49–60.
Mameli, M., Halbout, B., Creton, C., Engblom, D., Parkitna, J., Spanagel, R., et al. (2009). Cocaine-evoked synaptic plasticity: Persistence in the VTA triggers adaptations in the NAc. *Nature Neuroscience, 12*(8), 1036–1041.
Manbeck, K., Shelley, D., Schmidt, C., & Harris, A. (2014). Effects of oxytocin on nicotine withdrawal in rats. *Pharmacology, Biochemistry, and Behavior, 116*, 84–89.
Mansvelder, H., Mertz, M., & Role, L. (2009). Nicotinic modulation of synaptic transmission and plasticity in cortico-limbic circuits. *Seminars in Cell & Developmental Biology, 20*(4), 432–440.
Manzella, F., Maloney, S., & Taylor, G. (2015). Smoking in schizophrenic patients: A critique of the self-medication hypothesis. *World Journal of Psychiatry, 5*(1), 35.
Martínez-Téllez, R., Hernández-Torres, E., Gamboa, C., & Flores, G. (2009). Prenatal stress alters spine density and dendritic length of nucleus accumbens and hippocampus neurons in rat offspring. *Synapse, 63*(9), 794–804.
Mellen, J., & Sevenich MacPhee, M. (2001). Philosophy of environmental enrichment: Past, present, and future. *Zoo Biology, 20*(3), 211–226.
Miller, M. A., Bershad, A., King, A. C., Lee, R., & Wit, H. D. (2016). Intranasal oxytocin dampens cue-elicited cigarette craving in daily smokers. *Behavioural Pharmacology, 27*(8), 697–703.
Mineur, Y., Fote, G., Blakeman, S., Cahuzac, E., Newbold, S., & Picciotto, M. (2015). Multiple nicotinic acetylcholine receptor subtypes in the mouse amygdala regulate affective behaviors and response to social stress. *Neuropsychopharmacology, 41*(6), 1579–1587.
Murray, R. P., Johnston, J. J., Dolce, J. J., Lee, W. W., & O'Hara, P. (1995). Social support for smoking cessation and abstinence: The Lung Health Study. *Addictive Behaviors, 20*(2), 159–170.
Niaura, R., Shadel, W., Britt, D., & Abrams, D. (2002). Response to social stress, urge to smoke, and smoking cessation. *Addictive Behaviors, 27*(2), 241–250.
O'Brien, L., Albert, D., Chein, J., & Steinberg, L. (2011). Adolescents prefer more immediate rewards when in the presence of their peers. *Journal of Research on Adolescence, 21*(4), 747–753.
Ogunwale, M., Li, M., Ramakrishnam Raju, M., Chen, Y., Nantz, M., Conklin, D., et al. (2017). Aldehyde detection in electronic cigarette aerosols. *ACS Omega, 2*(3), 1207–1214.
Pallanti, S., Quercioli, L., & Hollander, E. (2004). Social anxiety in outpatients with schizophrenia: A relevant cause of disability. *American Journal of Psychiatry, 161*(1), 53–58.
Peartree, N., Hatch, K., Goenaga, J., Dado, N., Molla, H., Dufwenberg, M., et al. (2017). Social context has differential effects on acquisition of nicotine self-administration in male and female rats. *Psychopharmacology, 234*(12), 1815–1828.

Peretti-Watel, P., L'Haridon, O., & Seror, V. (2013). Time preferences, socioeconomic status and smokers' behaviour, attitudes and risk awareness. *European Journal of Public Health*, *23*(5), 783–788.

Persoskie, A., O'Brien, E., Nguyen, A., & Tworek, C. (2017). Measuring youth beliefs about the harms of e-cigarettes and smokeless tobacco compared to cigarettes. *Addictive Behaviors*, *70*, 7–13.

Peters, J., Kalivas, P., & Quirk, G. (2009). Extinction circuits for fear and addiction overlap in prefrontal cortex. *Learning & Memory*, *16*(5), 279–288.

Peters, J., LaLumiere, R., & Kalivas, P. (2008). Infralimbic prefrontal cortex is responsible for inhibiting cocaine seeking in extinguished rats. *Journal of Neuroscience*, *28*(23), 6046–6053.

Picciotto, M., Zoli, M., Rimondini, R., Léna, C., Marubio, L., Pich, E., et al. (1998). Acetylcholine receptors containing the β2 subunit are involved in the reinforcing properties of nicotine. *Nature*, *391*(6663), 173–177.

Reitzel, L., Businelle, M., Kendzor, D., Li, Y., Cao, Y., Castro, Y., et al. (2011). Subjective social status predicts long-term smoking abstinence. *BMC Public Health*, *11*(1), 135.

Reynolds, B., Patak, M., Shroff, P., Penfold, R. B., Melanko, S., & Duhig, A. M. (2007). Laboratory and self-report assessments of impulsive behavior in adolescent daily smokers and nonsmokers. *Experimental and Clinical Psychopharmacology*, *15*(3), 264.

Robinson, T., & Kolb, B. (1999). Alterations in the morphology of dendrites and dendritic spines in the nucleus accumbens and prefrontal cortex following repeated treatment with amphetamine or cocaine. *European Journal of Neuroscience*, *11*(5), 1598–1604.

Russo, S., Dietz, D., Dumitriu, D., Morrison, J., Malenka, R., & Nestler, E. (2010). The addicted synapse: Mechanisms of synaptic and structural plasticity in nucleus accumbens. *Trends in Neurosciences*, *33*(6), 267–276.

Saal, D., Dong, Y., Bonci, A., & Malenka, R. (2003). Drugs of abuse and stress trigger a common synaptic adaptation in dopamine neurons. *Neuron*, *37*(4), 577–582.

Scalici, F., & Schulz, P. (2017). Parents' and peers' normative influence on adolescents' smoking: Results from a Swiss-Italian sample of middle schools students. *Substance Abuse Treatment, Prevention, and Policy*, *12*(5).

Schane, R. E., Glantz, S. A., & Ling, P. M. (2009). Nondaily and social smoking: An increasingly prevalent pattern. *Archives of Internal Medicine*, *169*(19), 1742 1744.

Schoenbaum, G., Roesch, M., & Stalnaker, T. (2006). Orbitofrontal cortex, decision-making and drug addiction. *Trends in Neurosciences*, *29*(2), 116–124.

Segovia, G., Arco, A., & Mora, F. (2009). Environmental enrichment, prefrontal cortex, stress, and aging of the brain. *Journal of Neural Transmission*, *116*(8), 1007–1016.

Settles, R. E., Fischer, S., Cyders, M. A., Combs, J. L., Gunn, R. L., & Smith, G. T. (2012). Negative urgency: A personality predictor of externalizing behavior characterized by neuroticism, low conscientiousness, and disagreeableness. *Journal of Abnormal Psychology*, *121*(1), 160.

Shen, H., Moussawi, K., Zhou, W., Toda, S., & Kalivas, P. (2011). Heroin relapse requires long-term potentiation-like plasticity mediated by NMDA2b-containing receptors. *Proceedings of the National Academy of Sciences of the United States of America*, *108*(48), 19407–19412.

Simons-Morton, B., & Chen, R. (2006). Over time relationships between early adolescent and peer substance use. *Addictive Behaviors*, *31*(7), 1211–1223.

Simons-Morton, B., Lerner, N., & Singer, J. (2005). The observed effects of teenage passengers on the risky driving behavior of teenage drivers. *Accident; Analysis and Prevention*, *37*(6), 973–982.

Smith, R. (2002). Effects of cigarette smoking and nicotine nasal spray on psychiatric symptoms and cognition in schizophrenia. *Neuropsychopharmacology*, *27*(3), 479–497.

Solinas, M., Chauvet, C., Thiriet, N., El Rawas, R., & Jaber, M. (2008). Reversal of cocaine addiction by environmental enrichment. *Proceedings of the National Academy of Sciences of the United States of America*, *105*(44), 17145–17150.
Song, A. V., & Ling, P. M. (2011). Social smoking among young adults: Investigation of intentions and attempts to quit. *American Journal of Public Health*, *101*(7), 1291–1296.
Sonntag, H., Wittchen, H. U., Höfler, M., Kessler, R. C., & Stein, M. B. (2000). Are social fears and DSM-IV social anxiety disorder associated with smoking and nicotine dependence in adolescents and young adults? *European Psychiatry*, *15*(1), 67–74.
Spear, L. (2000). The adolescent brain and age-related behavioral manifestations. *Neuroscience & Biobehavioral Reviews*, *24*(4), 417–463.
Spillane, N. S., Smith, G. T., & Kahler, C. W. (2010). Impulsivity-like traits and smoking behavior in college students. *Addictive Behaviors*, *35*(7), 700–705.
Spinella, M. (2004). Neurobehavioral correlates of impulsivity: Evidence of prefrontal involvement. *International Journal of Neuroscience*, *114*(1), 95–104.
Stairs, D., & Bardo, M. (2009). Neurobehavioral effects of environmental enrichment and drug abuse vulnerability. *Pharmacology, Biochemistry, and Behavior*, *92*(3), 377–382.
Stewart, W., & Roth, P. (2001). Risk propensity differences between entrepreneurs and managers: A meta-analytic review. *Journal of Applied Psychology*, *86*(1), 145–153.
Strickland, J., & Smith, M. (2015). Animal models of social contact and drug self-administration. *Pharmacology, Biochemistry, and Behavior*, *136*, 47–54.
Thiel, K., Sanabria, F., & Neisewander, J. (2009). Synergistic interaction between nicotine and social rewards in adolescent male rats. *Psychopharmacology*, *204*(3), 391–402.
Walden, B., McGue, M., Iacono, W., Burt, S., & Elkins, I. (2004). Identifying shared environmental contributions to early substance use: The respective roles of peers and parents. *Journal of Abnormal Psychology*, *113*(3), 440–450.
Wang, T., Han, W., & Chen, H. (2016). Socially acquired nicotine self-administration with an aversive flavor cue in adolescent female rats. *Psychopharmacology*, *233*(10), 1837–1844.
Waters, K., Harris, K., Hall, S., Nazir, N., & Waigandt, A. (2006). Characteristics of social smoking among college students. *Journal of American College Health*, *55*(3), 133–139.
West, P., Sweeting, H., & Ecob, R. (1999). Family and friends influences on the uptake of regular smoking from mid-adolescence to early adulthood. *Addiction*, *94*(9), 1397–1411.
Wood, D., Buse, J., Wellman, C., & Rebec, G. (2005). Differential environmental exposure alters NMDA but not AMPA receptor subunit expression in nucleus accumbens core and shell. *Brain Research*, *1042*(2), 176–183.
Yang, J., Hou, C., Ma, N., Liu, J., Zhang, Y., Zhou, J., et al. (2007). Enriched environment treatment restores impaired hippocampal synaptic plasticity and cognitive deficits induced by prenatal chronic stress. *Neurobiology of Learning and Memory*, *87*(2), 257–263.
Zajonc, R. B. (1965). Social facilitation. *Science*, *149*, 269–274.
Zapolski, T. C., Cyders, M. A., & Smith, G. T. (2009). Positive urgency predicts illegal drug use and risky sexual behavior. *Psychology of Addictive Behaviors*, *23*(2), 348.
Zernig, G., Kummer, K., & Prast, J. (2013). Dyadic social interaction as an alternative reward to cocaine. *Frontiers in Psychiatry*, *4*(article 100).
Zhao-Shea, R., DeGroot, S., Liu, L., Vallaster, M., Pang, X., Su, Q., et al. (2015). Erratum: Corrigendum: Increased CRF signalling in a ventral tegmental area-interpeduncular nucleus-medial habenula circuit induces anxiety during nicotine withdrawal. *Nature Communications*, *6*(1), 7625.
Zhong, C., Du, C., Hancock, M., Mertz, M., Talmage, D., & Role, L. (2008). Presynaptic type III neuregulin 1 is required for sustained enhancement of hippocampal transmission by nicotine and for axonal targeting of 7 nicotinic acetylcholine receptors. *Journal of Neuroscience*, *28*(37), 9111–9116.

CHAPTER TWO

Assessing Social Alcohol Drinking in Rodent Models: Are We There Yet?

Andrey E. Ryabinin[1], Andre T. Walcott
Department of Behavioral Neuroscience, Oregon Health & Science University, Portland, OR, United States
[1]Corresponding author: e-mail address: ryabinin@ohsu.edu

Contents

Abstract

Pathological and social consequences associated with excessive alcohol consumption and dependence constitute a serious concern for human well-being. The success of preventative and therapeutic strategies for alcohol use disorder depends on the development of appropriate animal models of alcohol consumption. Alcohol consumption in humans typically occurs in social settings. In contrast, the vast majority of animal models investigate alcohol drinking in socially isolated animals. A number of rodent model studies have attempted to assess drinking of individual animals within social setting. These studies identified varied directions of effects of social environment on alcohol consumption. However, the interpretation of these results is strongly hampered by technical issues associated with the existing approaches to assess drinking of individual animals within each group. This review analyzes the existing literature mostly focusing on the interpretation caveats of performed studies and highlights some of the new directions in these studies.

International Review of Neurobiology, Volume 140
ISSN 0074-7742
https://doi.org/10.1016/bs.irn.2018.07.002

1. INTRODUCTION

Despite widespread casual consumption of alcoholic beverages, alcohol use disorder remains a tremendous health burden in human society. Excessive alcohol use accounts for about 6% of all deaths and is a leading risk factor for premature death and disability among people between the ages of 15 and 49 (National Institute on Alcohol Abuse and Alcoholism, 2017). In addition to health risks, alcohol affects social interactions, contributing to dissatisfaction in social relationships and promoting violence. On the other hand, alcohol consumption is greatly influenced by social factors. These influences include nonspecific effects of loneliness and social stress, as well as effects of peer pressure and social support from specific individuals or groups.

While human social attachments are influenced by human expectations, social institutions and economic factors, these factors interact with physiological effects of alcohol. In turn, human social behaviors have biological substrates. Since biological substrates are conserved between humans and other animals, understanding biological effects of alcohol on social behavior in laboratory animals would benefit our understanding of alcohol use disorder.

There is a precedent in other fields of preclinical neuroscience indicating that studies on social behaviors in rodents are helpful for understanding human social behaviors and psychiatric disorders. For example, experiments in prairie voles have identified oxytocin and vasopressin receptors as critical in regulation of specific long-term attachments (Insel & Hulihan, 1995; Lim & Young, 2004; Liu, Curtis, & Wang, 2001; Pitkow et al., 2001; Walum et al., 2012, 2008; Winslow, Hastings, Carter, Harbaugh, & Insel, 1993). Subsequent epidemiological studies using this knowledge have confirmed the association of polymorphisms in these receptors with problems in social attachment in humans, and the first clinical trials using oxytocin to treat autism spectrum disorders have been initiated (Auyeung et al., 2015; Chen, Barth, Johnson, Gotlib, & Johnson, 2011). Studies on social aspects of alcoholism are awaiting similar developments.

Preclinical studies using rodent models have left no doubt that social environment interacts with alcohol consumption. There have been already a substantial number of studies investigating either the effects of alcohol on social behaviors or effects of social environment on alcohol consumption. There have also been studies investigating effects of various other factors on alcohol drinking in social settings, as opposed to traditional rodent models investigating alcohol drinking in isolation. Our search for such studies was performed in Pubmed followed by analysis of references within the identified publications. We identified 33 published studies investigating alcohol consumption in socially housed animals (Table 1). It is difficult to design precise keywords that would capture all studies assessing social drinking in animals. Therefore, we apologize if we have inadvertently missed any studies. In our analysis of the literature we focused only on studies in which alcohol consumption is voluntary. By voluntary we mean that the animal had a choice whether to drink an alcohol solution or not (in most cases, the so called two-bottle choice paradigm). Therefore, we did not include studies in which alcohol is the only available fluid at any point of the study because thirst is an extremely strong motivation important for animal survival and can, in our opinion, override other motivations, including those associated with alcohol consumption. We also ignored publications in which animals are socially housed at some point during the study, but consumption is measured only when the animal is in isolation. There is a large number of such studies but we believe that there is no way of telling whether such prior experience or "probe trials" can accurately reflect an animal's consumption in social settings.

2. OVERVIEW OF EXISTING METHODS AND RESULTS

Over the years since the first studies trying to assess alcohol consumption in social settings, researchers have used several approaches that can be classified in four categories: (1) studies in which several animals are housed in a cage and total consumption per cage is measured; (2) studies in which animals are group-housed in cages but are separated from each other by a mesh or perforated partition; (3) studies in which several animals are housed in a cage, and the individual intake is estimated based on video recording; (4) studies in which several animals are housed per cage, and individual intake is estimated based on radio frequency tracking. We will discuss each of these approaches in the following sections.

As a general conclusion from these studies, social housing can either increase or inhibit alcohol consumption in laboratory animals, and some of the reviewed studies did not find significant effects (Table 1). The lack of significance could be due to its low effect size or due to difficulty to capture an effect in a particular model. Therefore, it seems to be more important to focus on identified effects than on lack of differences. With this idea in

Table 1 Summary of Studies Analyzing Animal Drinking in Social Conditions

Type	Species	Strain	Sex	Age	Simplified Direction of Effect	References
Total	Mouse	C3H-derived	Male	Adu	ND	Rodgers and Thiessen (1964)
Total	Mouse	129/JxBalb/cJ F1	Male	Adu	Iso > Soc	Brown (1968)
Total	Golden Hamster	N/A	Female	Adu	Iso > Soc	Slighter (1970)
Total	Rat	Long–Evans	Male	Adu	Iso > Soc	Deatherage (1972)
Total	Rat	Charles River albino	Male	Adu	NT	Heminway and Furumoto (1972))
Total	Rat	Long–Evans derived	Both	Adu	Iso > Soc	Kulkosky, Zellner, Hyson, and Riley (1980)
Total	Rat	Sprague-Dawley	Male	Adu	Iso < Soc	Weisinger, Denton, and Osborne (1989)
Total	Rat	Wistar	Male	Ado and Adu	NT	Wolffgramm and Heyne (1991)
Total	Rat	Wistar	Male	Ado and Adu	Iso > Soc	Juárez and Vázquez-Cortés (2003)
Total	Rat	P and NP	Male	Adu	Iso > Soc	Ehlers, Walker, Pian, Roth, and Slawecki (2007)
Total	Prairie vole	N/A	Both	Adu	ND	Anacker et al. (2014)
Mesh	Rat	Long–Evans	Male	Adu	Iso > Soc	Tomie, Lewis, Curiotto, and Pohorecky (2007)
Mesh	Prairie vole	N/A	Both	Adu	Iso > Soc	Anacker, Loftis, and Ryabinin (2011)
Mesh	Prairie vole	N/A	Both	Adu	ND	Hostetler, Anacker, Loftis, and Ryabinin (2012)
Mesh	Prairie vole and Meadow vole	N/A	Both	Adu	Iso < Soc	Anacker, Ahern, Young, and Ryabinin (2012)
Mesh	Rat	Wistar	Male	Ado	ND	Palm and Nylander (2014)
Mesh	Prairie vole	N/A	Both	Adu	Iso > Soc	Anacker and Ryabinin (2013)

Mesh	Mouse	CD-1	Both	Adu	Iso < Soc Iso > Soc	Tomie, DeFuria, Jones, Edwards, and Yu (2014)
Mesh	Mouse	CD-1	Both	Adu	Iso < Soc	Tomie, Samuel, Sprung, Malul, and Yu (2015)
Mesh	Prairie vole	N/A	Male	Adu	Iso < Soc	Walcott and Ryabinin (2017)
Total and mesh	Rat	Wistar	Male	Adu	Iso > Soc Mesh < Soc	Wolffgramm (1990)
Total and mesh	Prairie vole	N/A	Both	Adu	Iso < Soc	Anacker, Loftis, Kaur, and Ryabinin (2011)
Total and mesh	Mouse	CD-1	Both	Adu	Iso < Soc	Tomie et al. (2015)
Total and mesh	Mouse	Several	Both	Ado	Iso > Soc	Panksepp, Rodriguez, and Ryabinin (2017)
Video	Mouse	C57BL/6J	Both	Ado and Adu	Iso < Soc	Logue, Chein, Gould, Holliday, and Steinberg (2014)
Video	Rat	Sprague-Dawley	Both	Ado and Adu	Iso < Soc Iso > Soc	Varlinskaya, Truxell, and Spear (2015)
Radio	Rat	Long–Evans	Male	Adu	Soc > Iso	Pohorecky (2006)
Radio	Rat	Long–Evans	Male	Adu	Iso > Soc	Pohorecky (2008)
Radio	Rat	Long–Evans	Male	Adu	Soc > Iso	Pohorecky (2010)
Radio	Mouse	C57BL/6	Both	Adu	NT	Smutek et al. (2014)
Radio	Mouse	C57BL/6-derived	Female	Adu	NT	Stefaniuk et al. (2017)
Radio	Mouse	C57BL/6NTac	Male	Adu	NT	Thomsen et al. (2017)
Radio and total	Mouse	C57BL/6	Male	Adu	Iso > Soc	Holgate, Garcia, Chatterjee, and Bartlett (2017)

Type in left column indicates the approach to measure drinking (see text). Adu, adult; Ado, adolescent; Iso, isolate housing; Soc, social housing; Mesh, housing in cages separated by mesh. Direction of significant difference (> or <). ND, no difference detected. NT, not tested. We chose to indicate direction of effect if any difference was detected in the study, even if it was detected in only one or two conditions within each study.

mind, both directions of effects have been observed across each of the four types of approaches. In a few instances, both directions have been observed within a single study. It is clear that in each particular case, the direction is dependent on specific conditions, such as species, sex, age, proximity between other animals and alcohol intake levels (ethanol dose) in each study. This dichotomy of effects does not contradict human experiences: peer pressure can lead to excessive drinking, whereas social support can prevent excessive consumption. Thus, for each approach to measure alcohol intake, a model for social facilitation or a model of social inhibition of drinking can be developed. However, understanding of precise mechanisms regulating alcohol intake in such models requires precise monitoring of individual animal's behavior within the settings. Unfortunately, our analysis of existing models identified obvious and hidden pitfalls, making understanding of such individual behavior difficult. Below we will highlight advantages and limitations of each approach.

3. TOTAL INTAKE PER CAGE MODEL

The simplest approach to assessing consumption of alcohol, or any fluid, is by housing several animals per cage and measuring total intake of fluid (Fig. 1A). Intake of individual animals can then be estimated by dividing this total intake by the number of animals per cage. Most studies using this approach detected higher intake in isolated animals versus group-housed rodents. Some of these studies should be taken with caution because they were obtained in rodent strains that consume very small amounts of alcohol. Increased alcohol in isolation in animals consuming low amounts of alcohol could reflect not increased reinforcing effects of alcohol, but decreased avoidance of alcohol. Such models would not be very informative for mechanisms of excessive alcohol use. On the other hand, at least one study performed this analysis simultaneously in selectively-bred high preferring and in low preferring rats, and observed increased alcohol consumption in isolated animals of both lines (Ehlers et al., 2007). Thus, the direction of effect is not dependent on the amount of alcohol consumed.

Isolation is a well-known stressor for rodents. The higher drinking in isolated rodents could model the association of loneliness and depression with excessive alcohol consumption. On the other hand, decreased drinking in social condition could model social support interfering with excessive consumption. However, lower levels of intake in group settings could also simply reflect the fact that group-housed rodents will interfere with each

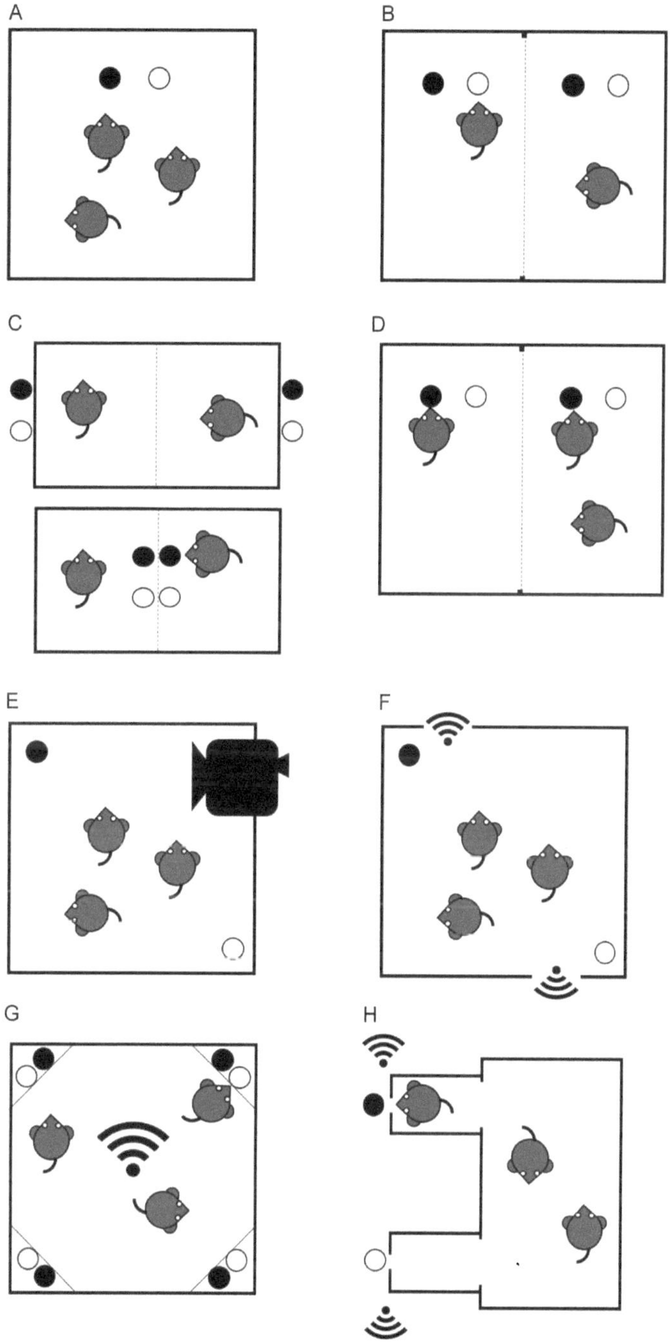

Fig. 1 See legend on next page.

other's drinking. Theoretically, this should not affect high and low drinking rodents equally as there should be less interference in low drinking lines. Nevertheless, the interference effects could be quite complex as rodents exhibit complex social hierarchies and they could be affecting drinking in a yet difficult to predict manner. The difference in intake between isolated versus socially housed animals could be due to differences in consumption in only some of the individuals within the socially-housed group. The discussed approach does not allow to assess drinking of individual rodents within the socially housed group.

One additional, at first glance simple, concern across practically all drinking studies in rodents is the difficulty in estimating the leaking of the bottles. It is customary and advisable to have an additional cage with no animals during alcohol consumption studies to assess whether shaking of the housing rack or evaporation interfered with measures of consumption. Ball bearings placed in the drinking spouts can decrease the general leaking from the bottle but are not obstacles if this leaking is performed intentionally by the animal. This consideration seems to be even more important for the discussed studies as either social or individual housing could affect measurements of intake differently (for example, when animals playing with the spouts or hitting them during social interactions). This behavior could be also different for different fluids. For example, animals could be interacting with the ethanol-containing fluid without drinking depending on whether they are housed alone or being distracted from this interaction by a cagemate.

A theoretical solution to the problem with leaking bottles would be to confirm alcohol intakes by blood alcohol levels. However, alcohol elimination rates are fast in smaller rodents, and assessing blood alcohol levels typically does not provide sufficient quantification sensitivity: it allows confirmation that some alcohol consumption has occurred but does not allow to distinguish smaller differences between groups.

While the majority of studies analyzing total intake found lower intake in group-housed rodents versus isolated animals, one study showed an opposite

Fig. 1 Schematic illustrations for various approaches to measure alcohol drinking during social housing in rodents. (A) Total intake per cage paradigm. (B) Mesh-divided cages. (C) Asocial (*top*) and social (*bottom*) placement of drinking cylinders in mesh-divided cages. (D) Combination of the total intake and mesh-divided cages approach. (E) Video-tracking of socially-housed animals. (F) Radio-frequency tracking of socially-housed animals. (G) Intellicage. (H) HM-2 system. *Black circles* illustrate alcohol-containing drinking vessels. *Open circles* illustrate water-containing drinking vessels. Cages are drawn not to scale.

effect (Weisinger et al., 1989). Thus, it is possible to observe increased intake in socially housed animals, suggesting that interference cannot explain decreased drinking in socially housed animals in the majority of studies. Increased drinking is reminiscent of peer pressure effects facilitating excessive alcohol consumption. This study was performed in a within-subjects design: one type of housing followed another. Thus, drinking in one set of conditions could affect drinking in subsequent conditions. In addition, it appears to be even more important to compare the changes in individual members of the group rather than total intake to estimate whether the different changes occur in different individuals (for example, in dominant versus subordinate animals). This is why a large number of experimenters used mesh housing to assess individual drinking in socially housed rodents.

4. MESH-SEPARATED COMPARTMENTS MODEL

In this model, a cage is divided into compartments (most frequently in two) separated by a mesh or perforated plastic wall (Fig. 1B). An individual animal is placed in each compartment. Each compartment is provided with its own food and drinking vessels. Therefore, there is no interference or competition between animals in their access for alcohol, water and food, and individual consumption can be assessed. On the other hand, the partition does not affect visual, auditory and olfactory communication between the animals. It is clear that the interactions between the animals are not completely natural (for example, a threat of attack from the other animal is never realized). Nevertheless, these conditions are more social than isolate housing.

Interestingly, the view that individual housing is completely asocial has also been recently challenged by studies showing that animals can communicate their state of anxiety and pain via olfactory signals (Laviola et al., 2017; Smith, Hostetler, Heinricher, & Ryabinin, 2016; Smith, Walcott, Heinricher, & Ryabinin, 2017; Walcott, Smith, Loftis, & Ryabinin, 2018). Such olfactory signals can regulate behavior between separate cages within the room. Thus, even individual housing cannot be considered completely asocial, unless each cage is housed with each own ventilated system, a difficult to achieve condition. Regardless, of this complication, communication between animals separated by a mesh is more complete than between animals sitting in different cages. Therefore, in most publications, mesh separation is referred to as social housing as opposed to isolate housing in separate cages.

Summarizing literature using this approach, we observed that researchers detected more varied outcomes in the directions of difference in alcohol consumption between mesh-separated rodents and isolate-housed animals. Again, different directions have been observed within each species. Moreover, both directions of differences have been observed within the same study using a single strain of mice, depending on sex and social housing conditions (Tomie et al., 2014). Even further, studies in prairie voles show that individual animals within the same sex and housing condition can exhibit different direction of effects on alcohol drinking when paired with another animal (Anacker, Loftis, & Ryabinin, 2011). Specifically, while the majority of prairie voles exhibiting high drinking in isolation decreased their drinking when housed with low drinking voles, some low drinking voles increased their drinking (Anacker, Loftis, & Ryabinin, 2011; Anacker & Ryabinin, 2013). Thus, similar to studies measuring total consumption per cage, both directions of effects could model varied effects of social conditions on alcohol abuse in humans.

Despite these encouraging observations giving credit to attempts to model the varied human conditions, studies using mesh housing can also be hard to interpret. It is obvious that the interactions between animals separated by a mesh are incomplete and, therefore, not completely natural. There is an opinion that these conditions are still less stressful than isolate conditions. For example, for prairie voles, which are extremely sensitive to social isolation, it is typical that recovery from surgery is performed in mesh-divided cages to reduce the stress of social isolation and improve healing. In agreement with this idea, unpublished studies in the Ryabinin laboratory showed that prairie voles are highly interactive if placed together after being housed singly in separate cages, but do not appear to show increased interactions if placed together after being housed in mesh-divided cages. This observation suggests that mesh separation does not present a significant obstacle for social interaction. However, more recent studies show that prairie voles are capable of consolation behavior, and that grooming of the stressed vole by its partner reduces stress in prairie voles (Burkett et al., 2016; Smith & Wang, 2014). Housing animals in mesh divided cages prevents such behavior and therefore can be considered stressful. Therefore, an effect of housing in mesh-divided cages on drinking can occur not because of an influence on drinking by its "neighbor," but because of increased stress. In fact, even though mice have not been yet demonstrated to show consolation behavior, some studies suggest higher levels of stress in mesh divided cages than in isolate animals (Rettich, Käsermann, Pelczar, Bürki, & Arras, 2006; Van Loo et al., 2007).

To avoid this complication, one study using prairie voles and meadow vole modified the mesh-divided cages such that the drinking cylinders were placed on either the mesh side, facing the other animal (social condition), or on the opposite side of the cage (asocial condition) (Anacker et al., 2012). Both prairie and meadow voles demonstrated higher intake of alcohol in the social condition than in the asocial condition, a finding suggesting facilitation of drinking in the social context (Fig. 1C). However, the effect was highly significant for alcohol intake, but did not reach statistical significance for alcohol preference, and there was also significantly higher total fluid intake (alcohol and water) in the social condition versus asocial condition. Therefore, one cannot exclude that the apparent greater alcohol consumption on the "social" side of the cage is partly due to higher likelihood of spillage of fluids on this side. Thus, this modification of mesh housing has its own caveats.

Taking together these studies, assessing alcohol consumption in mesh-divided cages allows better control of alcohol intake by individual animals within group, but is complicated by other methodological difficulties.

5. PARALLEL USE OF TOTAL INTAKE MODEL AND MESH DIVIDED COMPARTMENTS

Trying to reconcile the two models above, some researchers have used both models at the same time. The idea is that the conclusion about the interaction of social environment and alcohol consumption would be stronger if both models result in the same effect.

In some of these studies, similar effects were observed between the two models. Thus, prairie voles housed in pairs, regardless whether they were separated by a mesh or not, showed higher alcohol preference than voles housed in isolation (Anacker, Loftis, Kaur, et al., 2011). On the other hand, isolated adolescent FVB/NJ mice consumed more alcohol or saccharin-supplemented alcohol than pair-housed mice and showed higher rates of licking at the alcohol spout than when they were housed separated by mesh. In support of the same mechanisms involved in increased drinking and licking in isolated animals, the increased licking in this study was observed at a particular time of the day, but not throughout the day, when differences in blood alcohol levels were detected between mesh-separated and isolate-housed mice (Panksepp et al., 2017).

In contrast, studies in CD-1 mice using a setup combining both approaches (Fig. 1D) show that the effects are highly dependent on what sex and what particular configuration of housing was used (Tomie et al., 2015). Moreover, the very first, to our knowledge, study using both models showed contrasting effects of the two housing conditions on alcohol consumption: male Sprague-Dawley rats consumed more alcohol when isolate-housed versus pair-housed rats, but pair-housed rats consumed more alcohol on average versus mesh-separated rats (Wolffgramm, 1990).

It needs to be noted that the number of studies combining using both housing methods is low, and they involved different species, strains, age of animals, alcohol drinking paradigms and very different levels of total alcohol intake, making them difficult to compare. However, the observation that both models do not always agree in the direction of effect indeed suggests that additional factors, besides social contact, regulate alcohol differently between both models, and that caution needs to be exerted when interpreting results from both types of approaches.

6. VIDEO TRACKING

An alternative approach to separating animals by a partition is to house them socially but mark the animals and track individual consumption using video equipment (Fig. 1E). Total amount of fluid consumed is measured and then divided by the time each individual animal spends at the spout. Two studies to date have used this approach. Logue et al. (2014) observed that male adolescent, but not female or adult, C57BL/6J mice consume more alcohol when housed in triads versus when isolate-housed. Varlinskaya et al. (2015) observed that adolescent and adult male Sprague-Dawley rats consumed more sweetener-supplemented alcohol when they were housed in groups versus when they were housed in isolation, but adolescent female rats consumed more alcohol in isolation versus group housing. Thus, these two studies highlighted the complex interactions between social environment and alcohol consumption across sex and age of animals, an observation in agreement with complex effects of social environment on alcohol drinking in humans.

Giving credit to these studies, however, we cannot avoid noting the potential complications in the methods. Thus, these studies are based on the assumption that time spent at the spout reflects the amount of alcohol consumed. This assumption has not been experimentally tested and it is possible that in some conditions spending more time at the spout does not equate with increased drinking. It is also not clear whether the videos can always allow to clearly distinguish between two individuals at the same spout. Finally, one cannot also assess the potential higher leaking of fluid from the spouts in one condition versus the other.

Despite these caveats, this approach is clearly in the right direction. More studies like these two are needed. However, an additional problem with this approach is that this method is labor intensive and, therefore, expensive. It takes time to carefully evaluate animal's behavior from video recordings, especially when they are group-housed. Theoretically, analysis of video recordings can be automated. It should be noted, however, that despite the recent expansion of computerized analysis of animal behavior, including social behavior, no existing commercial package allows automated tracking of consummatory behaviors in social settings. We hope that this is only an issue of time. In the meantime, individual analysis of such behavior recently became possible using radio frequency tracking.

7. RADIO FREQUENCY TRACKING

In this approach, animals are surgically implanted with a microchip, which allows tracking the individual animal's location using radio frequency identification. The identification of the animal allows to match its location with a measurement of fluid consumption.

To our knowledge, this approach in the alcohol research field has been pioneered by Dr. Larissa Pohorecky. In a series of studies, she analyzed effects of dominance and stress in isolate- and triad-housed male Long–Evans rats (Pohorecky, 2006, 2008, 2010). The animals were "microchipped" and their proximity to the drinking spout was analyzed (Fig. 1F). The total intake used to assess individual alcohol consumption was measured by assessing changes in fluid tension. Alcohol drinking was assessed in basal conditions and following exposure to several types of stresses. These studies detected complex interactions between social hierarchy status and type of stressor in their effects on alcohol consumption. Depending on these factors, social environment could either promote or decrease alcohol intake.

Despite the clear technological breakthroughs achieved by these initial studies, they had also many caveats. First, it is not clear whether the spatial resolution is sufficient to distinguish between two animals when they were both are in proximity of the spout. Second, the intakes in these studies were very low questioning whether these studies measured reinforcing properties or simply avoidance of alcohol solutions. Third, animals were housed in cages with wire-mesh bottoms. These are very stressful housing conditions, which by themselves could affect animal behavior. Fourth, effects of social condition on leaking of the bottles could not be estimated. Nevertheless, these pioneering studies provided an excellent example for future advancement of this approach.

A relatively recent development to this approach was due to the development and commercial availability of the so-called Intellicage (NewBehavior AG). This highly sophisticated computerized housing system allows individual tracking of animals across a relatively large enclosure where access to up to eight drinking or feeding spouts can be individually programmed and the whole system can be expanded with additional housing and interaction enclosures (Fig. 1G). This system has only begun to be used in alcohol research. An advantage of the system is also that a large number of individuals can be tested simultaneously in one cage. One theoretical concern with this system is that an enriched environment provided by such housing conditions decreases the intake of alcohol, and the resulting low level of intake would be not informative for understanding excessive alcohol drinking. Indeed, such low intake has been observed under basal conditions in two of the three studies (Holgate et al., 2017; Stefaniuk et al., 2017). In fact, intake of alcohol is also difficult to assess in studies using Intellicage because individual drinking is evaluated using a lickometer system. Number of licks is known to correlate with fluid intake, but the existence of such correlation does not preclude the possibility that the number of licks varies depending on the context of consummatory behavior. Nevertheless, intermittent access to alcohol clearly enhanced alcohol consumption in these cages and thus should allow to model excessive alcohol consumption (Smutek et al., 2014).

One recent study using the Intellicage showed dramatically lower basal intake of alcohol in group-housed C57BL/6 mice compared to single housing (Holgate et al., 2017). However, the housing conditions were different between these conditions because Intellicage can be considered an enriched environment. To partially overcome this problem an additional group of mice was tested in parallel, in which housing was social but in the same type of cages as isolated animals. These group-housed mice showed intermediate levels of alcohol consumption between high intake in isolate-housed mice and low level in Intellicage group-housed. This result suggests that it is the social condition that contributes to regulation of alcohol drinking. However, the correct control for this effect would be to use the Intellicage with single-housed animals. This important control is difficult to perform because this housing system is very expensive and most laboratories cannot afford to have several Intellicage systems. The current studies have not yet used the advanced features of this system allowing analysis of complex behaviors, for example social hierarchies, in group housed-animals. Hopefully this system will be more available in the future allowing to study such interactions. However, even with this perspective, the difficulty to take into account potential leaking of bottles and estimate intake of each fluid casts doubts that this system will be the final word in analysis of drinking in social conditions.

The most recent development in this field is the so-called HM-2 system (MBRose; Fig. 1H). This system has been originally developed for obesity research. In the standard system, a community cage contains two channels, one connected to a food source and another connected to the fluid source. In the version of the system modified for studies of alcohol consumption, both channels are connected to fluid sources. One channel can be used for an ethanol solution and the other—for water, thereby allowing two-bottle choice studies. Both fluid sources are placed on a high precision balance. The balance registers the displacement of weight only when the animal enters the channel, and the channel only allows entry of a single animal. There is also a tray under the spout collecting the spilled fluid, and weight is tared when a photoelement indicates that the animal is not next to the channel. Thus, this is the first system that takes into account spillage of fluid.

Only one published study used this system for alcohol research to date (Thomsen et al., 2017). This study investigated the effects of a glucagon-like peptide 1 receptor agonist on alcohol consumption in male socially housed C57BL/6 mice. As could be feared based on previous studies in socially housed mice showing lower alcohol consumption, the baseline intake of alcohol in these conditions was low, as was alcohol preference. However, during reintroduction of alcohol following alcohol deprivation the intake and preference dramatically increased, and the researchers were able to detect an effect of the agonist. The advanced features of the system allowed the researchers to demonstrate that the effect of the agonist was primarily due to decreased drinking bout frequency and not bout size (Thomsen et al., 2017).

The HM-2 system has not yet been tested to evaluate effects of social housing. The main obstacle for such a study is that it would be relatively expensive to use each cage to investigate isolate-housed animals. As only a single study using HM-2 was performed so far, several questions still remain unanswered: how channels allowing entry of a single animal at a time affect alcohol consumption and whether some individual animals within the social group could occupy and prevent other animals from entering the channel. Since high intakes have been obtained in this system, these problems are unlikely but they still need to be experimentally addressed. Taken together, however, this system is an innovative way to tackle the social aspects of alcohol consumption.

8. CONCLUSIONS

Overall, it is clear that most researchers analyzing alcohol consumption in socially housed animals realized from the beginning the important complications that can arise in the interpretation of such complex behaviors. Different researchers chose to ignore one or the other complication in favor of their chosen approach. With the development of sophisticated modern technologies of computerized image analysis and sensitivity of radio frequency detection, an optimal solution to tracking consumption in individual animals within group settings will be found. This is not an easy task as the optimal solution should allow to integrate the analysis of consummatory behaviors with analysis of the entire behavioral repertoire of the animal in these settings. We hope that these advances will be achieved in the near future.

Our review primarily focused on the caveats of technical details involved in analysis of drinking in socially housed animals because these details still prevent mechanistic understanding of these behaviors. Therefore, the overview of results was relatively simple: mostly stating whether alcohol intake was higher in isolate or in social conditions. The interactions between social environment and alcohol's effects are much more complex than increased or decreased consumption. However, understanding these complex interactions is impossible without taking care of the multitude of highlighted technical complications. As appropriate models and technologies are developed we will be able to assess the mechanisms of these interactions. Understanding these mechanisms will contribute to development of rational pharmacotherapies, and prevention and treatment strategies for alcohol use disorder.

REFERENCES

Anacker, A. M. J., Ahern, T. H., Hostetler, C. M., Dufour, B. D., Smith, M. L., Cocking, D. L., et al. (2014). Drinking alcohol has sex-dependent effects on pair bond formation in prairie voles. *Proceedings of the National Academy of Sciences of the United States of America*, *111*(16), 6052–6057. https://doi.org/10.1073/pnas.1320879111.

Anacker, A. M. J., Ahern, T. H., Young, L. J., & Ryabinin, A. E. (2012). The role of early life experience and species differences in alcohol intake in microtine rodents. *PLoS One*, 7(6), e39753. https://doi.org/10.1371/journal.pone.0039753.

Anacker, A. M. J., Loftis, J. M., Kaur, S., & Ryabinin, A. E. (2011). Prairie voles as a novel model of socially facilitated excessive drinking. *Addictive Behaviors, 16*(1), 92–107. https://doi.org/10.1111/j.1369-1600.2010.00234.x.

Anacker, A. M. J., Loftis, J. M., & Ryabinin, A. E. (2011). Alcohol intake in prairie voles is influenced by the drinking level of a peer. *Alcoholism: Clinical and Experimental Research, 35*(10), 1884–1890. https://doi.org/10.1111/j.1530-0277.2011.01533.x.

Anacker, A. M. J., & Ryabinin, A. E. (2013). Identification of subpopulations of prairie voles differentially susceptible to peer influence to decrease high alcohol intake. *Frontiers in Pharmacology, 4,* 1–12. https://doi.org/10.3389/fphar.2013.00084.

Auyeung, B., Lombardo, M. V., Heinrichs, M., Chakrabarti, B., Sule, A., Deakin, J. B., et al. (2015). Oxytocin increases eye contact during a real-time, naturalistic social interaction in males with and without autism. *Translational Psychiatry, 5*(2), e507. https://doi.org/10.1038/tp.2014.146.

Brown, R. V. (1968). Effects of stress on voluntary alcohol consumption in mice. *Quarterly Journal of Studies on Alcohol, 29,* 49–53.

Burkett, J. P., Andari, E., Johnson, Z. V., Curry, D. C., de Waal, F. B. M., & Young, L. J. (2016). Oxytocin-dependent consolation behavior in rodents. *Science (New York, N.Y.), 351*(6271), 375–378. https://doi.org/10.1126/science.aac4785.

Chen, F. S., Barth, M. E., Johnson, S. L., Gotlib, I. H., & Johnson, S. C. (2011). Oxytocin receptor (OXTR) polymorphisms and attachment in human infants. *Frontiers in Psychology, 2,* 200. https://doi.org/10.3389/fpsyg.2011.00200.

Deatherage, G. (1972). Effects of housing density on alcohol intake in the rat. *Physiology & Behavior, 9*(1), 55–57.

Ehlers, C. L., Walker, B. M., Pian, J. P., Roth, J. L., & Slawecki, C. J. (2007). Increased alcohol drinking in isolate-housed alcohol-preferring rats. *Behavioral Neuroscience, 121*(1), 111–119. https://doi.org/10.1037/0735-7044.121.1.111.

Heminway, D. A., & Furumoto, L. (1972). Population density and alcohol consumption in the rat. *Quarterly Journal of Studies on Alcohol, 33*(3), 794–799.

Holgate, J. Y., Garcia, H., Chatterjee, S., & Bartlett, S. E. (2017). Social and environmental enrichment has different effects on ethanol and sucrose consumption in mice. *Brain and Behavior,* 7(8), e00767. https://doi.org/10.1002/brb3.767.

Hostetler, C. M., Anacker, A. M. J., Loftis, J. M., & Ryabinin, A. E. (2012). Social housing and alcohol drinking in male-female pairs of prairie voles (Microtus ochrogaster). *Psychopharmacology, 224*(1), 121–132. https://doi.org/10.1007/s00213-012-2836-4.

Insel, T. R., & Hulihan, T. J. (1995). A gender-specific mechanism for pair bonding: Oxytocin and partner preference formation in monogamous voles. *Behavioral Neuroscience, 109*(4), 782–789.

Juárez, J., & Vázquez-Cortés, C. (2003). Alcohol intake in social housing and in isolation before puberty and its effects on voluntary alcohol consumption in adulthood. *Developmental Psychobiology, 43*(3), 200–207. https://doi.org/10.1002/dev.10133.

Kulkosky, P. J., Zellner, D. A., Hyson, R. L., & Riley, A. L. (1980). Ethanol consumption of rats in individual, group, and colonial housing conditions. *Physiological Psychology, 8*(1), 56–60.

Laviola, G., Zoratto, F., Ingiosi, D., Carito, V., Huzard, D., Fiore, M., et al. (2017). Low empathy-like behaviour in male mice associates with impaired sociability, emotional memory, physiological stress reactivity and variations in neurobiological regulations. *PLoS One, 12*(12), e0188907. https://doi.org/10.1371/journal.pone.0188907.

Lim, M. M., & Young, L. J. (2004). Vasopressin-dependent neural circuits underlying pair bond formation in the monogamous prairie vole. *Neuroscience, 125*(1), 35–45. https://doi.org/10.1016/j.neuroscience.2003.12.008.

Liu, Y., Curtis, J. T., & Wang, Z. (2001). Vasopressin in the lateral septum regulates pair bond formation in male prairie voles (Microtus ochrogaster). *Behavioral Neuroscience, 115*(4), 910–919.

Logue, S., Chein, J., Gould, T., Holliday, E., & Steinberg, L. (2014). Adolescent mice, unlike adults, consume more alcohol in the presence of peers than alone. *Developmental Science, 17*(1), 79–85. https://doi.org/10.1111/desc.12101.

National Institute on Alcohol Abuse and Alcoholism. (2017). *Alcohol Facts and Statistics.* Retrieved April 25, 2018, from https://www.niaaa.nih.gov/alcohol-health/overview-alcohol-consumption/alcohol-facts-and-statistics.

Palm, S., & Nylander, I. (2014). Alcohol-induced changes in opioid peptide levels in adolescent rats are dependent on housing conditions. *Alcoholism: Clinical and Experimental Research, 38*(12), 2978–2987. https://doi.org/10.1111/acer.12586.

Panksepp, J. B., Rodriguez, E. D., & Ryabinin, A. E. (2017). Sweetened ethanol drinking during social isolation: Enhanced intake, resistance to genetic heterogeneity and the emergence of a distinctive drinking pattern in adolescent mice. *Genes, Brain, and Behavior, 16*(3), 369–383. https://doi.org/10.1111/gbb.12346.

Pitkow, L. J., Sharer, C. A., Ren, X., Insel, T. R., Terwilliger, E. F., & Young, L. J. (2001). Facilitation of affiliation and pair-bond formation by vasopressin receptor gene transfer into the ventral forebrain of a monogamous vole. *The Journal of Neuroscience: The Official Journal of the Society for Neuroscience, 21*(18), 7392–7396.

Pohorecky, L. A. (2006). Housing and rank status of male Long-Evans rats modify ethanol's effect on open-field behaviors. *Psychopharmacology, 185*(3), 289–297. https://doi.org/10.1007/s00213-005-0257-3.

Pohorecky, L. A. (2008). Psychosocial stress and chronic ethanol ingestion in male rats: Effects on elevated plus maze behavior and ultrasonic vocalizations. *Physiology & Behavior, 94*(3), 432–447. https://doi.org/10.1016/j.physbeh.2008.02.010.

Pohorecky, L. A. (2010). Acute novel stressors modify ethanol intake of psychosocially stressed rats. *Pharmacology, Biochemistry, and Behavior, 95*(4), 390–400. https://doi.org/10.1016/j.pbb.2010.02.017.

Rettich, A., Käsermann, H. P., Pelczar, P., Bürki, K., & Arras, M. (2006). The physiological and behavioral impact of sensory contact among unfamiliar adult mice in the laboratory. *Journal of Applied Animal Welfare Science: JAAWS, 9*(4), 277–288. https://doi.org/10.1207/s15327604jaws0904_3.

Rodgers, D. A., & Thiessen, D. D. (1964). Effects of population density on adrenal size, behavioral arousal, and alcohol preference of inbred mice. *Quarterly Journal of Studies on Alcohol, 25*, 240–247.

Slighter, R. G. (1970). Alcohol selection and position selection in hamsters caged singly and in groups. *Quarterly Journal of Studies on Alcohol, 31*(1), 20–27.

Smith, M. L., Hostetler, C. M., Heinricher, M. M., & Ryabinin, A. E. (2016). Social transfer of pain in mice. *Science Advances, 2*(10), e1600855. https://doi.org/10.1126/sciadv.1600855.

Smith, M. L., Walcott, A. T., Heinricher, M. M., & Ryabinin, A. E. (2017). Anterior cingulate cortex contributes to alcohol withdrawal-induced and socially transferred hyperalgesia. *eNeuro, 4*(4). ENEURO.0087–17.2017. https://doi.org/10.1523/ENEURO.0087-17.2017.

Smith, A. S., & Wang, Z. (2014). Hypothalamic oxytocin mediates social buffering of the stress response. *Biological Psychiatry, 76*(4), 281–288. https://doi.org/10.1016/j.biopsych.2013.09.017.

Smutek, M., Turbasa, M., Sikora, M., Piechota, M., Zajdel, J., Przewlocki, R., et al. (2014). A model of alcohol drinking under an intermittent access schedule using group-housed mice. *PLoS One, 9*(5), e96787. https://doi.org/10.1371/journal.pone.0096787.

Stefaniuk, M., Beroun, A., Lebitko, T., Markina, O., Leski, S., Meyza, K., et al. (2017). Matrix metalloproteinase-9 and synaptic plasticity in the central amygdala in control of alcohol-seeking behavior. *Biological Psychiatry, 81*(11), 907–917. https://doi.org/10.1016/j.biopsych.2016.12.026.

Thomsen, M., Dencker, D., Wörtwein, G., Weikop, P., Egecioglu, E., Jerlhag, E., et al. (2017). The glucagon-like peptide 1 receptor agonist Exendin-4 decreases relapse-like drinking in socially housed mice. *Pharmacology, Biochemistry, and Behavior, 160*, 14–20. https://doi.org/10.1016/j.pbb.2017.07.014.

Tomie, A., DeFuria, A. A., Jones, H. A., Edwards, S. D., & Yu, L. (2014). Effects of Cagemate Gender and the Cagemate's access to ethanol on ethanol and water intake of the proximal male or the proximal female CD-1 mouse. *Alcohol (Fayetteville, N.Y.), 48*(1), 73–82. https://doi.org/10.1016/j.alcohol.2013.10.006.

Tomie, A., Lewis, K., Curiotto, J., & Pohorecky, L. A. (2007). Intermittent exposure to a social stimulus enhances ethanol drinking in rats. *Pharmacology, Biochemistry, and Behavior, 87*(3), 341–348. https://doi.org/10.1016/j.pbb.2007.05.008.

Tomie, A., Samuel, A. G., Sprung, D. M., Malul, Y., & Yu, L. (2015). Effects of number of cagemates on home cage ethanol drinking during proximal cagemate drinking (PCD) procedures in male and female CD-1 mice. *Progress in Neuro-Psychopharmacology & Biological Psychiatry, 57*, 1–10. https://doi.org/10.1016/j.pnpbp.2014.10.001.

Van Loo, P. L. P., Kuin, N., Sommer, R., Avsaroglu, H., Pham, T., & Baumans, V. (2007). Impact of "living apart together" on postoperative recovery of mice compared with social and individual housing. *Laboratory Animals, 41*(4), 441–455. https://doi.org/10.1258/002367707782314328.

Varlinskaya, E. I., Truxell, E. M., & Spear, L. P. (2015). Ethanol intake under social circumstances or alone in Sprague-Dawley rats: Impact of age, sex, social activity, and social anxiety-like behavior. *Alcoholism: Clinical and Experimental Research, 39*(1), 117–125. https://doi.org/10.1111/acer.12604.

Walcott, A. T., & Ryabinin, A. E. (2017). Alcohol's effects on pair-bond maintenance in male prairie voles. *Frontiers in Psychiatry, 8*, 226. https://doi.org/10.3389/fpsyt.2017.00226.

Walcott, A. T., Smith, M. L., Loftis, J. M., & Ryabinin, A. E. (2018). Social transfer of alcohol withdrawal-induced hyperalgesia in female prairie voles. *Social Neuroscience, 105*(6), 1–8. https://doi.org/10.1080/17470919.2018.1456957.

Walum, H., Lichtenstein, P., Neiderhiser, J. M., Reiss, D., Ganiban, J. M., Spotts, E. L., et al. (2012). Variation in the oxytocin receptor gene is associated with pair-bonding and social behavior. *Biological Psychiatry, 71*(5), 419–426. https://doi.org/10.1016/j.biopsych.2011.09.002.

Walum, H., Westberg, L., Henningsson, S., Neiderhiser, J. M., Reiss, D., Igl, W., et al. (2008). Genetic variation in the vasopressin receptor 1a gene (AVPR1A) associates with pair-bonding behavior in humans. *Proceedings of the National Academy of Sciences of the United States of America, 105*(37), 14153–14156. https://doi.org/10.1073/pnas.0803081105.

Weisinger, R. S., Denton, D. A., & Osborne, P. G. (1989). Voluntary ethanol intake of individually- or pair-housed rats: Effect of ACTH or dexamethasone treatment. *Pharmacology, Biochemistry, and Behavior, 33*(2), 335–341.

Winslow, J. T., Hastings, N., Carter, C. S., Harbaugh, C. R., & Insel, T. R. (1993). A role for central vasopressin in pair bonding in monogamous prairie voles. *Nature, 365*(6446), 545–548. https://doi.org/10.1038/365545a0.

Wolffgramm, J. (1990). Free choice ethanol intake of laboratory rats under different social conditions. *Psychopharmacology, 101*(2), 233–239.

Wolffgramm, J., & Heyne, A. (1991). Social behavior, dominance, and social deprivation of rats determine drug choice. *Pharmacology, Biochemistry, and Behavior, 38*(2), 389–399.

CHAPTER THREE

Social Factors in Ethanol Sensitization

Rosana Camarini[1], Priscila Marianno, Mariana Rae
Department of Pharmacology, Institute of Biomedical Sciences, University of São Paulo, SP, Brazil
[1]Corresponding author: e-mail address: camarini@icb.usp.br

Contents

Abstract

Behavioral sensitization is a neuroadaptive process characterized by an increase in a particular behavior after repeated exposure to drugs or other stimuli, such as stress. Sensitization can also be extended to neurochemical and neuroendocrine sensitization. Several factors can influence sensitization to the effects of ethanol. For instance, stress is an important component in addiction that can strengthen ethanol-induced behaviors. In animal models, stressful situations can be induced by alterations in social aspects of the animal's environment, such as maternal separation, social conflicts, and housing conditions. Social conflict models involve acute, chronic or intermittent interaction of an animal to a conspecific and can occur at any stage of life, including preweaning, adolescence or adulthood. These events can influence ethanol-induced behavioral sensitization in different ways, such as increases in locomotion, drug reward, and drug-taking behaviors. On the other hand, environmental enrichment can produce a protective phenotype against drug-related behaviors. In this chapter, we discuss findings regarding consequences of social stress and environmental enrichment on sensitization to ethanol.

International Review of Neurobiology, Volume 140
ISSN 0074-7742
https://doi.org/10.1016/bs.irn.2018.07.003

1. INTRODUCTION

Drug addiction is a multi-dimensional disorder, which comprises complex neuromodulation involved with different stages of this disease (Koob & Volkow, 2010). Due to its complexity, developing animal models that incorporate most addiction characteristics in humans is problematic. Behavioral studies of addiction aim to explore components of this disorder, such as reward, reinforcement, relapse, aversion, drug-seeking/taking, impulsivity, tolerance and sensitization (Sanchis-Segura & Spanagel, 2006).

Behavioral sensitization is a neuroadaptive process characterized by an increase in a particular behavior after repeated administration of a drug, or other stimulus such as stress. Pivotal research has demonstrated that ventral tegmental area (VTA) in the midbrain, nucleus accumbens (NAcc) and medial prefrontal cortex (mPFC) are critical brain regions in this phenomenon. Specifically, dopaminergic projections from the VTA and glutamatergic projections from the mPFC to the NAcc are involved in the initiation or development of behavioral sensitization, while the NAcc is an essential neural substrate for its expression (Steketee & Kalivas, 2011). It is clear, though, that other neurotransmitters (e.g., GABA) and other brain regions (e.g., hippocampus, amygdala) can also mediate this neuroplastic event. Initiation or development of behavioral sensitization refers to the initial episodes of this process, while expression involves long-lasting neural changes that emerge as a consequence of induction, even after a long period of drug withdrawal or abstinence.

In general, locomotor activity has received most attention in behavioral sensitization studies, but sensitization can also occur in other behaviors such as conditioned place preference (CPP), drug-seeking, and drug-taking. CPP is a paradigm used to study drugs or social rewarding and aversive effects. Nonetheless, sensitization can also include neurochemical and neuroendocrine sensitization, in terms of increases in drug-induced neurotransmitter release or secretion of adrenocorticotropic hormone or corticosterone, respectively.

2. FACTORS INFLUENCING ETHANOL-INDUCED BEHAVIORAL SENSITIZATION

Locomotor sensitization in response to psychostimulants and morphine was first demonstrated in the 1970s (Shuster, Webster, & Yu, 1975; Wallach & Gershon, 1971), and it was first described for ethanol in 1980 by Masur and Boerngen (1980). In this study, female Swiss mice were treated with ethanol at different doses [0, 1.0, 2.5, or 3.5 g/kg; intraperitoneally (i.p.)] and locomotor activity was immediately assessed for 60 min on 30 consecutive days. A gradual increase in locomotor activity was observed after repeated administration of the 2.5 and 3.5 g/kg doses, indicating that ethanol, albeit not a pure psychostimulant, can induce locomotor behavioral sensitization. Interestingly, this pivotal study was conducted in female mice, which has subsequently been shown to be more sensitive to ethanol-induced behavioral sensitization than males (Masur & Boerngen, 1980). Later on, Masur, Oliveira de Souza, and Zwicker (1986) demonstrated the low sensitivity of rats to develop ethanol sensitization.

In 1993, Robinson and Berridge proposed the incentive sensitization theory of addiction, which was based on a gradual sensitization of the activity of motivational circuits in the brain, causing an incentive salience or "drug wanting" (Robinson & Berridge, 1993). A learned (Pavlovian) component added to this theory predicts that not only the drug but also drug-associated cues become hyper salient (Robinson & Berridge, 2000). This phenomenon could explain why drug-context pairing induces a more reliable sensitization than non-conditioned sensitization. Considering this important facet in behavioral sensitization, other tests that require drug-cue associations, such as CPP, have been used to complement the understanding of underlying mechanisms of cue-induced drug responses.

A causal relationship between behavioral sensitization to ethanol and subsequent greater alcohol consumption is still under study; however, there is evidence of an overlap in the neurocircuitry of both sensitization and relapse (Steketee & Kalivas, 2011). Specifically, C57BL/6J (B6) and DBA/2J (D2) mice exhibit marked differences in response to ethanol. While B6 mice are recognized for their preference for ethanol consumption and reduced tendency to develop ethanol sensitization, D2 mice show opposite characteristics toward ethanol effects (Camarini & Hodge, 2004; Phillips, Dickinson, & Burkhart-Kasch, 1994). Associations between sensitization and subsequent ethanol intake have been investigated in these two inbred mice. For example, B6 mice consumed more ethanol after behavioral sensitization compared with non-sensitized mice (Camarini & Hodge, 2004; Lessov, Palmer, Quick, & Phillips, 2001). Moreover, after induction of behavioral sensitization, D2 mice consumed more ethanol than animals pretreated with acute saline or with ethanol (Camarini & Hodge, 2004). Interestingly, with regard to social factors, D2 mice show lower sociability levels and lower social novelty preference than B6 mice (Moy et al., 2008).

Several factors can influence behavioral sensitization to ethanol (for a review, see Camarini & Pautassi, 2016). Some of these factors are related to the experimental protocol, such as dose and period of treatment. Other factors include the choice of the experimental subject, such as species, strain, age and sex. Another very important factor includes environmental influences, with emphasis on stressful stimuli.

Adolescence corresponds to a transition phase between childhood and adulthood with marked widespread neuronal changes and hormonal organization that characterize a shift to an emotional, social and cognitive refinement, important for the social development of an adult individual. These changes place adolescence in a period particularly sensitive to the consequences of all types of stressors, with focus on stresses of social origin such as bullying or isolation stress, especially considering that adolescents spend a significant amount of time with their peers instead of adults. In animal studies, a well-established social stress model is the social defeat (SD) model. In general, adolescents are more sensitive to stressful stimuli, and secrete higher levels of glucocorticoids in response to stressors (McCormick, Robarts, Kopeikina, & Kelsey, 2005; Romeo, 2010), yet they are less sensitive to ethanol locomotor sensitization as compared to adults (Carrara-Nascimento, Griffin, Pastrello, Olive, & Camarini, 2011; Faria et al., 2008). Interestingly, adolescent mice are also less sensitive to ethanol-induced CPP than adult mice, however, once they are pre-exposed to stress or ethanol, younger mice express CPP, suggesting a sensitization of the rewarding effect of ethanol dependent on previous stimuli-induced adaptation (Carrara-Nascimento, Olive, & Camarini, 2014; Song et al., 2007).

Concerning ethanol-dependent social behaviors, early adolescent mice are more sensitive to ethanol-induced facilitation of social activity at low doses of ethanol (0.5 and 0.75 g/kg) and less sensitive to ethanol-induced inhibition of social activity as compared to mid- and late-adolescent mice at higher ethanol doses (1.0–1.75 g/kg) (Varlinskaya & Spear, 2004).

As for sex differences, variations in brain monoamine and amino acid levels, circulating gonadal hormones (see Anker & Carroll, 2011, Becker & Hu, 2008, for review) and metabolizing enzymes activities (Quintanilla, Tampier, Sapag, Gerdtzen, & Israel, 2007) may be responsible for the differences between male and female rodents and humans in their responses to drugs of abuse. Female rodents show a more reliable and robust behavioral sensitization compared to males (Becker & Hu, 2008), including sensitization to ethanol (Didone, Quoilin, Tirelli, & Quertemont, 2008). Housing conditions and sex may be determining factors in finding a positive correlation between locomotor sensitization and ethanol consumption, since only female adolescent rhesus macaques tested in a social group setting showed such an association (Schwandt, Higley, Suomi, Heilig, & Barr, 2008). In a study screening adolescent rats for their sensitivity to ethanol-induced locomotor stimulation, it was found that adolescent female rats selected for being more sensitive to those effects show greater ethanol self-administration than their low-responder counterparts (Acevedo, Molina, Nizhnikov, Spear, & Pautassi, 2010).

3. BEHAVIORAL SENSITIZATION AND STRESS

Stressful experiences trigger activation of the hypothalamic–pituitary–adrenal (HPA) axis, increasing the release of corticotropin-releasing factor (CRF) from neurons of hypothalamic paraventricular nucleus into the blood vessels of the pituitary gland. CRF controls the release of the adrenocorticotrophic hormone (ACTH) in the anterior pituitary gland that, in turn, stimulates the synthesis and secretion of glucocorticoids, such as corticosterone in rodents and cortisol in humans, by the adrenal cortex (Herman & Cullinan, 1997). Exposure to a stressor and the subsequent release of glucocorticoids can interact with the mesolimbic dopaminergic system and facilitate dopamine release contributing to the reinforcing effects of drugs of abuse and subsequent drug self-administration (Piazza et al., 1993; Piazza & Le Moal, 1998; Sinha, 2008).

Stress is an important component in addiction that can strengthen ethanol-induced behaviors. Repeated exposure to stress results in greater locomotor response to ethanol (Santos-Rocha et al., 2018), a phenomenon known as cross-sensitization. Stress-induced neuroadaptive processes are thought to mediate enduring changes in mesolimbic systems capable of altering ethanol responsiveness and precipitating a greater response to this drug, demonstrating common long-term neurobiological changes induced by repeated stress and ethanol. Chronic unpredictable stress can increase locomotor activity in response to an acute ethanol injection in both adolescent and adult mice, known as stress-alcohol cross-sensitization (Santos-Rocha et al., 2018), suggesting that HPA axis activation can modulate ethanol behavioral sensitization (for a review, see Phillips, Roberts, & Lessov, 1997).

Alcohol consumption in rodents is also influenced by stress, as a history of exposure to stressful events can induce an escalation of voluntary ethanol intake, especially when stress exposure occurs early in development (Becker, Lopez, & Doremus-Fitzwater, 2011; Hwa, Holly, DeBold, & Miczek, 2016), although there are also reports of decreased or unchanged alcohol intake after stress (Becker et al., 2011). It has been suggested that there is a relationship between stress and the anxiolytic effects of ethanol seeking/consumption (Vengeliene et al., 2003); however, this is not the only motivation for alcohol drinking after stress. Ethanol itself activates HPA axis, and this dysregulation can also be responsible for influencing alcohol intake (for a review, see Becker et al., 2011).

Recognizing that are several factors that influence ethanol behavioral sensitization, we now discuss the influence of social factors by focusing on modeling psychosocial stress, including maternal separation, social defeat, and housing conditions, and, when pertinent, their intertwining with other factors such as age and sex.

3.1 Psychosocial Stress

Stressful social experiences are well documented to contribute to human psychopathology, such as depression, anxiety, and social phobia (Bjorkqvist, 2001; Kessler, 1997), and has long been associated with increased vulnerability to drug addiction (for a review, see Uhart & Wand, 2009). In animal models, alterations in social aspects of the environment can induce stressful situations leading to anxiety- and even depression-like behaviors. Social stress usually involves conflicts among individuals, but social isolation can also predispose the individual to anxiety/depression-related behaviors, physical health problems, and drug-seeking behavior.

Social stress models involve acute, chronic or intermittent interaction of an animal with a conspecific (Blanchard, McKittrick, & Blanchard, 2001). Social interactions can be positive (in case of maternal care) or negative (related to dominance and aggression, especially in males). Disrupting positive interactions or exposing animals to negative interactions can be considered socially stressful. These stressful events can occur at any stage of animal life, as preweaning, adolescence or adulthood, and influence behavioral sensitization in different ways. Even a single exposure to a social stressor can enhance adrenal glucocorticoids release (Koolhaas et al., 2011), and affect ethanol reinforcement and drug-taking behaviors (for a review, see Edwards, Little, Richardson, & Vendruscolo, 2015).

3.2 Maternal Separation

Maternal care is essential to the health of offspring (Fig. 1) and influences long-term neuroendocrine and behavioral functions. Variations in the

Fig. 1 The figure shows the mother crouching over the pups in a lactation posture. Maternal behaviors allow the pups to feel safe and protected. Disrupting this bond can compromise the behavior outcome regarding drug addiction, including behavioral sensitization.

quality and quantity of maternal care may be a critical aspect for offspring health, since reductions in this behavior seems to increase the risk of disturbances in postnatal development (Coutellier, 2012; Franks, Curley, & Champagne, 2011). Thus, maternal separation (MS), a form of early-life stress (Levine, 2005), has been shown to be an important risk factor of multiple psychiatric diseases. Animals that are separated from their mothers during a critical developmental window show various behavioral and psychological disturbances, such as anxious behavior, higher reactivity to stress, aggression, antisocial behavior, and seeking behavior for drugs such as alcohol, amphetamine, morphine and cocaine, both during adolescence and adulthood (Huot, Thrivikraman, Meaney, & Plotsky, 2001; Levine, 2005; Moffett et al., 2007; Vazquez et al., 2006; Zimmerberg & Sageser, 2011).

Different protocols of maternal separation have been described. Generally, pups are removed from the dam and maintained with the littermates during brief (e.g., 15 min) or prolonged (e.g., 180, 240 or 360 min) periods over a number of early postnatal days. In another approach, the pups can be isolated from their littermates, dam and nest (neonatal isolation) (for a review, see Moffett et al., 2007). Despite variations in experimental protocols, such as the timing and duration, the separation generally occurs in the first 2 weeks of pup life, when the HPA axis is under development. Although this period has been called "stress hyporesponsive period," there are reports that the paraventricular nucleus of hypothalamus and extra-hypothalamic brain regions are actually hyper-responsive to stressors during ontogenetic development (Levine, 1994; Sapolsky & Meaney, 1986; Vazquez et al., 2006). Thus, early life events, such as adverse experiences and inadequate maternal care, influence brain regions that regulate behavioral and neuroendocrine responses to stress (Levine, 1957, 2005; Sánchez, Ladd, & Plotsky, 2001), producing long-term consequences.

The maternal separation and consequent disruption in mother–infant interaction seems to produce neurochemical and behavioral changes in the offspring that are observed at different stages of life. For example, greater ethanol preference and consumption during adolescence and adulthood have been reported after maternal separation, suggesting a sensitized effect to both behaviors triggered by previous stressful experiences (Cruz, Quadros, Planeta, & Miczek, 2008; García-Gutiérrez et al., 2016; Huot et al., 2001; Song et al., 2007).

Exposure to early-life stress can also alter the locomotor stimulant effects induced by ethanol. One study evaluated the effects of a single maternal separation during 240 min in 15-day old male Wistar WKAH/HOK rat. Pups were separated in paired or isolated conditions and after maternal separation received an intragastric administration (i.g.) of ethanol (2.5 g/kg). The analyses of the locomotor activity revealed that isolated pups treated with ethanol showed hyperlocomotion compared to water-controls and to those paired with a littermate. Ethanol-induced locomotor activity was potentiated in social isolated pups and this effect was blocked by the CRF1 receptor antagonist (Arias et al., 2010). Several researches have implicated CRF1 signaling pathways in acute and repeated alcohol-related behaviors outcome, including psychomotor sensitization to ethanol (Pastor et al., 2008).

Another study analyzed the effects of repeated maternal separation [4 h/day from postnatal days (PND) 1 to 14] on ethanol-induced locomotor activity in 15-day old male and female Wistar rat pups. Maternal isolated pups exhibited significantly greater locomotor activity than controls, indicating that chronic maternal separation affects ethanol sensitivity in infancy (Fernández et al., 2014). However, Arias, Revillo, and Spear (2012) demonstrated that isolation from the dams and littermates for 240 min in a different period (PND 15–18) did not induce locomotor sensitization in rats repeatedly treated with 2.5 g/kg ethanol.

Maternal separation also seems to produce gender-dependent effects regarding ethanol-induced behavioral sensitization, since this stressful event facilitates the development of ethanol sensitization only in female mice. Few studies evaluated the effects of maternal separation on locomotor sensitization to ethanol. In Kawakami, Quadros, Takahashi, and Suchecki (2007), mice were separated from the dam for 15 or 180 min, from PND 2 to 14 and tested to ethanol sensitization when adults, using a 5-days injection protocol of ethanol administration. During adulthood, mice were treated with 2.2 g/kg ethanol (i.p.) every other day for 10 days. Although the phenomenon observed has been discrete, behavioral sensitization was more evident in females than in males. In females, long maternal separation induced a faster development of behavioral sensitization to ethanol compared to brief maternal separation. In conclusion, female mice seem to be more sensible to develop ethanol sensitization when exposed to long maternal separation (Kawakami et al., 2007). However, this effect was not confirmed using a more prolonged ethanol treatment (daily injections of 2.2 g/kg ethanol for 21 days). The potentiating effect of maternal separation on ethanol-induced behavioral sensitization in females was only verified when the ethanol treatment regimen did not induce a robust sensitization, based on a longer treatment, which may have masked the effects obtained with a weak sensitization observed in the previous study of Kawakami et al. (2007) (Kawakami, Quadros, & Suchecki, 2016).

3.3 Social Defeat

The SD stress model is based on the resident-intruder paradigm, in which the resident animal will attack the intruder in its home cage. This aggressive behavior is generated by territoriality and can be exacerbated by the presence of a female. In this animal model, an unknown intruder animal is transferred into the home cage of a resident for a short period of time, and generally results in a submissive behavior exhibition of the intruder caused by the aggressive conspecific (Golden, Covington, Berton, & Russo, 2011; Miczek, 1979; Miczek, Thompson, & Shuster, 1982). Variations in the SD stress paradigm include the number of confronts (single or repeated), intensity (defined by the number of attacks) and duration of exposure to the aggressor (Blanchard et al., 2001; Beery & Kaufer, 2015; Norman et al., 2015).

Acute and chronic SD stress result in increased activation of the HPA axis, and is associated with elevated plasmatic corticosterone levels (Buwalda et al., 1999; Covington & Miczek, 2005; Keeney et al., 2006). An important aspect of this model is that elevated corticosterone responses were similar in the first and last episodes of social defeat, suggesting non-adaptation in HPA axis in this stress model (Covington & Miczek, 2005; Norman et al., 2015). Thus, rodents do not habituate to repeated episodes of this type of stress.

It is well demonstrated that repeated exposure to this model of social stress produces depressive-like behavior (for a review, see Hollis & Kabbaj, 2014), with signs of anxiety-like behavior, deficits in social interaction, social avoidance behavior and anhedonia, in rodents (Favoretto, Macedo, & Quadros, 2017; Hollis, Wang, Dietz, Gunjan, & Kabba, 2010; Iñiguez et al., 2014; Macedo et al., 2018).

In addition to the effects on depressive-like behavior, social stress can contribute to the vulnerability to drug use (for a review, see Miczek, Yap, & Covington, 2008). The consequences of SD stress and its protocol variations on sensitizing behavioral responses to ethanol are described in Norman et al. (2015). While a mild stress (15 bites) had no effect on alcohol drinking compared to non-stressed control CFW mice, the moderate stress group (30 bites) showed escalated ethanol self-administration. Moreover, the exposure to social defeat during ethanol abstinence also facilitated the development of "alcohol deprivation effect" (Funk, Vohra, & Lê, 2004), defined as a temporary increase in voluntary intake following a period of forced ethanol deprivation (Sinclair, 1972). Moreover, social stress can induce a sensitized response to the rewarding effects of ethanol, once defeated mice showed an enhanced preference to the ethanol-paired compartment in the CPP test (Macedo et al., 2018).

Although repeated social defeat induces cross-sensitization to psychostimulants (Covington & Miczek, 2001; Nikulina, Covington, Ganschow, Hammer, & Miczek, 2004; Yap et al., 2015), the effects of SD stress on ethanol locomotor sensitization has generated negative results. In a 10-day defeat protocol, with 5-min session of confrontation every other day, Swiss mice did not show locomotor stimulation after receiving an i.p. injection of 2.2 g/kg ethanol at 10 days after the last defeat. Thus, a history of SD stress did not induce a cross-sensitized response to ethanol (Macedo et al., 2018). Similar results were described by Norman et al. (2015), in which mice were defeated during 10 consecutive days and were classified as receiving moderate or mild stress depending on the number of attacks. Neither moderate nor mild stress modified the locomotor activity of the mice in response to 2.0 g/kg ethanol (i.p.).

It is also relevant pointing out the importance of temporal parameters to drug access after the stress exposure. Increased ethanol intake is only detected following a temporal gap between the stress and ethanol availability, but not immediately after exposure to stress (Caldwell & Riccio, 2010), suggesting a delayed effect of social stress on ethanol drinking. Indeed, defeated mice that previously acquired ethanol intake showed decreased ethanol intake when the drug was offered immediately after stress (Funk, Harding, Juzytsch, & Lê, 2005; Norman et al., 2015; van Erp & Miczek, 2001), but increased consumption, 20 days after defeat (Norman et al., 2015). Thus, escalation of ethanol drinking induced by social defeat emerge only after a specified period of exposure to stress, suggesting a neural plasticity induced by the aggressive stress that triggers drug-taking behavior.

3.4 Housing Conditions

Some species such as rodents, primates and humans are characterized as social species and have evolved to seek companionship for both comfort and protection (Silk, 2007). Therefore, social interaction and hierarchy play an important role in behavior and wellbeing for such animals. Rodents are preferentially kept in cages in pair or group housing with recommended space to avoid isolation and overcrowding. All primary enclosures should provide a safe and healthy environment, allowing animals to form the social structure and hierarchy as close as that found in nature. Inappropriate housing conditions can lead to stress and ultimately affect behavior (Arndt et al., 2009).

The effects of housing conditions on several behaviors seems to be age and sex-dependent. For example, females are less responsive to overcrowding compared to males and a possible explanation is that dominant vs submissive relation is almost absent in female groups. This type of dynamics is often related to high stress levels (Brown & Grunberg, 1995). Regarding age-related differences, adolescents seem more reactive to housing effects compared to adults, possibly due to the fact that during this phase, animals are more playful and willing to interact socially compared to adults (Laviola, Adriani, Morley-Fletcher, & Terranova, 2002).

Adolescent rats housed in overcrowded cages (nine animals per cage) for 27 days have higher circulating corticosterone levels than adults (Uarquin, Meyer, Cardenas, & Rojas, 2016). This is particularly relevant for addiction, since stress is a relevant factor to alcohol intake and relapse. Social isolation for 40 days during early development (starting at weaning; PND 21), produced higher voluntary alcohol consumption compared to group-housed control in adult male and female C57BL/6J mice. However, social isolation during adulthood (late isolation; PND 60) did not increase ethanol intake, suggesting that the timing of chronic social isolation can be a relevant factor for later effects on drug self-administration (Lopez, Doremus-Fitzwater, & Becker, 2011). Similar results were described in rats, where isolated Long-Evans rats showed a pronounced ethanol intake compared to group-housed rats when they were isolated at PND 21, but not at PND 65 (Schenk, Gorman, & Amit, 1990). Both isolated Fawn-Hooded and Wistar rats consumed more ethanol at high concentrations (16%, v/v), but not at lower concentrations (2%, 4% and 8%, v/v) as compared to grouped-reared rats in the two-bottle choice test (Hall, Huang, Fong, Pert, & Linnoila, 1998). Deehan, Cain, and Kiefer (2007) studied groups of Long-Evans rats housed under one of three different conditions (isolation condition, social condition and enriched condition for 90 days). Afterward, rats were trained to lever press for 10% ethanol using the sucrose/ethanol fading procedure. Rats in the isolation condition showed increased active lever responding for 10% ethanol when compared to the other groups. Effectively, in both studies, rats were isolated from PND 21. However, social isolation increased ethanol intake in rats selectively bred for high or low preference for ethanol (P and NP rats, respectively) even when the rats were isolated at PND 43–50. In this study, isolation enhanced forced ethanol intake in both P and NP rats, but increased ethanol drinking in a limited-access paradigm only in P rats (Ehlers, Walker, Pian, Roth, & Slawecki, 2007), suggesting a genetic component in the effects of social isolation on ethanol drinking.

In another study, Long-Evans rats were isolated at PND 21–25 and were subsequently exposed to ethanol using different ethanol drinking paradigms: access to 10% ethanol solutions as the sole liquid source, two-bottle choice between 10% ethanol and water, and operant self-administration using a sucrose-fading procedure. Social isolation showed to increase ethanol intake in all those multiple ethanol drinking paradigms (McCool & Chappell, 2009).

Social isolation rearing (starting on PND 28) in Long-Evans rats is associated with increased release of dopamine and norepinephrine in the NAcc in response to an acute 2.0 g/kg ethanol challenge administered in adulthood (Karkhanis, Locke, McCool, Weiner, & Jones, 2014). This study suggests a dopaminergic/noradrenergic neurochemical sensitization in the NAcc, a brain region known to have a critical role in the actions of ethanol. Interestingly, social isolation reduced dopamine D2 receptor binding density in the NAcc (Lodge & Lawrence, 2003), which could represent a neuroplasticity response to the sensitization of dopamine release elicited by social isolation and/or ethanol. Therefore, disrupting social interaction either by isolation or overcrowding, especially during development, has a high impact on behavioral and neurochemical sensitization responses, and can determine the outcome of how these animals will behave in the future.

It is important to mention that social isolation did not modify ethanol consumption in Fawn-Hooded rats socially isolated at PND 21 (Lodge & Lawrence, 2003) and in Sprague-Dawley CD rats housed individually starting at PND 25 (Pisu et al., 2011). Differences in the duration for which the animals were exposed to ethanol, duration of social isolation or strain may have contributed to the apparent discrepancies found in these studies.

Unfortunately, only a very limited number of studies have focused on the impacts of housing conditions on locomotor sensitization. A study by Araujo et al. (2005) addressed specifically the effects of different housing conditions on locomotor sensitization to ethanol. Female Swiss mice were housed in standard (5 animals/cage), isolated (1 animal/cage) and overcrowded (12 animals/cage) conditions. Although no differences in locomotor activity among groups were found after an acute ethanol injection, animals in overcrowded conditions displayed higher locomotor sensitization following repeated 1.8 g/kg ethanol administration compared to control and isolated groups. Thus, stressful housing conditions can potentiate ethanol sensitization. In fact, mice housed in overcrowded conditions display higher anxiety-like behaviors compared to individually housed mice (Reiss, Wolter-Sutter, Krezel, & Ouagazzal, 2007).

Even though stress responsiveness could be a plausible explanation for the findings in Araujo's report, it is important to point out that female mice were used in that study. As discussed previously, females tend to be less responsive to overcrowding when compared to male mice, at least in regard to HPA axis reactivity. An interesting hypothesis discussed in this study is that animals could also have been influenced by their counterparts' reaction to housing conditions and/or ethanol.

To test this hypothesis, Araujo et al. (2006) designed an experiment in which all female mice in one of the home cages were injected with 1.8 g/kg ethanol (i.p.) during 21 consecutive days, while in an another home cage, only half of the animals were administered with the drug. In this latter cage, the other half received saline. The mice kept in home cages in which all the counterparts received ethanol showed higher behavioral sensitization compared to those animals maintained in heterogeneous cages. This provides evidence of behavioral influences among counterparts regarding sensitization of drug responses.

These studies support the notion that housing conditions can affect behavioral outcome related to drug addiction, and that these changes are age- and sex-dependent. Also, these studies suggest that stress, social interaction and cage mates' behaviors are potential mediators of these aforementioned behaviors. Nevertheless, additional studies are needed to further elucidate the underlying mechanisms of different housing conditions on addiction-related behaviors in general, and more specifically ethanol-induced behavioral sensitization.

4. BEHAVIORAL SENSITIZATION AND ENVIRONMENTAL ENRICHMENT

In contrast to the role of psychosocial stressors increasing risk of alcohol relapse, protective factors as a positive environment associated with psychotherapy, behavior therapy and relapse prevention are effective approaches of intervention to achieve abstinence. Moreover, mindfulness, meditation and exercise are also alternatives modalities that have becoming more prevalent in addiction treatment (Lynch, Peterson, Sanchez, Abel, & Smith, 2013; Zgierska et al., 2009). In animal models, enriched social environment, environmental enrichment (EE) and voluntary exercise have demonstrated to have positive results in cognitive function, depressive- and anxiety-like behaviors, and addiction-related responses (Simpson & Kelly, 2011).

EE was first categorized by Hebb in the 1940s, following his observations that rats raised as pets had improved behavioral responses when compared to their peers kept under laboratory conditions. In this context, EE is defined as "a combination of inanimate and social stimulation" (Rosenzweig, Bennett, Hebert, & Morimoto, 1978; Van Praag, Kempermann, & Gage, 2000).

As an attempt to mimic the physical and social complexity found in nature, but in a laboratory setting, animals are kept in larger cages with objects for exploration (cubes, toys, tunnels, houses, nesting material) and may have access to voluntary exercise (ladders, running wheels), as shown in Fig. 2. This setting enables animals to have increased and more complex social interactions. Therefore, EE can be considered as a controlled environment that allows animals to be exposed to novelty, voluntary exercise and social interaction. Studies have attempted to study each individual variable on its own, but have failed to reproduce the same behavioral outcomes as EE, suggesting that the effects of EE are most likely due to the combination of all factors mentioned above (Van Praag et al., 2000). For example, the inclusion of a running wheel in EE in order to promote physical exercise

Fig. 2 Example of a cage used for environmental enrichment. Cages are larger than standard cages and usually contain ladders, tunnels, houses, toys and running wheels. This type of environment offers novel stimuli, voluntary physical exercise and positive social interaction, which has been shown to promote a protective phenotype against drug-induced behavioral sensitization.

and adult neurogenesis is debatable. While there is evidence that exercise is critical to increase hippocampal neurogenesis in male (Mustroph et al., 2012) and female C57BL/6J mice (Kobilo et al., 2011), other studies report that hippocampal neurogenesis is increased in mice exposed to EE without a running wheel, although these two forms of environmental manipulations (EE and running wheel) differentially influence some proliferating granule cells in the subgranular zone (SGZ) of the dentate gyrus (Steiner, Zurborg, Hörster, Fabel, & Kempermann, 2008).

Exposure of animal to an enriched environment causes various changes in neuronal and behavioral output. Most relevant, previous studies have showed that animals living in enriched conditions have higher levels of various aspects of neuroplasticity including adult neurogenesis (Brown et al., 2003;

Segovia, Yagüe, García-Verdugo, & Mora, 2006), dendritic branching, spine density and synaptic formation (Green & Greenough, 1986; Green, Greenough, & Schlumpf, 1983) and gliogenesis in hippocampus and prefrontal cortex (Simpson & Kelly, 2011; Van Praag et al., 2000).

EE has also been shown to cause molecular, neurochemical and behavioral changes, such as increases in CREB expression in the hippocampus of mice (Williams et al., 2001) and in neurotrophic factor levels, especially brain-derived neurotrophic factor (BDNF), which is highly associated with hippocampal neurogenesis and contributes to learning and memory (Rossi et al., 2006). Enriched animals show better performance in memory tasks (Lambert, Fernandez, & Frick, 2005), lower anxiety-like behavior and stress responses (Novaes et al., 2017) and improved social recognition memory (Toyoshima, Yamada, Sugita, & Ichitani, 2018), as compared to animals kept under standard conditions.

Due to these aforementioned positive effects, a number of studies have been performed to better understand the effects of EE on pathologies and disorders, with the hope that changes in the environment could lead to improvement in these conditions. These include Huntington's disease (Van Dellen, Blakemore, Deacon, York, & Hannan, 2000), Alzheimer's disease (Arendash et al., 2004), stroke (Risedal et al., 2002), schizophrenia (Mcomish et al., 2008), depression (Llorens-Martín, Tejeda, & Trejo, 2011) and even cancer (Cao et al., 2010). Extensive work has also been done to explore the effects of EE on drug addiction (for a review, see Stairs & Bardo, 2009), as a promising tool in preventing and treating addiction for different drugs. In humans, EE could be introduced by occupations and lifestyles that stimulate cognition, promote physical exercise and active participation in cultural events, and engaging in leisure activities (e.g., reading, painting, etc.).

The protective actions of EE on addiction-related behaviors in animals have been demonstrated by reversal of CPP and behavioral sensitization to cocaine (Solinas, Chauvet, Thiriet, El Rawas, & Jaber, 2008), decreases in self-administration of amphetamine (Bardo, Klebaur, Valone, & Deaton, 2001), as well as blockade of CPP and decreased behavioral sensitization to morphine (Xu, Hou, Gao, He, & Zhang, 2007). More recently, Sikora et al. (2018) described reductions in drug-seeking behavior for heroin, nicotine and methamphetamine in animals exposed to EE after self-administration training for the aforementioned drugs. Regarding ethanol, it has been reported that EE decreased alcohol intake after acute stress exposure (Marianno, Abrahao, & Camarini, 2017) and prevented and reversed alcohol behavioral sensitization (Rueda, Teixeira, Yonamine, & Camarini, 2012). On the other hand, some reports found no change or even increase in drug reward following enriched environment exposure. For instance, Thiriet et al. (2011) showed that EE was unable to reduce CPP for methamphetamine in rats. Higher place preference magnitudes following conditioning with cocaine (Green et al., 2010) and ethanol (Pautassi et al., 2017; Rae et al., 2018) have been reported. Rockman, Gibson, and Benarroch (1989) found that a long period of enrichment induced a higher voluntary ethanol intake than social isolation.

The seemingly controversy in effects of EE on drug-induced behaviors can be at least partially explained by the numerous differences in procedures used by individual laboratories. As reviewed by Simpson and Kelly (2011), age, gender, animal strain, toys, cage size, number of animals per cage and duration are some of the elements that can affect EE outcome. Moreover, different procedures employed (i.e., CPP, sensitization, self-administration) measure different aspects of drug addiction, so it is likely that the environmental influence on behaviors varies on the basis of different drug classes as well as specific addiction-related behaviors (Bardo & Bevins, 2000).

Specifically regarding alcohol sensitization, only one paper to our knowledge has addressed this matter. Rueda et al. (2012) described a protective effect of EE in preventing and reversing alcohol-induced behavioral sensitization. Briefly, male Swiss mice were kept under standard or enriched conditions for 17 days prior to and during the 15 days of alcohol sensitization. EE prevented mice to develop ethanol sensitization (see Fig. 3). In a different protocol, mice were exposed to the EE during 1 week after ethanol behavioral sensitization had developed, which was enough to reverse ethanol-induced behavioral sensitization. This study was effective in describing not only a protective effect of EE on the development of ethanol behavioral sensitization, but also a possible tool in reversing sensitization once acquired.

Although the mechanisms underlying EE effects on ethanol-induced locomotor behavioral sensitization are still unknown, one possible explanation is its effects on levels of the neurotrophin BDNF. In Rueda et al. (2012), EE decreased BDNF levels in the prefrontal cortex, a result consistent with a down-regulation of BDNF–TrkB signaling in the medial prefrontal cortex of rats exhibiting reduced cocaine-induced sensitization (Lu, Cheng, Lim, Khoshnevisrad, & Poo, 2010). Further studies are needed to specifically explore this.

In addition, studies on the relevance of BDNF to EE effects on ethanol-induced CPP have also been performed. For example, Pautassi et al. (2017)

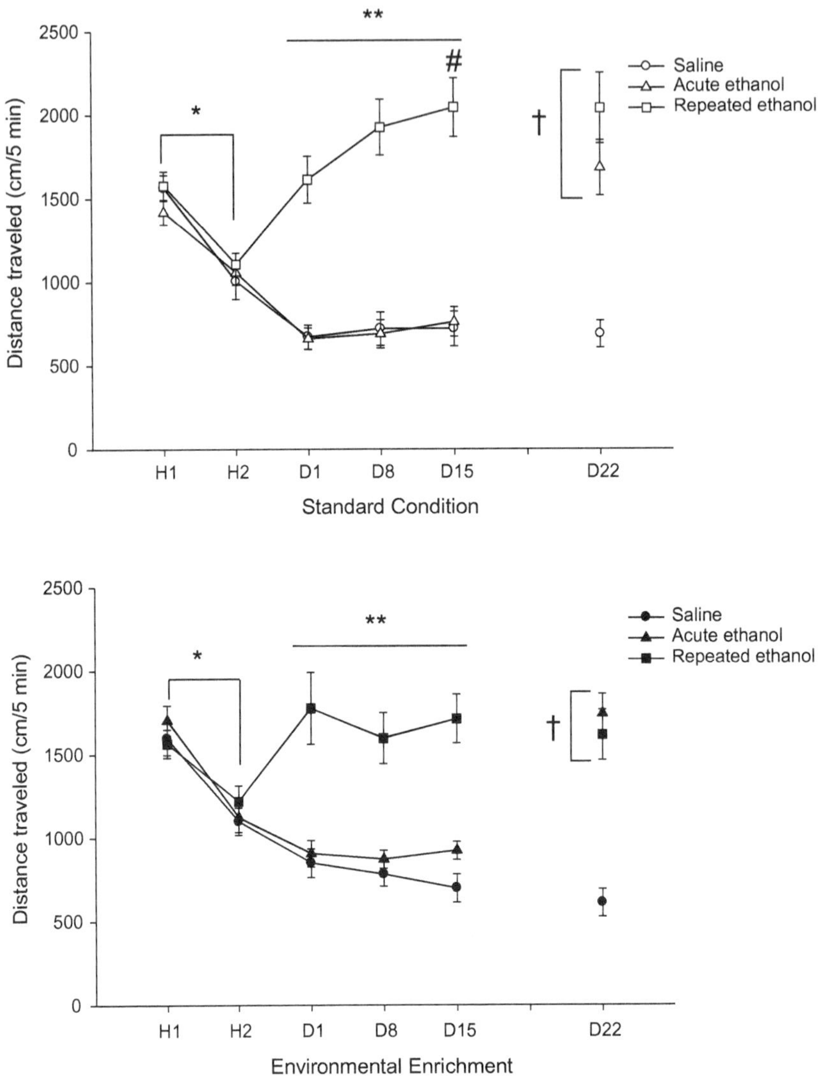

Fig. 3 Effects of EE on behavioral sensitization to ethanol in control (A) and environmentally enriched (B) groups. Control mice displayed a gradual increase in locomotor activity following 1.8 g/kg ethanol injections. However, animals kept under enriched conditions did not display locomotor sensitization. *Authorized adaptation from Rueda, A. V. L., Teixeira, A. M. A., Yonamine, M., & Camarini, R. (2012). Environmental enrichment blocks ethanol-induced locomotor sensitization and decreases BDNF levels in the prefrontal cortex in mice.* Addiction Biology, *17(4), 736–745. https://doi.org/10.1111/j.1369-1600.2011.00408.x.*

showed that adolescent mice exposed to EE exhibited a significant enhancement of CPP to ethanol, reflecting a sensitization of ethanol-induced conditioned reward. Moreover, they also displayed higher BDNF levels in the prefrontal cortex than controls reared in standard conditions (Pautassi et al., 2017). Taken together, the data suggests that BDNF levels in the prefrontal cortex may play an important role in the sensitization of locomotor responses to ethanol.

5. CONCLUSIONS

Behavioral sensitization is a key tool to study neuroplasticity induced by repeated drug exposure. A number of social factors, including psychosocial stress and environmental and housing conditions, can influence the outcome of animals' development to drug-induced sensitization. These studies provide valuable information that helps understand the sensitized incentive salience component of addiction processes and how social factors can exert a pivotal influence over this phenomenon. Moreover, they provide a basis for establishing prevention or treatment approaches for alcoholism. Although extensive work has been done regarding on ethanol-induced behavioral sensitization and its modulation by social factors, additional studies are necessary to better elucidate the complex mechanisms underlying the interactions between negative/positive social stimuli and sensitization of responses to alcohol.

REFERENCES

Acevedo, M. B., Molina, J. C., Nizhnikov, M. E., Spear, N. E., & Pautassi, R. M. (2010). High ethanol dose during early adolescence induces locomotor activation and increases subsequent ethanol intake during late adolescence. *Developmental Psychobiology*, *52*(5), 424–440. https://doi.org/10.1002/dev.20444.

Anker, J. J., & Carroll, M. E. (2011). Females are more vulnerable to drug abuse than males: Evidence from preclinical studies and the role of ovarian hormones. *Current Topics in Behavioral Neuroscience*, *8*, 73–96. https://doi.org/10.1007/7854_2010_93.

Araujo, N. P., Camarini, R., Souza-Formigoni, M. L., Carvalho, R. C., Abílio, V. C., Silva, R. H., et al. (2005). The importance of housing conditions on behavioral sensitization and tolerance to ethanol. *Pharmacology, Biochemistry, and Behavior*, *82*(1), 40–45. https://doi.org/10.1016/j.pbb.2005.07.009.

Araujo, N. P., Fukushiro, D. F., Cunha, J. L., Levin, R., Chinen, C. C., Carvalho, R. C., et al. (2006). Drug-induced home cage conspecifics' behavior can potentiate behavioral sensitization in mice. *Pharmacology, Biochemistry, and Behavior*, *84*(1), 142–147. https://doi.org/10.1016/j.pbb.2006.04.019.

Arendash, G. W., Garcia, M. F., Costa, D. A., Cracchiolo, J. R., Wefes, I. M., & Potter, H. (2004). Environmental enrichment improves cognition in aged Alzheimer's transgenic mice despite stable β-amyloid deposition. *Neuroreport*, *15*(11), 1751–1754. https://doi.org/10.1097/01.wnr.0000137183.68847.4e.

Arias, C., Revillo, D. A., & Spear, N. E. (2012). Chronic tolerance to the locomotor stimulating effect of ethanol in preweanling rats as a function of social stress. *Alcohol*, *46*(3), 245–252. https://doi.org/10.1016/j.alcohol.2011.09.025.

Arias, C., Solari, A. C., Mlewski, E. C., Miller, S., Haymal, B., Spear, N. E., et al. (2010). Social isolation and stress related hormones modulate the stimulating effect of ethanol in preweanling rats. *Behavioural Brain Research*, *211*(1), 64–70. https://doi.org/10.1016/j.bbr.2010.03.010.

Arndt, S. S., Laarakker, M. C., van Lith, H. A., van der Staay, F. J., Gieling, E., Salomons, A. R., et al. (2009). Individual housing of mice—Impact on behaviour and stress responses. *Physiology & Behavior*, *97*(3–4), 385–393. https://doi.org/10.1016/j.physbeh.2009.03.008.

Bardo, M. T., & Bevins, R. A. (2000). Conditioned place preference: What does it add to our preclinical understanding of drug reward? *Psychopharmacology*, *153*(1), 31–43. https://doi.org/10.1007/s002130000569.

Bardo, M. T., Klebaur, J. E., Valone, J. M., & Deaton, C. (2001). Environmental enrichment decreases intravenous self-administration of amphetamine in female and male rats. *Psychopharmacology*, *155*(3), 278–284. https://doi.org/10.1007/s0022130100720.

Becker, J. B., & Hu, M. (2008). Sex differences in drug abuse. *Frontiers in Neuroendocrinology*, *29*(1), 36–47. https://doi.org/10.1016/j.yfrne.2007.07.003.

Becker, H. C., Lopez, M. F., & Doremus-Fitzwater, T. L. (2011). Effects of stress on alcohol drinking: A review of animal studies. *Psychopharmacology*, *218*(1), 131–156. https://doi.org/10.1007/s00213-011-2443-9.

Beery, A. K., & Kaufer, D. (2015). Stress, social behavior, and resilience: Insights from rodents. *Neurobiology of Stress*, *1*, 116–127. https://doi.org/10.1016/j.ynstr.2014.10.004.

Bjorkqvist, K. (2001). Social defeat as a stressor in humans. *Physiology & Behavior*, *73*(3), 435–442. https://doi.org/10.1016/S0031-9384(01)00490-5.

Blanchard, R. J., McKittrick, C. R., & Blanchard, D. C. (2001). Animal models of social stress: Effects on behavior and brain neurochemical systems. *Physiology & Behavior*, *73*(3), 261–271. https://doi.org/10.1016/S0031-9384(01)00449-8.

Brown, J., Cooper-Kuhn, C. M., Kempermann, G., van Praag, H., Winkler, J., Gage, F. H., et al. (2003). Enriched environment and physical activity stimulate hippocampal but not olfactory bulb neurogenesis. *European Journal of Neuroscience*, *17*(10), 2042–2046. https://doi.org/10.1046/j.1460-9568.2003.02647.x.

Brown, K. J., & Grunberg, N. E. (1995). Effects of housing on male and female rats: Crowding stresses male but calm females. *Physiology & Behavior*, *58*(6), 1085–1089. https://doi.org/10.1016/0031-9384(95)02043-8.

Buwalda, B., de Boer, S. F., Schmidt, E. D., Felszeghy, K., Nyakas, C., Sgoifo, A., et al. (1999). Long-lasting deficient dexamethasone suppression of hypothalamic-pituitary-adrenocortical activation following peripheral CRF challenge in socially defeated rats. *Journal of Neuroendocrinology*, *11*(7), 513–520. https://doi.org/10.1046/j.1365-2826.1999.00350.x.

Caldwell, E. E., & Riccio, D. C. (2010). Alcohol self-administration in rats: Modulation by temporal parameters related to repeated mild social defeat stress. *Alcohol*, *44*(3), 265–274. https://doi.org/10.1016/j.alcohol.2010.02.012.

Camarini, R., & Hodge, C. W. (2004). Ethanol preexposure increases ethanol self-administration in C57BL/6J and DBA/2J mice. *Pharmacology, Biochemistry, and Behavior*, *79*(4), 623–632. https://doi.org/10.1016/j.pbb.2004.09.012.

Camarini, R., & Pautassi, R. M. (2016). Behavioral sensitization to ethanol: Neural basis and factors that influence its acquisition and expression. *Brain Research Bulletin*, *125*, 53–78. https://doi.org/10.1016/j.brainresbull.2016.04.006.

Cao, L., Liu, X., Lin, E. J., Wang, C., Choi, E. Y., Riban, V., et al. (2010). Environmental and genetic activation of a brain-adipocyte BDNF/leptin axis causes cancer remission and inhibition. *Cell*, *142*(1), 52–64. https://doi.org/10.1016/j.cell.2010.05.029.

Carrara-Nascimento, P. F., Griffin, W. C., 3rd, Pastrello, D. M., Olive, M. F., & Camarini, R. (2011). Changes in extracellular levels of glutamate in the nucleus accumbens after ethanol-induced behavioral sensitization in adolescent and adult mice. *Alcohol*, *45*(5), 451–460. https://doi.org/10.1016/j.alcohol.2011.01.002.

Carrara-Nascimento, P. F., Olive, M. F., & Camarini, R. (2014). Ethanol pre-exposure during adolescence or adulthood increases ethanol intake but ethanol-induced conditioned place preference is enhanced only when pre-exposure occurs in adolescence. *Developmental Psychobiology*, *56*(1), 36–48. https://doi.org/10.1002/dev.21089.

Coutellier, L. (2012). Rodent models of early environment effects on offspring development and susceptibility to neurological diseases in adulthood. *Translational Neuroscience*, *3*(3), 258–262. https://doi.org/10.2478/s13380-012-0034-9.

Covington, H. E., 3rd, & Miczek, K. A. (2001). Repeated social-defeat stress, cocaine or morphine. Effects on behavioral sensitization and intravenous cocaine self-administration "binges" *Psychopharmacology*, *158*(4), 388–398. https://doi.org/10.1007/s002130100858.

Covington, H. E., 3rd, & Miczek, K. A. (2005). Intense cocaine self-administration after episodic social defeat stress, but not after aggressive behavior: Dissociation from corticosterone activation. *Psychopharmacology*, *183*(3), 331–340. https://doi.org/10.1007/s00213-005-0190-5.

Cruz, F. C., Quadros, I. M., Planeta, C. d. S., & Miczek, K. A. (2008). Maternal separation stress in male mice: Long-term increases in alcohol intake. *Psychopharmacology*, *201*(3), 459–468. https://doi.org/10.1007/s00213-008-1307-4.

Deehan, G. A., Jr., Cain, M. E., & Kiefer, S. W. (2007). Differential rearing conditions alter operant responding for ethanol in outbred rats. *Alcoholism, Clinical and Experimental Research*, *31*, 1692–1698.

Didone, V., Quoilin, C., Tirelli, E., & Quertemont, E. (2008). Parametric analysis of the development and expression of ethanol-induced behavioral sensitization in female Swiss mice: Effects of dose, injection schedule, and test context. *Psychopharmacology*, *201*(2), 249–260. https://doi.org/10.1007/s00213-008-1266-9.

Edwards, S., Little, H. J., Richardson, H. N., & Vendruscolo, L. F. (2015). Divergent regulation of distinct glucocorticoid systems in alcohol dependence. *Alcohol*, *49*(8), 811–816. https://doi.org/10.1016/j.alcohol.2015.04.004.

Ehlers, C. L., Walker, B. M., Pian, J. P., Roth, J. L., & Slawecki, C. J. (2007). Increased alcohol drinking in isolate-housed alcohol-preferring rats. *Behavioral Neuroscience*, *121*(1), 111–119. https://doi.org/10.1037/0735-7044.121.1.111.

Faria, R. R., Lima Rueda, A. V., Sayuri, C., Soares, S. L., Malta, M. B., Carrara-Nascimento, P. F., et al. (2008). Environmental modulation of ethanol-induced locomotor activity: Correlation with neuronal activity in distinct brain regions of adolescent and adult Swiss mice. *Brain Research*, *1239*, 127–140. https://doi.org/10.1016/j.brainres.2008.08.056.

Favoretto, C. A., Macedo, G. C., & Quadros, I. M. (2017). Effects of ethanol on social avoidance induced by chronic social defeat stress in mice. *Stress*, *20*(1), 68–74. https://doi.org/10.1080/10253890.2017.1280667.

Fernández, M., Fabio, M. C., Nizhnikov, M. E., Spear, N. E., Abate, P., & Pautassi, R. M. (2014). Maternal isolation during the first two postnatal weeks affects novelty-induced responses and sensitivity to ethanol-induced locomotor activity during infancy. *Developmental Psychobiology*, *56*(5), 1070–1082. https://doi.org/10.1002/dev.21192.

Franks, B., Curley, J. P., & Champagne, F. A. (2011). Measuring variations in maternal behavior: Relevance for studies of mood and anxiety. In: Gould, T. D. (Ed.), *Mood and Anxiety Related Phenotypes in Mice: Vol. II,* (pp. 209–224). New York: Springer.

Funk, D., Harding, S., Juzytsch, W., & Lê, A. D. (2005). Effects of unconditioned and conditioned social defeat on alcohol self-administration and reinstatement of alcohol seeking in rats. *Psychopharmacology, 183*(3), 341–349. https://doi.org/10.1007/s00213-005-0194-1.

Funk, D., Vohra, S., & Lê, A. D. (2004). Influence of stressors on the rewarding effects of alcohol in Wistar rats: Studies with alcohol deprivation and place conditioning. *Psychopharmacology, 176*(1), 82–87. https://doi.org/10.1007/s00213-004-1859-x.

García-Gutiérrez, M. S., Navarrete, F., Aracil, A., Bartoll, A., Martínez-Gras, I., Lanciego, J. L., et al. (2016). Increased vulnerability to ethanol consumption in adolescent maternal separated mice. *Addiction Biology, 21*(4), 847–858. https://doi.org/10.1111/adb.12266.

Golden, S. A., Covington, H. E., 3rd, Berton, O., & Russo, S. J. (2011). A standardized protocol for repeated social defeat stress in mice. *Nature Protocols, 6*(8), 1183–1191. https://doi.org/10.1038/nprot.2011.361.

Green, T. A., Alibhai, I. N., Roybal, C. N., Winstanley, C. A., Theobald, D. E., Birnbaum, S. G., et al. (2010). Environmental enrichment produces a behavioral phenotype mediated by low cyclic adenosine monophosphate response element binding (CREB) activity in the nucleus accumbens. *Biological Psychiatry, 67*(1), 28–35. https://doi.org/10.1016/j.biopsych.2009.06.022.

Green, E. J., & Greenough, W. T. (1986). Altered synaptic transmission in dentate gyrus of rats reared in complex environments: Evidence from hippocampal slices maintained in vitro. *Journal of Neurophysiology, 55*(4), 739–750. https://doi.org/10.1152/jn.1986.55.4.739.

Green, E. J., Greenough, W. T., & Schlumpf, B. E. (1983). Effects of complex or isolated environments on cortical dendrites of middle-aged rats. *Brain Research, 264*(2), 233–240. https://doi.org/10.1016/0006-8993(83)90821-1.

Hall, F. S., Huang, S., Fong, G. W., Pert, A., & Linnoila, M. (1998). Effects of isolation-rearing on voluntary consumption of ethanol, sucrose and saccharin solutions in Fawn Hooded and Wistar rats. *Psychopharmacology, 139*(3), 210–216.

Herman, J. P., & Cullinan, W. E. (1997). Neurocircuitry of stress: Central control of the hypothalamo-pituitary-adrenocortical axis. *Trends in Neurosciences, 20*(2), 78–84. https://doi.org/10.1016/S0166-2236(96)10069-2.

Hollis, F., & Kabbaj, M. (2014). Social defeat as an animal model for depression. *ILAR Journal, 55*(2), 221–232. https://doi.org/10.1093/ilar/ilu002.

Hollis, F., Wang, H., Dietz, D., Gunjan, A., & Kabba, J. M. (2010). The effects of repeated social defeat on long-term depressive-like behavior and short-term histone modifications in the hippocampus in male Sprague-Dawley rats. *Psychopharmacology, 211*(1), 69–77. https://doi.org/10.1007/s00213-010-1869-9.

Huot, R. L., Thrivikraman, K. V., Meaney, M. J., & Plotsky, P. M. (2001). Development of adult ethanol preference and anxiety as a consequence of neonatal maternal separation in Long Evans rats and reversal with antidepressant treatment. *Psychopharmacology, 158*(4), 366–373. https://doi.org/10.1007/s002130100701.

Hwa, L. S., Holly, E. N., DeBold, J. F., & Miczek, K. A. (2016). Social stress-escalated intermittent alcohol drinking: Modulation by CRF-R1 in the ventral tegmental area and accumbal dopamine in mice. *Psychopharmacology, 233*(4), 681–690. https://doi.org/10.1007/s00213-015-4144-2.

Iñiguez, S. D., Riggs, L. M., Nieto, S. J., Dayrit, G., Zamora, N. N., Shawhan, K. L., et al. (2014). Social defeat stress induces a depression-like phenotype in adolescent male c57BL/6 mice. *Stress, 17*(3), 247–255. https://doi.org/10.3109/10253890.2014.910650.

Karkhanis, A. N., Locke, J. L., McCool, B. A., Weiner, J. L., & Jones, S. R. (2014). Social isolation rearing increases nucleus accumbens dopamine and norepinephrine responses to acute ethanol in adulthood. *Alcoholism, Clinical and Experimental Research, 38*(11), 2770–2779. https://doi.org/10.1111/acer.12555.
Kawakami, S. E., Quadros, I. M., & Suchecki, D. (2016). Naltrexone prevents in males and attenuates in females the expression of behavioral sensitization to ethanol regardless of maternal separation. *Frontiers in Endocrinology, 7*, 135. https://doi.org/10.3389/fendo.2016.00135.
Kawakami, S. E., Quadros, I. M., Takahashi, S., & Suchecki, D. (2007). Long maternal separation accelerates behavioural sensitization to ethanol in female, but not in male mice. *Behavioural Brain Research, 184*(2), 109–116. https://doi.org/10.1016/j.bbr.2007.06.023.
Keeney, A., Jessop, D. S., Harbuz, M. S., Marsden, C. A., Hogg, S., & Blackburn-Munro, R. E. (2006). Differential effects of acute and chronic social defeat stress on hypothalamic-pituitary-adrenal axis function and hippocampal serotonin release in mice. *Journal of Neuroendocrinology, 18*(5), 330–338. https://doi.org/10.1111/j.1365-2826.2006.01422.x.
Kessler, R. C. (1997). The effects of stressful life events on depression. *Annual Review of Psychology, 48*, 191–214. https://doi.org/10.1146/annurev.psych.48.1.191.
Kobilo, T., Liu, Q. R., Gandhi, K., Mughal, M., Shaham, Y., & van Praag, H. (2011). Running is the neurogenic and neurotrophic stimulus in environmental enrichment. *Learning Memory, 18*(9), 605–609. https://doi.org/10.1101/lm.2283011.
Koob, G. F., & Volkow, N. D. (2010). Neurocircuitry of addiction. *Neuropsychopharmacology, 35*(4), 217–238. https://doi.org/10.1038/npp.2010.4.
Koolhaas, J. M., Bartolomucci, A., Buwalda, B., de Boer, S. F., Flügge, G., Korte, S. M., et al. (2011). Stress revisited: A critical evaluation of the stress concept. *Neuroscience and Biobehavioral Reviews, 35*(5), 1291–1301. https://doi.org/10.1016/j.neubiorev.2011.02.003.
Lambert, T. J., Fernandez, S. M., & Frick, K. M. (2005). Different types of environmental enrichment have discrepant effects on spatial memory and synaptophysin levels in female mice. *Neurobiology of Learning and Memory, 83*(3), 206–216. https://doi.org/10.1016/j.nlm.2004.12.001.
Laviola, G., Adriani, W., Morley-Fletcher, S., & Terranova, M. L. (2002). Peculiar response of adolescent mice to acute and chronic stress and to amphetamine: Evidence of sex differences. *Behavioral Brain Research, 130*(1–2), 117–125. https://doi.org/10.1016/S0166-4328(01)00420-X.
Lessov, C. N., Palmer, A. A., Quick, E. A., & Phillips, T. J. (2001). Voluntary ethanol drinking in C57BL/6J and DBA/2J mice before and after sensitization to the locomotor stimulant effects of ethanol. *Psychopharmacology, 155*(1), 91–99. https://doi.org/10.1007/s002130100699.
Levine, S. (1957). Infantile experience and resistance to physiological stress. *Science, 126*(3270), 405. https://doi.org/10.1126/science.126.3270.405.
Levine, S. (1994). The ontogeny of the hypothalamic-pituitary-adrenal axis. The influence of maternal factors. *Annals of the New York Academy of Sciences, 746*, 275–288. https://doi.org/10.1111/j.1749-6632.1994.tb39245.x.
Levine, S. (2005). Developmental determinants of sensitivity and resistance to stress. *Psychoneuroendocrinology, 30*(10), 939–946. https://doi.org/10.1016/j.psyneuen.2005.03.013.
Llorens-Martín, M., Tejeda, G. S., & Trejo, J. L. (2011). Antidepressant and proneurogenic influence of environmental enrichment in mice: Protective effects vs recovery. *Neuropsychopharmacology, 36*(12), 2460–2468. https://doi.org/10.1038/npp.2011.134.

Lodge, D. J., & Lawrence, A. J. (2003). The effect of isolation rearing on volitional ethanol consumption and central CCK/dopamine systems in Fawn Hooded rats. *Behavioural Brain Research, 141*, 113–122. https://doi.org/10.1016/S0166-4328(02)00328-5.

Lopez, M. F., Doremus-Fitzwater, T. L., & Becker, H. C. (2011). Chronic social isolation and chronic variable stress during early development induce later elevated ethanol intake in adult C57BL/6J mice. *Alcohol, 45*, 355–364. https://doi.org/10.1016/j.alcohol.2010.08.017.

Lu, H., Cheng, P. L., Lim, B. K., Khoshnevisrad, N., & Poo, M. M. (2010). Elevated BDNF after cocaine withdrawal facilitates LTP in medial prefrontal cortex by suppressing GABA inhibition. *Neuron, 67*(5), 821–833. https://doi.org/10.1016/j.neuron.2010.08.012.

Lynch, W. J., Peterson, A. B., Sanchez, V., Abel, J., & Smith, M. A. (2013). Exercise as a novel treatment for drug addiction: A neurobiological and stage-dependent hypothesis. *Neuroscience and Biobehavioral Reviews, 37*(8), 1622–1644. https://doi.org/10.1016/j.neubiorev.2013.06.011.

Macedo, G. C., Morita, G. M., Domingues, L. P., Favoretto, C. A., Suchecki, D., & Quadros, I. M. H. (2018). Consequences of continuous social defeat stress on anxiety- and depressive-like behaviors and ethanol reward in mice. *Hormones and Behavior, 97*, 154–161. https://doi.org/10.1016/j.yhbeh.2017.10.007.

Marianno, P., Abrahao, K. P., & Camarini, R. (2017). Environmental enrichment blunts ethanol consumption after restraint stress in C57BL/6 mice. *PLoS One, 12*(1), e0170317. https://doi.org/10.1371/journal.pone.0170317.

Masur, J., & Boerngen, R. (1980). The excitatory component of ethanol in mice: A chronic study. *Pharmacology, Biochemistry, and Behavior, 13*(6), 777–780.

Masur, J., Oliveira de Souza, M. L., & Zwicker, A. P. (1986). The excitatory effect of ethanol: Absence in rats, no tolerance and increased sensitivity in mice. *Pharmacology, Biochemistry, and Behavior, 24*(5), 1225–1228.

McCool, B. A., & Chappell, A. M. (2009). Early social isolation in male Long-Evans rats alters both appetitive and consummatory behaviors expressed during operant ethanol self-administration. *Alcoholism, Clinical and Experimental Research, 33*(2), 273–282. https://doi.org/10.1111/j.1530-0277.2008.00830.x.

McCormick, C. M., Robarts, D., Kopeikina, K., & Kelsey, J. E. (2005). Long-lasting, sex- and age-specific effects of social stressors on corticosterone responses to restraint and on locomotor responses to psychostimulants in rats. *Hormones and Behavior, 48*(1), 64–74. https://doi.org/10.1016/j.yhbeh.2005.01.008.

Mcomish, C. E., Burrows, E., Howard, M., Scarr, E., Kim, D., Shin, H. S., et al. (2008). Phospholipase C-beta1 knockout mice exhibit endophenotypes modeling schizophrenia which are rescued by environmental enrichment and clozapine administration. *Molecular Psychiatry, 13*(7), 661–672. https://doi.org/10.1038/sj.mp.4002046.

Miczek, K. A. (1979). A new test for aggression in rats without aversive stimulation: Differential effects of d-amphetamine and cocaine. *Psychopharmacology, 60*(3), 253–259.

Miczek, K. A., Thompson, M. L., & Shuster, L. (1982). Opioid-like analgesia in defeated mice. *Science, 215*(4539), 1520–1522. https://doi.org/10.1126/science.7199758.

Miczek, K. A., Yap, J. J., & Covington, H. E., 3rd (2008). Social stress, therapeutics and drug abuse: Preclinical models of escalated and depressed intake. *Pharmacology & Therapeutics, 120*(2), 102–128. https://doi.org/10.1016/j.pharmthera.2008.07.006.

Moffett, M. C., Vicentic, A., Kozel, M., Plotsky, P., Francis, D. D., & Kuhar, M. J. (2007). Maternal separation alters drug intake patterns in adulthood in rats. *Biochemical Pharmacology, 73*(3), 321–330. https://doi.org/10.1016/j.bcp.2006.08.003.

Moy, S. S., Nadler, J. J., Young, N. B., Nonneman, R. J., Segall, S. K., Andrade, G. M., et al. (2008). Social approach and repetitive behavior in eleven inbred mouse strains. *Behavioural Brain Research, 191*(1), 118–129. https://doi.org/10.1016/j.bbr.2008.03.015.

Mustroph, M. L., Chen, S., Desai, S. C., Cay, E. B., DeYoung, E. K., & Rhodes, J. S. (2012). Aerobic exercise is the critical variable in an enriched environment that increases hippocampal neurogenesis and water maze learning in male C57BL/6J mice. *Neuroscience, 219*, 62–71. https://doi.org/10.1016/j.neuroscience.2012.06.007.

Nikulina, E. M., Covington, H. E., 3rd, Ganschow, L., Hammer, R. P., Jr., & Miczek, K. A. (2004). Long-term behavioral and neuronal cross-sensitization to amphetamine induced by repeated brief social defeat stress: Fos in the ventral tegmental area and amygdala. *Neuroscience, 123*(4), 857–865. https://doi.org/10.1016/j.neuroscience.2003.10.029.

Norman, K. J., Seiden, J. A., Klickstein, J. A., Han, X., Hwa, L. S., DeBold, J. F., et al. (2015). Social stress and escalated drug self-administration in mice I. Alcohol and corticosterone. *Psychopharmacology, 232*(6), 991–1001. https://doi.org/10.1007/s00213-014-3733-9.

Novaes, L. S., Dos Santos, N. B., Batalhote, R. F. P., Malta, M. B., Camarini, R., Scavone, C., et al. (2017). Environmental enrichment protects against stress-induced anxiety: Role of glucocorticoid receptor, ERK, and CREB signaling in the basolateral amygdala. *Neuropharmacology, 113*, 457–466. Pt. A. https://doi.org/10.1016/j.neuropharm.2016.10.026.

Pastor, R., McKinnon, C. S., Scibelli, A. C., Burkhart-Kasch, S., Reed, C., Ryabinin, A. E., et al. (2008). Corticotropin-releasing factor-1 receptor involvement in behavioral neuroadaptation to ethanol: A urocortin1-independent mechanism. *Proceedings of the National Academy of Sciences of the United States of America, 105*(26), 9070–9075. https://doi.org/10.1073/pnas.0710181105.

Pautassi, R. M., Suárez, A. B., Hoffmann, L. B., Rueda, A. V., Rae, M., Marianno, P., et al. (2017). Effects of environmental enrichment upon ethanol-induced conditioned place preference and pre-frontal BDNF levels in adolescent and adult mice. *Scientific Reports*, 7(1), 8574. https://doi.org/10.1038/s41598-017-08795-0.

Phillips, T. J., Dickinson, S., & Burkhart-Kasch, S. (1994). Behavioral sensitization to drug stimulant effects in C57BL/6J and DBA/2J inbred mice. *Behavioral Neuroscience, 108*(4), 789–803. https://doi.org/10.1037/0735-7044.108.4.789.

Phillips, T. J., Roberts, A. J., & Lessov, C. N. (1997). Behavioral sensitization to ethanol: Genetics and the effects of stress. *Pharmacology, Biochemistry, and Behavior, 57*(3), 487–493. https://doi.org/10.1016/S0091-3057(96)00448-0.

Piazza, P. V., Deroche, V., Deminière, J. M., Maccari, S., Le Moal, M., & Simon, H. (1993). Corticosterone in the range of stress-induced levels possesses reinforcing properties: Implications for sensation-seeking behaviors. *Proceedings of the National Academy of Sciences of the United States of America, 90*(24), 11738–11742. https://doi.org/10.1073/pnas.90.24.11738.

Piazza, P. V., & Le Moal, M. (1998). The role of stress in drug self-administration. *Trends in Pharmacological Sciences, 19*(2), 67–74. https://doi.org/10.1016/S0165-6147(97)01115-2.

Pisu, M. G., Mostallino, M. C., Dore, R., Maciocco, E., Secci, P. P., & Serra, M. (2011). Effects of voluntary ethanol consumption on emotional state and stress responsiveness in socially isolated rats. *European Neuropsychopharmacology, 21*(5), 414–425. https://doi.org/10.1016/j.euroneuro.2010.07.006.

Quintanilla, M. E., Tampier, L., Sapag, A., Gerdtzen, Z., & Israel, Y. (2007). Sex differences, alcohol dehydrogenase, acetaldehyde burst, and aversion to ethanol in the rat: A systems perspective. *American Journal of Physiology: Endocrinology and Metabolism, 293*(2), E531–E537. https://doi.org/10.1152/ajpendo.00187.2007.

Rae, M. B., Zanos, P., Georgiou, P., Chivers, P., Bailey, A., & Camarini, R. (2018). Environmental enrichment enhances alcohol reward via an oxytocinergic-dependent mechanism. *Neuropharmacology, 138*, 267–274.

Reiss, D., Wolter-Sutter, A., Krezel, W., & Ouagazzal, A. M. (2007). Effects of social crowding on emotionality and expression of hippocampal nociceptin/orphanin FQ system transcripts in mice. *Behavioral Brain Research, 184*(2), 167–173. https://doi.org/10.1016/j.bbr.2007.07.010.

Risedal, A., Mattsson, B., Dahlqvist, P., Nordborg, C., Olsson, T., & Johansson, B. B. (2002). Environmental influences on functional outcome after a cortical infarct in the rat. *Brain Research Bulletin, 58*(3), 315–321. https://doi.org/10.1016/S0361-9230(02)00796-7.

Robinson, T. E., & Berridge, K. C. (1993). The neural basis of drug craving: An incentive-sensitization theory of addiction. *Brain Research. Brain Research Reviews, 18*(3), 247–291.

Robinson, T. E., & Berridge, K. C. (2000). The psychology and neurobiology of addiction: An incentive-sensitization view. *Addiction, 95*(8), 91–117, Suppl. 2.

Rockman, G. E., Gibson, J. E., & Benarroch, A. (1989). Effects of environmental enrichment on voluntary ethanol intake in rats. *Pharmacology, Biochemistry, and Behavior, 34*, 689–693.

Romeo, R. D. (2010). Pubertal maturation and programming of hypothalamic-pituitary-adrenal reactivity. *Frontiers in Neuroendocrinology, 31*(2), 232–240. https://doi.org/10.1016/j.yfrne.2010.02.004.

Rosenzweig, M. R., Bennett, E. L., Hebert, M., & Morimoto, H. (1978). Social grouping cannot account for cerebral effects of enriched environments. *Brain Research, 153*(3), 563–576. https://doi.org/10.1016/0006-8993(78)90340-2.

Rossi, C., Angelucci, A., Costantin, L., Braschi, C., Mazzantini, M., Babbini, F., et al. (2006). Brain-derived neurotrophic factor (BDNF) is required for the enhancement of hippocampal neurogenesis following environmental enrichment. *European Journal of Neuroscience, 24*(7), 1850–1856. https://doi.org/10.1111/j.1460-9568.2006.05059.x.

Rueda, A. V. L., Teixeira, A. M. A., Yonamine, M., & Camarini, R. (2012). Environmental enrichment blocks ethanol-induced locomotor sensitization and decreases BDNF levels in the prefrontal cortex in mice. *Addiction Biology, 17*(4), 736–745. https://doi.org/10.1111/j.1369-1600.2011.00408.x.

Sánchez, M. M., Ladd, C. O., & Plotsky, P. M. (2001). Early adverse experience as a developmental risk factor for later psychopathology: Evidence from rodent and primate models. *Development and Psychopathology, 13*(3), 419–449.

Sanchis-Segura, C., & Spanagel, R. (2006). Behavioural assessment of drug reinforcement and addictive features in rodents: An overview. *Addiction Biology, 11*(1), 2–28. https://doi.org/10.1111/j.1369-1600.2006.00012.x.

Santos-Rocha, J. B., Rae, M., Teixeira, A. M. A., Teixeira, S. A., Munhoz, C. D., Muscará, M. N., et al. (2018). Involvement of neuronal nitric oxide synthase in cross-sensitization between chronic unpredictable stress and ethanol in adolescent and adult mice. *Alcohol, 68*, 71–79. https://doi.org/10.1016/j.alcohol.2017.10.004.

Sapolsky, R. M., & Meaney, M. J. (1986). Maturation of the adrenocortical stress response: Neuroendocrine control mechanisms and the stress hyporesponsive period. *Brain Research, 396*(1), 64–76. https://doi.org/10.1016/0165-0173(86)90010-X.

Schenk, S., Gorman, K., & Amit, Z. (1990). Age-dependent effects of isolation on the self-administration of ethanol in laboratory rats. *Alcohol, 7*, 321–326.

Schwandt, M. L., Higley, J. D., Suomi, S. J., Heilig, M., & Barr, C. S. (2008). Rapid tolerance and locomotor sensitization in ethanol-naïve adolescent rhesus macaques. *Alcoholism: Clinical and Experimental Research, 32*(7), 1217–1228. https://doi.org/10.1111/j.1530-0277.2008.00676.x.

Segovia, G., Yagüe, A. G., García-Verdugo, J. M., & Mora, F. (2006). Environmental enrichment promotes neurogenesis and changes the extracellular concentrations of glutamate and GABA in the hippocampus of aged rats. *Brain Research Bulletin, 70*(1), 8–14. https://doi.org/10.1016/j.brainresbull.2005.11.005.

Shuster, L., Webster, G. W., & Yu, G. (1975). Increased running response to morphine in morphine-pretreated mice. *Journal of Pharmacology and Experimental Therapeutics, 192*(1), 64–67.

Sikora, M., Nicolas, C., Istin, M., Jaafari, N., Thiriet, N., & Solinas, M. (2018). Generalization of effects of environmental enrichment on seeking for different classes of drugs of abuse. *Behavioral Brain Research, 341*, 109–113. https://doi.org/10.1016/j.bbr.2017.12.027.

Silk, J. B. (2007). The adaptive value of sociality in mammalian groups. *Philosophical Transactions of the Royal Society B: Biological Sciences, 362*(1480), 539–559. https://doi.org/10.1098/rstb.2006.1994.

Simpson, J., & Kelly, J. P. (2011). The impact of environmental enrichment in laboratory rats—Behavioural and neurochemical aspects. *Behavioural Brain Research, 222*(1), 246–264. https://doi.org/10.1016/j.bbr.2011.04.002.

Sinclair, J. D. (1972). The alcohol-deprivation effect. Influence of various factors. *Quarterly Journal of Studies on Alcohol, 33*(3), 769–782.

Sinha, R. (2008). Chronic stress, drug use, and vulnerability to addiction. *Annals of the New York Academy of Sciences, 1141*, 105–130. https://doi.org/10.1196/annals.1441.030.

Solinas, M., Chauvet, C., Thiriet, N., El Rawas, R., & Jaber, M. (2008). Reversal of cocaine addiction by environmental enrichment. *Proceedings of the National Academy of Sciences of the United States of America, 105*(44), 17145–17150. https://doi.org/10.1073/pnas.0806889105.

Song, M., Wang, X. Y., Zhao, M., Wang, X. Y., Zhai, H. F., & Lu, L. (2007). Role of stress in acquisition of alcohol-conditioned place preference in adolescent and adult mice. *Alcoholism: Clinical and Experimental Research, 31*(12), 2001–2005. https://doi.org/10.1111/j.1530-0277.2007.00522.x.

Stairs, D. J., & Bardo, M. T. (2009). Neurobehavioral effects of environmental enrichment and drug abuse vulnerability. *Pharmacology, Biochemistry, and Behavior, 92*(3), 377–382. https://doi.org/10.1016/j.pbb.2009.01.016.

Steiner, B., Zurborg, S., Hörster, H., Fabel, K., & Kempermann, G. (2008). Differential 24 h responsiveness of Prox1-expressing precursor cells in adult hippocampal neurogenesis to physical activity, environmental enrichment, and kainic acid-induced seizures. *Neuroscience, 154*, 521–529. https://doi.org/10.1016/j.neuroscience.2008.04.023.

Steketee, J. D., & Kalivas, P. W. (2011). Drug wanting: Behavioral sensitization and relapse to drug-seeking behavior. *Pharmacological Reviews, 63*(2), 348–365. https://doi.org/10.1124/pr.109.001933.

Thiriet, N., Gennequin, B., Lardeux, V., Chauvet, C., Decressac, M., Janet, T., et al. (2011). Environmental enrichment does not reduce the rewarding and neurotoxic effects of methamphetamine. *Neurotoxicity Research, 19*(1), 172–182. https://doi.org/10.1007/s12640-010-9158-2.

Toyoshima, M., Yamada, K., Sugita, M., & Ichitani, Y. (2018). Social enrichment improves social recognition memory in male rats. *Animal Cognition*, Epub ahead of print. https://doi.org/10.1007/s10071-018-1171-5.

Uarquin, D. G., Meyer, J. S., Cardenas, F. P., & Rojas, M. J. (2016). Effect of overcrowding on hair corticosterone concentrations in juvenile male Wistar rats. *Journal of the American associations for Laboratory Animal Sciences, 55*(6), 749–755.

Uhart, M., & Wand, G. S. (2009). Stress, alcohol and drug interaction: An update of human research. *Addiction Biology, 14*, 43–64. https://doi.org/10.1111/j.1369-1600.2008.00131.x.

Van Dellen, A., Blakemore, C., Deacon, R., York, D., & Hannan, A. J. (2000). Delaying onset of Huntington's in mice. *Nature, 404*(6779), 721–722. https://doi.org/10.1038/35008142.

van Erp, A. M., & Miczek, K. A. (2001). Persistent suppression of ethanol self-administration by brief social stress in rats and increased startle response as index of withdrawal. *Physiology & Behavior, 73*(3), 301–311. https://doi.org/10.1016/S0031-9384(01)00458-9.

Van Praag, H., Kempermann, G., & Gage, F. H. (2000). Neural consequences of environmental enrichment. *Nature Reviews. Neuroscience, 1*(3), 191–198. https://doi.org/10.1038/35044558.

Varlinskaya, E. I., & Spear, L. P. (2004). Changes in sensitivity to ethanol-induced social facilitation and social inhibition from early to late adolescence. *Annals of the New York Academy of Sciences, 1021*, 459–461. https://doi.org/10.1196/annals.1308.064.

Vazquez, V., Penit-Soria, J., Durand, C., Besson, M. J., Giros, B., & Daugé, V. (2006). Brief early handling increases morphine dependence in adult rats. *Behavioural Brain Research, 170*(2), 211–218. https://doi.org/10.1016/j.bbr.2006.02.0.

Vengeliene, V., Siegmund, S., Singer, M. V., Sinclair, J. D., Li, T. K., & Spanagel, R. (2003). A comparative study on alcohol-preferring rat lines: Effects of deprivation and stress phases on voluntary alcohol intake. *Alcoholism, Clinical and Experimental Research, 27*(7), 1048–1054. https://doi.org/10.1097/01.ALC.0000075829.81211.0C.

Wallach, M. B., & Gershon, S. (1971). Sensitization to amphetamines. *Psychopharmacology Bulletin,* 7(4), 30–31.

Williams, B. M., Luo, Y., Ward, C., Redd, K., Gibson, R., Kuczaj, S. A., et al. (2001). Environmental enrichment: Effects on spatial memory and hippocampal CREB immunoreactivity. *Physiology and Behavior, 73*(4), 649–658. https://doi.org/10.1016/S0031-9384(01)00543-1.

Xu, Z., Hou, B., Gao, Y., He, F., & Zhang, C. (2007). Effects of enriched environment on morphine-induced reward in mice. *Experimental Neurology, 204*(2), 714–719. https://doi.org/10.1016/j.expneurol.2006.12.027.

Yap, J. J., Chartoff, E. H., Holly, E. N., Potter, D. N., Carlezon, W. A., Jr., & Miczek, K. A. (2015). Social defeat stress-induced sensitization and escalated cocaine self-administration: The role of ERK signaling in the rat ventral tegmental area. *Psychopharmacology, 232*(9), 1555–1569. https://doi.org/10.1007/s00213-014-3796-7.

Zgierska, A., Rabago, D., Chawla, N., Kushner, K., Koehler, R., & Marlatt, A. (2009). Mindfulness meditation for substance use disorders: A systematic review. *Substance Abuse, 30*(4), 266–294. https://doi.org/10.1080/08897070903250019.

Zimmerberg, B., & Sageser, K. A. (2011). Comparison of two rodent models of maternal separation on juvenile social behavior. *Frontiers in Psychiatry, 2*, 39. https://doi.org/10.3389/fpsyt.2011.00039.

FURTHER READING

Fahlke, C., Hard, E., & Eriksson, C. J. (1997). Effects of early weaning and social isolation on subsequent alcohol intake in rats. *Alcohol, 14*(2), 175–180. https://doi.org/10.1016/S0741-8329(96)00141-3.

CHAPTER FOUR

Social Influences in Animal Models of Opiate Addiction

Seven E. Tomek, M. Foster Olive[1]
Department of Psychology, Arizona State University, Tempe, AZ, United States
[1]Corresponding author: e-mail address: foster.olive@asu.edu

Contents

Abstract

Opiate addiction has reached an epidemic prevalence in recent years, yet social influences on the use and abuse of opiates has been widely understudied. In particular, the neurobiological substrates of opiate addiction and their modulation by social influences are largely unknown, perhaps due to the lack of widespread incorporation of social variables into animal models of opiate addiction. As reviewed here, animal models such as oral and intravenous drug self-administration, conditioned place preference, behavioral sensitization, and the effects of various stressors, have been useful in identifying some of the neurochemical circuitry that mediate social influences on opiate addiction. However, it is clear from our review that newer paradigms that incorporate various social elements are greatly needed to provide more translational insights into the neurobiological basis of opiate addiction. These elements include social and environmental enrichment, presence of conspecifics, and procedures that require subjects to exert effort to engage in prosocial behavior. A wider implementation of social variables into animal models of opiate addiction will help inform neurobehavioral strategies to increase the efficacy of treatment.

International Review of Neurobiology, Volume 140
ISSN 0074-7742
https://doi.org/10.1016/bs.irn.2018.07.004

Opiates are a broad class of alkaloids that are widely used for their analgesic and anti-tussive properties. Naturally occurring opiates found in the latex of the opium poppy *Papaver somniferum* include codeine and morphine, the latter of which is acetylated to form the highly addictive opiate heroin (diacetylmorphine). Other opiate alkaloids, now manufactured almost entirely via synthetic routes, include oxycodone, hydromorphone, methadone, buprenorphine, and fentanyl, to name a few (Bond & Witton, 2017). The term "opiate" is often used in reference to these compounds, as is the term "opioid," though some clinicians and researchers prefer to use the term opiate to refer to alkaloids chemical compounds and the term opioid to refer to endogenous opioid systems. In this review we will attempt to maintain this distinction. Prior to reviewing information on the current opiate epidemic and rodent models of opiate addiction, we will first review the mammalian endogenous opioid systems.

1. ENDOGENOUS OPIOID SIGNALING

The endogenous opioid system is implicated in numerous normal and pathophysiological processes, including addiction, nociception, consummatory behaviors, reproduction, mood regulation, thermoregulation, respiration, and immune and peripheral organ function (Bodnar, 2013). This system is comprised of several families of neuropeptides, each derived from larger precursor proteins including pro-opiomelanocortin, proenkephalin, and prodynorphin. Following their synthesis, these precursor peptides are enzymatically cleaved to form multiple bioactive fragments, including various endorphins (i.e., β-endorphin), enkephalins (i.e., Met- and Leu-enkephalin), and dynorphin A. Additional members of the opioid peptide family include endomorphins, the precursor of which has remained elusive, nociceptin/orphanin FQ, nocistatin, morphiceptin, and others. These neuropeptides are released from dense core vesicles and bind to one or more widely distributed G-protein coupled receptors (GPCRs) including as μ, δ, κ, and opioid receptor-like 1 (ORL1) (Compton, Jones, & Baldwin, 2016; Maher, Martin, & Childers, 2005). Numerous variants of each of these receptors can arise from alternative splicing and posttranslational modifications, which alter their trafficking and ligand binding profiles (Stein, 2016). Other receptors such as sigma and ORL1 receptors have been proposed to bind opiate alkaloids, but their inclusion into the classical family of opioid receptors is often debated due to their structural differences and non-traditional binding profiles.

Opioid peptides as well as synthetic opiate analgesics such as morphine, hydromorphone, oxycodone, etc. exert their primarily pharmacological effects by activating one or more of the classic μ, δ, or κ receptors. These GPCRs utilize inhibitory G proteins such as G_i/G_o to inhibit cyclic adenosine monophosphate production and reduce neural activity via several different second and third messenger systems (Al-Hasani & Bruchas, 2011). While opioid receptor activation produces a general reduction in cellular activity, this can also lead to increased firing of various neuronal subpopulations such as midbrain dopaminergic neurons (Johnson & North, 1992) via disinhibition of local inhibitory GABAergic interneurons, resulting in increases in dopamine release in the ventral striatum and nucleus accumbens (Compton et al., 2016). While beyond the scope of the present chapter, chronic opiate-induced adaptations in receptor function and intracellular signaling can result in the development of tolerance and physiological dependence, leading to higher doses needed to achieve the same desired effects.

2. HISTORICAL ASPECTS OF SOCIAL INFLUENCES ON OPIATE USE

The use of opium dates back over 4000 years, as written accounts of its medical applications have been found in Asia Minor and surrounding areas, and many references to opium use can also be found in the writing of the ancient Greeks (Kapoor, 1995). From these regions, it is believed that opium use subsequently spread westward into Europe and eastward toward India, China, and other parts of Southeast Asia. However, in the 13th century, opium use in Europe was drastically reduced as of the result of Inquisition-stoked associations of the drug with Eastern religions. Nevertheless, the use of opium for both its euphorigenic and medical properties continued to become a widespread societal issue in countries such as China. Numerous legislative efforts were made to suppress the use of opium, yet its sale and trafficking became a source of international conflict that culminated in the First and Second Opium Wars between China and Britain (Goldberg & Latimer, 2014). Around the end of the Second Opium War in the 1860s, the United States was undergoing exponential growth as a result of industrialization, the lucrative allure of the California Gold Rush, and the expansion of the railroad industry. This rapid growth resulted in a significant labor shortage, and many impoverished Chinese sought work and financial stability through the promise of opportunity in the United States (Olive, 2017).

Ultimately, however, Chinese immigrants were ostracized for many reasons including the low cost of their labor, and forced to form their own subcommunities within the United States. This stigmatization of Chinese immigrants carried over to their use of opium, forcing them to confine their use into so-called "opium dens." While many factors played a role, it is plausible to argue that this demonization of Chinese immigrants and their opium use was one of the first widespread negative social influences on the use of opiate drugs. Ironically, however, it was these same opium dens that introduced opium to non-Chinese citizens (Goldberg & Latimer, 2014).

Other significant events in the development and spread of opiate use soon followed (Olive, 2017). One of the first was in 1805 when German pharmacist Friedrich Sertürner first isolated and characterized morphine as the primary active alkaloid component of opium. Subsequent work identified additional active alkaloids including codeine, thebaine, and papaverine. Not long after, the hypodermic syringe was developed which permitted easy administration of morphine and other opiates directly into the bloodstream or muscle. This invention came at a fortuitous time, as the American Civil War erupted the following decade, and morphine was found to be a potent pain reliever for wounded soldiers. Its widespread use as a battlefield analgesic resulting in a high incidence of morphine dependence in soldiers.

Despite its widespread use in this exclusively male population, recreational opiate use also became popular among upper class women (Courtwright, 1982). The post-Civil War spike in morphine dependence led scientists to explore the development of other analgesics, and it was soon discovered that boiling morphine in acetic acid produced diacetylmorphine. In the late 1800s, German pharmaceutical company Bayer commenced marketing of diacetylmorphine under the brand name of Heroin as an anti-tussive and treatment for tuberculosis. Remarkably, in the early 1900s, the American Medical Association approved the use of prescription heroin and promoted its dispensing in lieu of morphine.

3. THE CURRENT OPIATE EPIDEMIC

Currently, opiate use is at an all-time high (Rudd, 2016). Much of the current epidemic is being attributed to the dramatic increase in prescriptions of opiate analgesic (Ciccarone, Ondocsin, & Mars, 2017; Ostling et al., 2018). The Centers for Disease Control (CDC) reports that the number of issued prescriptions for opiates has quadrupled since 1999;

however, there has not been a parallel increase in reports of incidences of chronic pain (Department of Health, Services, & for Disease Control, 2017). A European study examining gender differences in opiate dependent adults reported no gender differences regarding the age of onset of regular opiate use; however, at up to 4 years of consistent use there are gender specific differences in the consequences of opiate addiction. Specifically, women were shown to exhibit more social issues and men experience more legal difficulties with opiate use, specifically looking at history of arrests (Hölscher et al., 2010).

Heroin is one of the most commonly used and abused opiates, and can be administered in several different ways, including smoking, oral intake, insufflation, or most frequently, intravenous injection. It is not uncommon for a new user to start out using heroin in a less invasive way such as smoking or insufflation, but begin injecting heroin to bolster the speed and intensity of their "high" (Bond & Witton, 2017). When consumed, heroin provides the user with feelings of intense euphoria, accompanied by warmth and flushing, limbs feeling heavy, and some undesirable effects such as dry mouth, nausea, and vomiting. These initial experiences are followed with feelings of lethargy and impaired mental functioning. Further contributing to the rapidly increasing rates of heroin use is its relative availability and low cost, as well as containing laced with fentanyl or other potent synthetic opiates (Ciccarone et al., 2017; Ostling et al., 2018). The addition of fentanyl or other potent opiates to the heroin supply has increased its addiction potential as well as its proclivity to induce fatal overdose (Rudd, 2016).

Since the end of the 20th century, heroin was the drug of choice for low-income, Caucasian males between the ages of 18–25 (Jones, Campopiano, Baldwin, & McCance-Katz, 2015). However, the ongoing increase of heroin use is now showing record rates in both males and females, and spanning the entire socioeconomic spectrum (Cicero & Ellis, 2017; Compton et al., 2016; Jones et al., 2015). In addition, the demographics of heroin use has transitioned from predominant use by low-income ethnic minorities in urban areas to use in suburban and rural regions (Cicero & Ellis, 2017). Like other abused substances, there is a cyclical pattern that occurs with opiate use, in which the drug is consumed, euphoria and intoxication ensue, tolerance is developed, higher amounts of the drug is ingested, resulting in symptoms of withdrawal when abstinent. These feelings of dysphoria and negative affect prompt drug seeking to stave off the physical and mental discomfort (Koob, 2008).

Opiate use disorder is characterized in the fifth edition of the American Psychiatric Association's *Diagnostic and Statistical Manual of Mental Disorders (DSM-5)* as meeting at least two or more criteria that include impaired control over drug use, compulsive use, continued use despite harm, craving, withdrawal symptoms, or persistent or recurrent social or interpersonal problems caused or made by the effects of opiates (APA, 2013). Many studies have aimed to evaluate intrapersonal problems and their role in substance use disorders; however, understanding the social influences on opiate addiction is highly complex because it incorporates a dynamic set of genetic, environmental, and social factors.

While it is widely accepted that a person's peer group or social network impacts their drug using behaviors (Buchanan & Latkin, 2008; Eitan, Emery, Bates, & Horrax, 2017; Luthar, Anton, Merikangas, & Rounsaville, 1992; Russell, Trudeau, & Leland, 2015), researchers continue to examine the directionality of social influence. In other words, it is unclear opiate addict self-selects into a peer group of people who use drugs, or if the peer group impacts the person's decision to begin or maintain use (Bohnert, Bradshaw, & Latkin, 2009). Results from a self-report study investigating this phenomenon found evidence that both scenarios are at play, but that change in drug use over time was not due to a change in their peers' drug-taking behavior, but a function of a changing social network (Bohnert et al., 2009). This is supported by findings that psychosocial factors including peer-group relationships, family issues, employment and social support can be used as predictors of long-term opiate use (Scherbaum & Specka, 2008). Cicero and Ellis (2017) recently reported in a meta-analysis of studies of prescription opiate abusers, many report that prescription opiates are used as a "safe" alternative to illicit opiates with less stringent legal ramifications for non-medical use and distribution. In addition, while "getting high" and alleviation of withdrawal are often the primary motivating factors in non-medical use of prescription opiates, other factors include coping with life stressors and self-medication of psychological and emotional issues (Cicero & Ellis, 2017).

Social networks and influences are contributing factors not only in the initial onset of the first use and development of opiate use addiction, but the escalation of non-injection to injection use (Koram et al., 2011). These investigators found that the transition from smoked or snorted heroin use to intravenous injection was associated with the users' perception of their peer's drug-related behaviors. Initial drug use is highly influenced by family and peers, and especially in young individuals the transition to injection drug use is socially sanctioned and condoned by friends (Sherman, Smith, Laney, & Strathdee, 2002). Moreover, the increased prevalence of heroin use has diminished its traditional social stigma (Cicero & Ellis, 2017).

Successful cessation of opiate use is also impacted by whether a recovering addict's social network consists of active drug users or drug-free persons (Buchanan & Latkin, 2008; Eitan et al., 2017). Individuals who successfully completed treatment and remained abstinent after treatment had reduced their drug-using network members significantly, compared to non-quitters (Buchanan & Latkin, 2008). Though directionality is an ongoing topic of study, neighborhood poverty is significantly correlated with current heroin use (Williams & Latkin, 2007), and an important aspect of maintaining or attenuating heroin addiction is the type and quality of social relationships. While this social aspect can offer a benefit to reducing drug use, it does not buffer against the detriments associated with neighborhood poverty. Williams and Latkin (2007) found that social support of friends and family significantly impacted ongoing heroin use, but only if the friends and family were a positive support. If social networks were prone to drug use, then drug use was reinforced and persisted. Results of another study examining social influences on the escalation of opiate use in adolescents found that increased exposure to drug-using peers led to a higher likelihood of initiation of drug use and greater use durations and frequency (Russell et al., 2015). In addition to interpersonal relations, psychosocial functioning is predictive of later drug use (Griffith, Knight, Joe, & Simpson, 1998). Other significant risk factors include age of first drug experimentation, sensation-seeking personality traits (Luthar et al., 1992), proclivity to depression, and low self-esteem (Griffith et al., 1998).

Drugs are often consumed in social settings, and much like the peer/network relationship and opiate dependence, the directionality of the social context and drug effects is unclear. The subjective effects of opiates (in this case) may be heightened by the presence of peers, or the valence of social interaction might be mediated by opiates (de Wit & Sayette, 2018). de Wit and Sayette (2018) examined the directionality of this phenomenon and found that opiates appeared to increase prosocial effects and reduce subjective responses to negative social perceptions. Little research is available on the effects of social context on mood responses to opiates; however, de Wit and Sayette (2018) discussed a study in which men self-administering heroin in a clinical setting initially were more social, but over the remaining 8 days of the study chose to spend more and more time alone. There currently very little research in humans that has examined this exact construct in controlled settings; however, there are preclinical studies in which the rewarding effects of opiates in laboratory animals are modulated by social enrichment and or social interaction as will be discussed below.

Familial support and peer relationships are an important component of an opiate addicts success in treatment and recovery (Griffith et al., 1998). Interestingly, individuals struggling with substance abuse have more contact with their families than non-drug users, even if their familial situation is dysfunctional or stressful (Griffith et al., 1998; Stanton & Shadish, 1997). More so, people with a history of dysfunctional family dynamics are significantly more likely to associate with deviant and/or drug-using peers (Griffith et al., 1998). The proximity of the family to the drug abusing individual, and their influence on the development and maintenance of the drug addiction via modeling, enabling, or stressors, make social and familial involvement in treatment necessary. A comprehensive meta-analysis of treatment outcomes for drug-addicted individuals revealed that clients who participated in therapy that incorporated family involvement used significantly less drugs after treatment than those who received therapy that did not include family participation (Stanton & Shadish, 1997). Social and familial networks are useful in recovery, but the quality of support is highly influential. For example, the higher the number of substance-abusers within the social network the higher the likelihood of relapse (Wasserman, Stewart, & Delucchi, 2001).

The most common methods of treatment for opiate dependence are maintenance pharmacotherapies (i.e., methadone, buprenorphine), or cognitive behavioral therapy, and most often, a combination of both therapy and drug maintenance. Methadone, a full μ opioid receptor agonist, has been widely used as a treatment for opiate dependence for nearly 40 years (Donny, Brasser, Bigelow, Stitzer, & Walsh, 2005). However, methadone has a very long elimination half-life which can lead to accumulation and overdose. Buprenorphine is a partial μ opioid receptor agonist, and while considered safer than methadone (lower risks of over dose and respiratory depression), it carries some abuse liability (Walsh et al., 2017). Methadone and buprenorphine aid in treatment by activating the same opioid receptors in the brain as the abused opiate, thus lessening cravings and attenuating withdrawal symptoms. Buprenorphine is often formulated with the nonspecific opioid receptor antagonist naloxone that is intended to deter abuse via opioid receptor blockade if taken intravenously, whereas if taken as intended via the oral route the effects of naloxone are largely negligible due to its poor oral bioavailability.

4. STANDARD RODENT MODELS OF OPIATE ADDICTION

4.1 Self-Administration

The animal self-administration model is widely considered to have the most face and predictive validity model of human drug addiction (Belin-Rauscent & Belin, 2011; Panlilio & Goldberg, 2007). Laboratory models of self-administration allow researchers to observe and quantify physiological, behavioral, and cognitive aspects of drug effects (Jones & Comer, 2013). This method allows researchers to design and develop studies that are illustrative of the human condition, in that the animals voluntarily determine if and how much drug they self-administer. The self-administration model is rooted in operant conditioning, which relies on the premise that when the animal receives an infusion of drug, its reinforcing properties increase the likelihood of the behavior that produced the drug delivery. Laboratory animals can self-administer both drug (heroin, cocaine, alcohol) and non-drug (sucrose, food) reinforcers through a variety of clinically relevant routes, including orally, intracranially, or intravenously (Belin-Rauscent & Belin, 2011). Often, delivery of the reinforcer is paired with a discrete cue such as a light, tone, or both. A major strength of the self-administration model is that it allows experimental subjects to consume opiates volitionally and intravenously, which is consistent with the most common method of heroin use in humans (Panlilio & Goldberg, 2007). In order to allow laboratory animals to self-administer opiates, first they need to be surgically implanted with an intravenous catheter. The catheter is usually inserted into the jugular vein and attached to a subcutaneous vascular port. Tubing contained within a spring tether then connects the port to a computer-controlled syringe pump. The animal can initiate a drug infusion by pressing a lever or inserting their nose into a nosepoke aperture (Watterson et al., 2014).

Studies that employ the self-administration paradigm often use operant reinforcement schedules. A common schedule is a fixed ratio (FR) schedule, in which a specific number of responses are required in order for the animal to receive the drug reinforcer. For example, in an FR1 schedule, a single lever press results in delivery of one reinforcer. Progressive ratio (PR) is another schedule of reinforcement, in which the number of lever presses needed to receive each successive reinforcer progressively increases (either arithmetically or exponentially). In a PR schedule, the session ends when the animal fails to respond for a specified time, which is considered the "breakpoint." This schedule is often used quantify motivation for the drug and is thus a measure of reinforcer strength or efficacy (Panlilio & Goldberg, 2007).

More complex schedules of reinforcement can be developed to assess other properties of substance abuse such as cue-induced drug seeking and reinstatement. The reinstatement model of self-administration is considered a useful method that translates to the phenomenon of relapse in addicted humans (Shaham, Shalev, Lu, De Wit, & Stewart, 2003). This model operates on the idea that after stable levels of drug self-administration are achieved, extinction of operant responding is achieved by withholding the drug reinforcer, and subsequently drug-seeking behavior can be reinstated as the result of brief exposure to the drug (drug priming), exposure to a stressor, or re-exposure to drug-paired cues (Belin-Rauscent & Belin, 2011; Namba, Tomek, Olive, Beckmann, & Gipson, 2018; Shaham et al., 2003).

Previous studies have established that laboratory animals willingly intravenously self-administer opiates including heroin, morphine, and oxycodone, and will continue to increase their intake if given unlimited access, while animals that have limited access (i.e., 3 h or less per day) fail to develop dependence and maintain significant levels of intake (Chen et al., 2006). Chen et al. (2006) proposed and tested an animal model of opiate dependence in which they observed escalating heroin self-administration in relation to sleep and food intake patterns. Their results indicated that if given the opportunity, rats will self-administer heroin to the point of dependence, here measured by tracking the reduced food and water intake in relation to the increase of heroin consumption. Interestingly, rats given the opportunity to self-administer heroin in long access (>12 h) daily sessions are more resistant to extinction with the drug is withheld (Badiani, Belin, Epstein, Calu, & Shaham, 2011). Rats that exhibit steady heroin intake show lower intracranial self-stimulation (ICSS) thresholds compared to rat with escalated intake patterns which show elevations in ICSS thresholds, indicative of a diminished impact of the effects of heroin on brain reward function and possibly reflective of opiate tolerance (Vendruscolo et al., 2011).

4.2 Conditioned Place Preference

Another popular paradigm utilized in modeling opiate addiction is the conditioned place preference (CPP) approach. While there are both biased and unbiased variations, a typical unbiased CPP experiment is conducted in a two compartment apparatus featuring environmental (i.e., tactile, visual) cues that are unique to each compartment and do not produce any innate preference for either (Beloate & Coolen, 2017). During conditioning, the animal repeatedly receives passive drug injections followed by confinement to one of the two compartments, allowing the formation of associations between drug effects and contextual cues, and the other compartment with a neutral substance such as saline. Following conditioning, animals are re-tested in a drug-free state, and if the animal chooses to spend more time in the drug-conditioned compartment, then the drug is interpreted to have a rewarding value. Critics of the CPP paradigm argue that the ability of a drug to produce CPP does not always correlate to drug addiction in humans, as both natural rewards and drugs that have low abuse potential in humans can produce CPP in rodents (Tzschentke, 2007). In addition, the number of drug exposures is relatively low in CPP procedures, rarely results in the development of physical dependence, and is passively administered by an experimenter. Conversely, conditioned place aversion (CPA) occurs when an aversive drug experience, such as a very high drug dose or during withdrawal, drives the animal to spend less time in the drug or withdrawal-conditioned compartment.

Extensive research has been performed on opiates using the CPP paradigm, in species ranging from invertebrates to fish to mammals. However, the bulk of the literature has focused primarily on rodent species (i.e., mice, rats, and prairie voles) (Tzschentke, 2007). These studies have shown repeatedly that morphine, heroin, buprenorphine, oxycodone, fentanyl and other opioid receptor agonists, or partial agonists induce CPP, with classical bell-shaped dose response relationships characterized by mid-range doses producing CPP, lower doses having no effects, and high doses producing sedation and/or CPA (Rech, Mokler, & Briggs, 2012; Tzschentke, 2007).

Opioid receptor antagonists such as naloxone are also capable of producing CPA, though some researchers have reported no effect of these treatments in the absence of prior opioid administration (Tzschentke, 2007). Interestingly, when animals conditioned with morphine receive naloxone injections before the CPP test session, they exhibit stronger CPP than animals receiving saline. Animals conditioned with morphine and naloxone simultaneously failed to develop CPP, suggesting that opioid antagonists block the acquisition but enhance the expression of morphine CPP (Neisewander, Pierce, & Bardo, 1990). Morphine CPP is also abolished by selectively blocking protein synthesis in the basolateral amygdala, hippocampus, or nucleus accumbens (Milekic, 2006; Pooriamehr, Sabahi, & Miladi-Gorji, 2017). When animals are prevented from synthesizing new proteins after conditioning with morphine and 24 h prior to testing, they are unable to consolidate the contextual drug-related memories from the conditioning sessions, presumably due to the necessity of newly synthetized proteins (i.e., glutamate receptors, structural proteins, etc.) for establishing contextual memories.

Housing conditions, which dictate the level of baseline social enrichment available to each animal, have a significant effect on the sensitivity of rodents toward developing CPP for various drugs including opiates. Rats reared in group-housing conditions acquired CPP in response to a wide range of heroin doses tested, while animals reared in isolation developed CPP only in response to the highest dose (Schenk, Hunt, Colle, & Amit, 1982). However, others have found that changes in the social climate between housing and conditioning contexts enhanced sensitivity to morphine CPP in adolescent mice (Kennedy, Panksepp, Runckel, & Lahvis, 2012). For example, mice reared in isolation but conditioned with peers developed CPP to lower doses of morphine compared to mice that were immersed in social groups during housing and conditioning. These inconsistent findings may be due to differences in species or experimental methodologies (i.e., testing apparatus, social interactions within/outside drug context, etc.). It is also important to note these socially induced changes in opiate reward sensitivity reflect only the relatively short period of CPP acquisition and do not necessarily generalize to other paradigms of long-term opiate use (Bozarth, Murray, & Wise, 1989; El Rawas, Thiriet, Lardeux, Jaber, & Solinas, 2009).

Interactions with drug-naïve animals has also been shown to slow the rate of acquisition of morphine CPP. Adolescent mice housed in peer groups that had previously been exposed to morphine acquired CPP to morphine more quickly and with lower doses than animals housed with drug-naïve animals (Bates, Emery, Wellman, & Eitan, 2014). Animals housed with drug-naïve conspecifics also extinguished CPP more rapidly, which reflects the ability of humans to achieve higher success at sobriety when avoiding contact with other drug users (Neisewander, Peartree, & Pentkowski, 2012). The consideration of social enrichment is important for creating an accurate model of the human condition of drug addiction, as major contributing factors to opiate addiction in adolescence are inadequate social support networks and social isolation (Bates et al., 2014; Kennedy et al., 2012).

The CPP paradigm can also be used to model relapse. Following standard conditioning and testing for a place preference, animals undergo extinction either by additional conditioning with only saline in both compartments, or by repeated placement into the testing apparatus, which induces a steady reduction in preference for the drug-paired compartment in the absence of additional drug conditioning sessions. Following extinction, animals are exposed to either stress or a priming injection of the original conditioning drug, which are sufficient to reinstate CPP, and thus serve as a potential model of relapse (Weiss, 2005). The CPP model of relapse has been adapted to incorporate social stressors, as these types of social interactions are a major contributor to the recidivism of recovering addicts in humans (Ribeiro Do Couto et al., 2006). Social stress in rodents can be induced via interaction with an aggressive male conspecific who is prone to fighting. When subjected to this stressful interaction after extinction training, CPP can be reinstated (Ribeiro Do Couto et al., 2006; Tzschentke, 2007). Social defeat can also prevent acquisition of morphine CPP when experienced a few days before testing, which suggests a decrease in sensitivity to the rewarding effects of opiates (Coventry, D'Aquila, Brain, & Willner, 1997). These data support the hypothesis that social stress can modulate opiate reward and promote relapse.

4.3 Behavioral Sensitization

Sensitization, also referred to as reverse tolerance, is a phenomenon in which following repeated exposure to the same dose of a particular drug of abuse, a specific behavioral, physiological, or cellular response increases, rather than decreases, over time. Such sensitized responses can persist for weeks or months. The most commonly used measured in preclinical addiction research is the sensitization to the locomotor activating effects of a particular drug, the most common type examined being psychostimulants (Belin-Rauscent & Belin, 2011). Behavioral sensitization can also model individual differences in drug sensitivity (Steketee & Kalivas, 2011). However, behavioral (i.e., locomotor) sensitization can also occur in rodents exposed to low doses of opiate drugs, as higher doses tend to be sedating and do produce hypolocomotion. Behavioral sensitization paradigms are relatively straightforward and economical to conduct, and are believed to model the lasting maladaptive plasticity in the brain following repeated drug intake that may result in increased sensitivity to the neurobiological effects of abused drugs that possibly contribute to increased propensity for intake, abuse, and relapse (Belin-Rauscent & Belin, 2011).

5. SUMMARY OF EXISTING LITERATURE ON SOCIAL INFLUENCES ON OPIATE ADDICTION BEHAVIORS IN RODENTS

5.1 Social and Environmental Enrichment

In preclinical research, attempts to understand the effects of the environment on the development and maintenance of opiate addiction often incorporate the paradigm of environmental enrichment. An enriched environment is one that contains objects in the home cage such as a running wheel, toys, tubes, and other additions as compared to the standard polycarbonate cage environment that a laboratory animal typically resides in (Xu, Hou, Gao, He, & Zhang, 2007). An environment can also be considered enriched if the laboratory animal subject is housed with other animals, as opposed to being housed in isolation. The benefits of an enriched environment appears to have protective effects from the consequences of opiate use, as discussed below.

An early study examined the role of social environment in the consumption of an oral morphine solution, where rats were raised either in social colonies vs in isolation (Alexander, Beyerstein, Hadaway, & Coambs, 1981). In addition to comparing the effects of initial environment (social vs. isolated), this study was conducted in a counterbalanced fashion such that half of each group was placed in the opposite condition 65 days into the study. Results indicated that rats initially housed in an enriched environment consumed less morphine than isolated rats. Further, rats living in a colony at time of testing consumed less morphine than isolated rats regardless of their prior environmental conditions (Alexander et al., 1981). These findings are among the earliest to demonstrate the influence of social housing conditions and environment enrichment on voluntary opiate intake in rodents.

The impact of social enrichment was further tested in a study that examined how quickly isolated rats learned to intravenously self-administer heroin as well as the amount of heroin self-administered compared to group housed rats (Bozarth et al., 1989). In this experiment, rats were either housed individually in cages that precluded any visual or tactile contact with other rats, or in groups of 10 in a large cage that allowed for ample social interaction. Rats were allowed to self-administer heroin in 2-h sessions for a total of 25 days (in five 5-day a week blocks). After each self-administration session, rats were returned to their respective isolated or group-housing conditions. Both groups of animals learned to self-administer heroin over the five 5-day blocks of 2 h/day heroin access, yet single-housed rats self-administered significantly more heroin during the second, third, and fourth blocks of 5-day periods time than socially housed animals. By the end of the study there were no significant differences in total heroin intake, yet isolated rats acquired self-administration more quickly (Bozarth et al., 1989). These observations suggest that social enrichment may affect the initial reinforcing effects of heroin and/or learning related to general operant conditioning. Another study examined differences between group housed and isolated rats in their sensitivity to the conditioned rewarding effects of heroin (0, 20, 40, and 80 μg/kg, s.c.) using the CPP paradigm (Schenk et al., 1982). Results of this study indicated that group housed rats showed conditioned heroin reward at all doses tested, whereas isolated rats only demonstrated rewarding effects at the highest dose. Thus, social interaction can alter the conditioned rewarding effects of opiates.

The positive effects of social enrichment are further supported by a study where socially isolated rats (both male and female) were shown to ingest a higher volume of a morphine-containing solution than their socially housed conspecifics (Raz & Berger, 2010). Another study that examined the potentially insulating effects of enrichment on the rewarding properties of opiates also showed changes in intake rates between environmentally enriched rats and those housed in standard conditions (Abadi, Miladi-Gorji, & Bigdeli, 2016). In this study, subjects were randomly assigned into enriched or standard environments, passively made dependent on morphine with twice daily injections (10 mg/kg, 12 h intervals) of morphine for 14 days, and then undergoing naloxone-precipitated withdrawal. During withdrawal, rats were given the option of consuming a morphine or sucrose-containing solution. Results showed that rats who had experienced environmental enrichment consumed significantly less morphine at a dose of 0.5 mg/mL morphine than rats housed under standard conditions.

In another study (El Rawas et al., 2009), mice were randomly assigned to either enriched or standard housing conditions. Mice were then tested for the development of a heroin or sucrose CPP, assessed for opiate-stimulated locomotor activity, and also underwent in vivo microdialysis procedures to examine heroin-induced changes in extracellular dopamine in the nucleus accumbens. Results show that the rewarding effects of heroin were reduced in mice from the enriched environment condition, but sucrose-conditioned mice showed no effect of environment (El Rawas et al., 2009). However, no effects of housing condition were observed on the locomotor activating effects of heroin nor the ability of heroin to elevate extracellular dopamine levels in the nucleus accumbens. These data indicate that environmental enrichment offers some protective benefits in regard to heroin (i.e., diminished conditioned rewarding effects), although locomotor and mesolimbic dopaminergic effects of heroin appear to be unaltered.

These studies show that an enriched environment has several beneficial outcomes with regards to opiate intake. There is also evidence that an enriched environment is potentially relevant for maintaining abstinence during treatment (Galaj, Manuszak, & Ranaldi, 2016; Peck, Galaj, Eshak, Newman, & Ranaldi, 2015). For example exposure of mice to environmental enrichment during withdrawal eliminates behavioral sensitization and CPP to cocaine (Solinas, Chauvet, Thiriet, Rawas, & Jaber, 2008). Other studies have investigated the role of an enriched environment in drug-seeking despite aversive consequences as well as when heroin self-administration has been attenuated (Galaj et al., 2016; Peck et al., 2015). In one study (Peck et al., 2015), researchers utilized a conflict paradigm, which consists of administering a series of electric shocks to the floor of an operant conditioning chamber immediately below the operandum (i.e., lever) immediately following response-contingent delivery of the drug, and increasing in intensity until the animal ceases all drug self-administration (Cooper, Barnea-Ygael, Levy, Shaham, & Zangen, 2007). Thus, in this model, subjects a presented with a conflict between self-administering a drug and receiving an electric shock, which is intended to mimic an aversive consequence of drug intake. Rats were first allowed to self-administer heroin for approximately 15 days before being randomly assigned to either an enriched or standard housing environment condition. After 2 days in their respective conditions, response-contingent shock delivery commenced with an escalating current until rats did not press the lever that previously resulted in drug delivery for 3 consecutive days. Results of this study showed that even though rats from both the enriched and standard environment conditions achieved abstinence following introduction of the electric shock, enriched rats reached the criteria for maintaining abstinence (3 consecutive days of an absence of lever presses) significantly faster than rats housed in a standard environment (Peck et al., 2015). These data support the idea that an enriched environment is potentially a useful tool in reducing the motivation to self-administer opiates, even under conditions of punishment and negative consequences that result from drug intake.

The promising and possibly insulating abilities of an enriched environment are further supported in another study by Galaj et al. (2016). In this study, single-housed rats were allowed to self-administer heroin for 15 days in 3 h daily sessions. Lever presses resulted in an infusion of heroin were paired with presentation of a light cue. After heroin acquisition, rats were randomly assigned into an enriched or standard housing condition. After 15 days in their new enriched environments, rats underwent extinction sessions for 30 days, in which they were put back into the self-administration apparatus but lever presses did not result in drug infusion or light cue presentation. The day after the last extinction session, reinstatement sessions commenced, where the light cue was non-contingently presented twice, and subsequent lever presses result in the light cue and syringe pump activation but without an actual infusion of heroin. Results showed that rats in the enriched environment conditions pressed the active lever during reinstatement significantly less than rats in the standard environment condition. Thus, cue-triggered heroin seeking can be diminished by environmental enrichment.

The benefits of environmental enrichment may extend into subsequent generations, as offspring of morphine-dependent rats consumed less morphine if their parents had been reared in an enriched environment as opposed to a standard environment (Pooriamehr et al., 2017). Also, when pregnant morphine-dependent mothers were allowed to exercise via swimming (Torabi, Pooriamehr, Bigdeli, & Miladi-Gorji, 2017) or wheel running (Haydari, Miladi-Gorji, Mokhtari, & Safari, 2014), their offspring consumed significantly less morphine than the offspring of the morphine-dependent sedentary mothers. Conversely, while social enrichment seems to be an insulating factor in some aspects of opiate addiction, there is also evidence that the loss of a social reward can result in a vulnerability to drug seeking (Beloate & Coolen, 2017). This phenomenon was studied in the context of pair bonding and social and environmental enrichment. Beloate and Coolen (2017) attribute the resulting drug-seeking behavior to changes in the mesocorticolimbic reward pathway, which may also depend on the age at which subjects are socially isolated.

Few studies have examined social influences on behavioral (locomotor) sensitization to opiates. In one study (Xu et al., 2007), standard environment and enriched environment housed mice were placed in a novel environment, and it was shown that the enriched mice demonstrated reduced baseline locomotor activity compared to the standard housed mice. These animals then received six daily injections of morphine followed by a challenge injection of morphine (10 mg/kg) after a 5-day drug-free interval. Acquisition of behavioral sensitization was observed in mice under either housing conditions, yet the magnitude of behavioral sensitization in the enriched environment condition was less pronounced than that in the standard environment condition. Mice in the environmentally enriched condition showed a significantly less robust behavioral response (i.e., expression of sensitization) than mice in the standard environment condition.

A similar study was performed with rats, in which adolescent and adult rats, either group housed or singly housed, were received repeated injections of morphine. The adolescent morphine enriched environment rats did not exhibit the enhanced locomotor response as compared to the saline only and saline enriched environment rats. This relationship was not found (to a statistically significant degree) in adult rats, demonstrating that adolescent rats were vulnerable to social influences on morphine sensitization, but adult rats were not (Hofford, Schul, Wellman, & Eitan, 2012). Another study encompassed both environmental factors and stress to investigate the influence of social crowding upon morphine-induced locomotor behavior. Each condition had four rats per cage, but in the social crowding condition, the cages were 50% smaller than the size of the standard cages. Results showed that the social overcrowding procedure could produce morphine sensitization in rats with higher baseline motor activity (Xigeng et al., 2004). These differences were only significant under stress conditions, which provides a translational value to the study, as human addiction is often related to and impacted by various stressors.

5.2 Stress

There is substantial evidence that stress plays a role in the development and maintenance of drug addiction in humans (Koob, 2008). Researchers utilize animal models of stress and addiction in order to investigate the neural and behavioral mechanisms involved. In order to induce stress in laboratory rodents, aversive conditions are employed such as footshock, restraint or tail pinch, or involve social or maternal deprivation (Conrad, Ortiz, & Judd, 2017; Lu, Shepard, Hall, & Shaham, 2003). Most stressors activate corticotrophin-releasing factor (CRF) containing systems in the brain (Koob, 2008), which mediates the hypothalamic activation of the hypothalamic–pituitary–adrenal (HPA) axis, which in turn facilitates peripheral hormonal, autonomic, and behavioral responses (Varghese et al., 2015). It is theorized that CRF systems are relevant to all stages of the addiction cycle but plays the biggest role in the withdrawal/negative affect stage (Zorrilla, Logrip, & Koob, 2014). Chronic abuse of addictive drugs ultimately leads to a state of stress and negative affect (Zorrilla et al., 2014). Alteration in the typical functioning of CRF has been implicated in the drive to abuse addictive substances and the susceptibility to relapse (Zorrilla et al., 2014). Maternal separation, as a stressor, is also implicated in overactivation of the HPA axis in adulthood, and evidence suggests it is associated with increased opiate self-administration in adulthood (Neisewander et al., 2012).

Another study similar to the conflict model described earlier examined the effects of stress on morphine self-administration via pairing of mild electric shocks with lever pressing for morphine. Each time a rat pressed a lever, it received a footshock, followed by the infusion of morphine. The shock duration was either 0.2 or 0.02 s, and when the rats experienced the shock of a longer duration, they increased their lever pressing until they reached lethal levels of morphine consumption (Lu et al., 2003). These lethal effects could have been due to the analgesic effects of morphine, whereby rats could have increased their self-administration in order to alleviate the pain associated with the shock, which may have resulted in overdose. However, the ability of a stressor to increase opiate self-administration is not necessarily specific to electric footshock stress, as another study that utilized food deprivation as a stressor and observed increased etonitazene self-administration only on days of food restriction (Carroll & Boe, 1982).

6. NEWER RODENT PARADIGMS FOR ASSESSING PROSOCIAL BEHAVIOR IN OPIATE ADDICTION

In an attempt to give insight into the neural mechanisms of impaired social function in opiate addiction, Tomek, Stegmann, and Olive (2018) utilized an established prosocial paradigm in which a rat will release or rescue a conspecific from a plastic restrainer instead of receiving food or other palatable rewards (Ben-Ami Bartal, Decety, & Mason, 2011) (see Fig. 1). After baseline rescuing behaviors were measured, rats were allowed to self-administer sucrose pellets (orally) or heroin (intravenously) in long access (6 h/day) sessions for 2 weeks. The day after the final day of heroin self-administration, rats were given the opportunity to choose between releasing their cagemate from the restraint or continuing to self-administer heroin or sucrose. Results showed that rats with a history of sucrose self-administration continued to release their cagemate, whereas rats who had a history of heroin self-administration did not continue to release their cagemate and continued to self-administer heroin (Tomek et al., 2018). These results are consistent with a study in which "rescuer" rats were given an injection of either the benzodiazepine midazolam, the beta-adrenergic antagonist nadolol, or saline (Ben-Ami Bartal et al., 2016). Rats administered nadolol continued to rescue

Fig. 1 Photograph of a "rescuer" rat investigating a modified plastic restrainer with a removable door, inside of which is located the rat's "trapped" cagemate. Using this paradigm, it is has been demonstrated that "rescuer" rats will exhibit prosocial behavior by releasing its cagemate from the restrainer in lieu of obtaining chocolate (Ben-Ami Bartal et al., 2011), but will fail to do for unfamiliar rats or in rats with a history of heroin self-administration (Tomek et al., 2018). Such paradigms may be of considerable use in modeling social influences on opiate intake as well as deficits in social behavior that result from long-term opiate use. *Photo courtesy of Elisha Fornwalt.*

their cagemate at the same rate as controls; however midazolam-administered rats released their conspecific less often, all the while still opening the restraint door for chocolate (Ben-Ami Bartal et al., 2016). The authors interpreted these results as a drug induced interruption or dysregulation social affective processing that appears necessary to motivate the rat to open the restraint door for its cagemate. Overall, these findings suggest that the rescue paradigm may represent a novel rodent model of impaired social function in opiate use disorders. Utilization of this model may lead to improved strategies in facilitating recovery and treatment.

7. LIMITATIONS OF CURRENT MODELS

While each of the animal models discussed above have various strengths, there are weaknesses as well. For example, in humans and rodents, stress can contribute to drug use and relapse; however, the types of stressor humans encounter are typically related to a relationship, socioeconomic status, major life event, or lack of social support, as opposed to the aversive physical stressors administered to rodents in a laboratory (Heilig, Epstein, Nader, & Shaham, 2016). These differences are necessary to reconcile if claims of translational effectiveness are being made, or goals of improving outcomes for drug-addicted individuals are going to be realized. In environmental enrichment studies, a major limitation is the control group, which most often consists of animals in isolation, which is also a stressor and therefore a potential confound (Beloate & Coolen, 2017; Solinas, Thiriet, Chauvet, & Jaber, 2010). Limitations of CPP include inconsistent experimental results that may be due to differences between species or experimental methodology, such as utilizing different CPP apparatus and social circumstances of the experimental subjects (Neisewander et al., 2012). It is also important to note these socially induced changes in opiate CPP sensitivity reflect only the relatively short period of CPP acquisition and do not necessarily generalize to other paradigms of long-term opiate use (Bozarth et al., 1989; El Rawas et al., 2009). Sensitization is also a useful model; however, it is not only contingent upon ongoing exposure to drugs of abuse, it is affected by learning, stress, and environmental circumstances (Robinson & Berridge, 2000), making it difficult to precisely identify the underlying mechanisms. While intravenous self-administration paradigms are considered the gold standard of rodent models of addiction and have the benefits of volitional drug intake and self-titration of desired amount ingested, they bear various limitations such as subject attrition due to loss of catheter patency, experience of postsurgical discomfort, and artificial drug-taking environments that usually lack the presence of conspecifics. There is thus a need for improved approaches with even more translational value.

8. CONCLUSIONS

Over the last 40 years, preclinical models of opiate addiction have provided significant insight into the behavioral, physiological, neurobiological, genetic, and environmental basis of opiate addiction. These models have aided researchers in making significant progress in understanding of the neurobiological mechanisms of addictive drugs. These models are, for the most part reliable, replicable, and capture various (but certainly not all) aspects of addiction in humans. Research on social influences of addiction has demonstrated that enriched environments and social circumstances are beneficial in the reduction of drug-taking and drug-seeking behaviors in both humans and animals (Abadi et al., 2016; Hofford et al., 2012; Kennedy et al., 2012; Raz & Berger, 2010; Russell et al., 2015). However, the fact remains that, even with the myriad of rodent models of opiate addiction at the disposal of researchers worldwide, highly effective treatment modalities and prevention of relapse still remain significant obstacles to obtaining any degree of success in preventing or treating opiate addiction.

REFERENCES

Abadi, A. H. A., Miladi-Gorji, H., & Bigdeli, I. (2016). Effect of environmental enrichment on physical and psychological dependence signs and voluntary morphine consumption in morphine-dependent and morphine-withdrawn rats. *Behavioural Pharmacology*, *27*(2–3), 270–278. https://doi.org/10.1097/FBP.0000000000000197.

Alexander, B. K., Beyerstein, B. L., Hadaway, P. F., & Coambs, R. B. (1981). Effect of early and later colony housing on oral ingestion of morphine in rats. *Pharmacology, Biochemistry, and Behavior*, *15*(4), 571–576. https://doi.org/10.1016/0091-3057(81)90211-2.

Al-Hasani, R., & Bruchas, M. R. (2011). Molecular mechanisms of opioid receptor-dependent signalling and behaviour. *Anesthesiology*, *115*(6), 1363–1381. https://doi.org/10.1097/ALN.0b013e318238bba6.Molecular.

American Psychiatric Association. (2013). *Diagnostic and statistical manual of mental disorders* (5th ed.). . Washington, DC: American Psychiatric Publishing.

Badiani, A., Belin, D., Epstein, D., Calu, D., & Shaham, D. (2011). Opiate versus psychostimulant addiction: The differences do matter. *Nature Reviews Neuroscience*, *12*(11), 685–700. https://doi.org/10.1038/nrn3104.

Bates, M. L. S., Emery, M. A., Wellman, P. J., & Eitan, S. (2014). Social housing conditions influence morphine dependence and the extinction of morphine place preference in adolescent mice. *Drug and Alcohol Dependence*, *142*, 283–289. https://doi.org/10.1016/j.drugalcdep.2014.06.036.

Belin-Rauscent, A., & Belin, A. (2011). Animal models of drug addiction. *Addictions—From Pathophysiology to Treatment*, *53*, 21–63. https://doi.org/10.1007/978-1-60761-934-5.

Beloate, L. N., & Coolen, L. M. (2017). Influences of social reward experience on behavioral responses to drugs of abuse: Review of shared and divergent neural plasticity mechanisms for sexual reward and drugs of abuse. *Neuroscience and Biobehavioral Reviews*, *83*, 356–372. https://doi.org/10.1016/j.neubiorev.2017.10.024.

Ben-Ami Bartal, I., Decety, J., & Mason, P. (2011). Empathy and pro-social behavior in rats. *Science*, *334*(60619), 1427–1430. http://www.jstor.org/stable/41352251.

Ben-Ami Bartal, I., Shan, H., Molasky, N. M. R., Murray, T. M., Williams, J. Z., Decety, J., et al. (2016). Anxiolytic treatment impairs helping behavior in rats. *Frontiers in Psychology*, 7, 850. https://doi.org/10.3389/fpsyg.2016.00850.

Bodnar, R. J. (2013). Endogenous opiates and behavior: 2012. *Peptides*, *50*, 55–95. https://doi.org/10.1016/j.peptides.2013.10.001.

Bohnert, A. S. B., Bradshaw, C. P., & Latkin, C. A. (2009). A social network perspective on heroin and cocaine use among adults: Evidence of bidirectional influences. *Addiction*, *104*(7), 1210–1218. https://doi.org/10.1111/j.1360-0443.2009.02615.x.

Bond, A. J., & Witton, J. (2017). Perspectives on the pharmacological treatment of heroin addiction. *Clinical Medicine Insights: Psychiatry*, *8*, 117955731773732. https://doi.org/10.1177/1179557317737322.

Bozarth, M. A., Murray, A., & Wise, R. A. (1989). Influence of housing conditions on the acquisition of intravenous heroin and cocaine self-administration in rats. *Pharmacology, Biochemistry, and Behavior*, *33*, 903–907. https://doi.org/10.1016/0091-3057(89)90490-5.

Buchanan, A. S., & Latkin, C. A. (2008). Drug use in the social networks of heroin and cocaine users before and after drug cessation. *Drug and Alcohol Dependence*, *96*(3), 286–289. https://doi.org/10.1016/j.drugalcdep.2008.03.008.

Carroll, M. E., & Boe, I. N. (1982). Increased intravenous drug self-administration during deprivation of other reinforcers. *Pharmacology, Biochemistry, and Behavior, 17*(3), 563–567. https://doi.org/10.1016/0091-3057(82)90319-7.

Centers for Disease Control and Prevention. (2017). *Annual surveillance report of drug-related risks and outcomes—United States, 2017. Surveillance special report 1.* U.S. Department of Health and Human Services. https://www.cdc.gov/drugoverdose/pdf/pubs/2017-cdc-drug-surveillance-report.pdf. Published August 31, 2017.

Chen, S. A., O'Dell, L. E., Hoefer, M. E., Greenwell, T. N., Zorrilla, E. P., & Koob, G. F. (2006). Unlimited access to heroin self-administration: Independent motivational markers of opiate dependence. *Neuropsychopharmacology, 31*(12), 2692–2707. https://doi.org/10.1038/sj.npp.1301008.

Ciccarone, D., Ondocsin, J., & Mars, S. G. (2017). Heroin uncertainties: Exploring users' perceptions of fentanyl-adulterated and -substituted "heroin". *International Journal of Drug Policy, 46*, 146–155. https://doi.org/10.1016/j.drugpo.2017.06.004.

Cicero, T. J., & Ellis, M. S. (2017). The prescription opioid epidemic: A review of qualitative studies on the progression from initial use to abuse. *Dialogues in Clinical Neuroscience, 19*(3), 259–269.

Compton, W. M., Jones, C. M., & Baldwin, G. T. (2016). Relationship between non-medical prescription-opioid use and heroin use. *New England Journal of Medicine, 374*(2), 154–163. https://doi.org/10.1056/NEJMra1508490.

Conrad, C. D., Ortiz, J. B., & Judd, J. M. (2017). Chronic stress and hippocampal dendritic complexity: Methodological and functional considerations. *Physiology and Behavior, 178*, 66–81. https://doi.org/10.1016/j.physbeh.2016.11.017.

Cooper, A., Barnea-Ygael, N., Levy, D., Shaham, Y., & Zangen, A. (2007). A conflict rat model of cue-induced relapse to cocaine seeking. *Psychopharmacology, 194*(1), 117–125. https://doi.org/10.1007/s00213-007-0827-7.

Courtwright, D. T. (1982). *Dark paradise: A history of opiate addiction in America.* Harvard University Press.

Coventry, T. L., D'Aquila, P. S., Brain, P., & Willner, P. (1997). Social influences on morphine conditioned place preference. *Behavioural Pharmacology, 8*(6), 575–584. https://doi.org/10.1097/00008877-199711000-00015.

de Wit, H., & Sayette, M. (2018). Considering the context: Social factors in responses to drugs in humans. *Psychopharmacology, 235*(4), 935–945. https://doi.org/10.1007/s00213-018-4854-3.

Donny, E. C., Brasser, S. M., Bigelow, G. E., Stitzer, M. L., & Walsh, S. L. (2005). Methadone doses of 100 mg or greater are more effective than lower doses at suppressing heroin self-administration in opioid-dependent volunteers. *Addiction, 100*(10), 1496–1509. https://doi.org/10.1111/j.1360-0443.2005.01232.x.

Eitan, S., Emery, M. A., Bates, M. L. S., & Horrax, C. (2017). Opioid addiction: Who are your real friends? *Neuroscience and Biobehavioral Reviews, 83*, 697–712. https://doi.org/10.1016/j.neubiorev.2017.05.017.

El Rawas, R., Thiriet, N., Lardeux, V., Jaber, M., & Solinas, M. (2009). Environmental enrichment decreases the rewarding but not the activating effects of heroin. *Psychopharmacology, 203*(3), 561–570. https://doi.org/10.1007/s00213-008-1402-6.

Galaj, E., Manuszak, M., & Ranaldi, R. (2016). Environmental enrichment as a potential intervention for heroin seeking. *Drug and Alcohol Dependence, 163*, 195–201. https://doi.org/10.1016/j.drugalcdep.2016.04.016.

Goldberg, W., & Latimer, D. (2014). *Flowers in the blood: The story of opium.* New York: Skyhorse Publishing.

Griffith, J., Knight, D., Joe, G., & Simpson, D. (1998). Implications of family and peer relations for treatment engagement and follow-up outcomes: An integrative model. *Psychology of Addictive Behaviors, 12*(2), 113–126. https://doi.org/10.1037/0893-164X.12.2.113.

Haydari, S., Miladi-Gorji, H., Mokhtari, A., & Safari, M. (2014). Effects of voluntary exercise on anxiety-like behavior and voluntary morphine consumption in rat pups borne from morphine-dependent mothers during pregnancy. *Neuroscience Letters*, *578*, 50–54. https://doi.org/10.1016/j.neulet.2014.06.026.

Heilig, M., Epstein, D. H., Nader, M. A., & Shaham, Y. (2016). Time to connect: Bringing social context into addiction neuroscience. *Nature Reviews. Neuroscience*, *17*(9), 592–599. https://doi.org/10.1038/nrn.2016.67.

Hofford, R. S., Schul, D. L., Wellman, P. J., & Eitan, S. (2012). Social influences on morphine sensitization in adolescent rats. *Addiction Biology*, *17*(3), 547–556. https://doi.org/10.1111/j.1369-1600.2011.00315.x.

Hölscher, F., Reissner, V., Di Furia, L., Room, R., Schifano, F., Stohler, R., et al. (2010). Differences between men and women in the course of opiate dependence: Is there a telescoping effect? *European Archives of Psychiatry and Clinical Neuroscience*, *260*(3), 235–241. https://doi.org/10.1007/s00406-009-0053-x.

Johnson, S. W., & North, R. A. (1992). Opioids excite dopamine neurons by hyperpolarization of local interneurons. *Journal of Neuroscience*, *12*(2), 483–488. https://doi.org/10.1523/JNEUROSCI.12-02-00483.1992.

Jones, C. M., Campopiano, M., Baldwin, G., & McCance-Katz, E. (2015). National and state treatment need and capacity for opioid agonist medication-assisted treatment. *American Journal of Public Health*, *105*(8), e55–e63. https://doi.org/10.2105/AJPH.2015.302664.

Jones, J. D., & Comer, S. D. (2013). A review of human drug self-administration procedures. *Behavioural Pharmacology*, *24*(5–6), 384–395. https://doi.org/10.1097/FBP.0b013e3283641c3d.

Kapoor, L. D. (1995). *Opium poppy: Botany, chemistry and pharmacology*. Binghamton, New York: Haworth Press.

Kennedy, B. C., Panksepp, J. B., Runckel, P. A., & Lahvis, G. P. (2012). Social influences on morphine-conditioned place preference in adolescent BALB/cJ and C57BL/6J mice. *Psychopharmacology*, *219*(3), 923–932. https://doi.org/10.1007/s00213-011-2421-2.

Koob, G. F. (2008). A role for brain stress systems in addiction. *Neuron*, *59*(1), 11–34. https://doi.org/10.1016/j.neuron.2008.06.012.

Koram, N., Liu, H., Li, J., Li, J., Luo, J., & Nield, J. (2011). Role of social network dimensions in the transition to injection drug use: Actions speak louder than words. *AIDS and Behavior*, *15*(7), 1579–1588. https://doi.org/10.1007/s10461-011-9930-1.

Lu, L., Shepard, J. D., Hall, F. S., & Shaham, Y. (2003). Effect of environmental stressors on opiate and psychostimulant reinforcement, reinstatement and discrimination in rats: A review. *Neuroscience and Biobehavioral Reviews*, *27*(5), 457–491. https://doi.org/10.1016/S0149-7634(03)00073-3.

Luthar, S. S., Anton, S. F., Merikangas, K. R., & Rounsaville, B. J. (1992). Vulnerability to drug abuse among opioid addicts' siblings: Individual, familial, and peer influences. *Comprehensive Psychiatry*, *33*(3), 190–196. https://doi.org/10.1016/0010-440X(92)90029-P.

Maher, C. E., Martin, T. J., & Childers, S. R. (2005). Mechanisms of mu opioid receptor/G-protein desensitization in brain by chronic heroin administration. *Life Sciences*, *77*(10), 1140–1154. https://doi.org/10.1016/j.lfs.2005.03.004.

Milekic, M. H. (2006). Persistent disruption of an established morphine conditioned place preference. *Journal of Neuroscience*, *26*(11), 3010–3020. https://doi.org/10.1523/JNEUROSCI.4818-05.2006.

Namba, M. D., Tomek, S. E., Olive, M. F., Beckmann, J. S., & Gipson, C. D. (2018). The winding road to relapse: Forging a new understanding of cue-induced reinstatement

models and their associated neural mechanisms. *Frontiers in Behavioral Neuroscience, 12*, 17. https://doi.org/10.3389/fnbeh.2018.00017.
Neisewander, J. L., Peartree, N. A., & Pentkowski, N. S. (2012). Emotional valence and context of social influences on drug abuse-related behavior in animal models of social stress and prosocial interaction. *Psychopharmacology, 224*(1), 33–56. https://doi.org/10.1007/s00213-012-2853-3.
Neisewander, J. L., Pierce, R. C., & Bardo, M. T. (1990). Naloxone enhances the expression of morphine-induced conditioned place preference. *Psychopharmacology, 100*(2), 201–205.
Olive, M. F. (2017). *Understanding drugs: Heroin*. New York: Chelsea House Publishers.
Ostling, P. S., Davidson, K. S., Anyama, B. O., Helander, E. M., Wyche, M. Q., & Kaye, A. D. (2018). America's opioid epidemic: A comprehensive review and look into the rising crisis. *Current Pain and Headache Reports, 22*(5), 32. https://doi.org/10.1007/s11916-018-0685-5.
Panlilio, L. V., & Goldberg, S. R. (2007). Self-administration of drugs in animals and humans as a model and an investigative tool. *Addiction, 102*(12), 1863–1870. https://doi.org/10.1111/j.1360-0443.2007.02011.x.
Peck, J. A., Galaj, E., Eshak, S., Newman, K. L., & Ranaldi, R. (2015). Environmental enrichment induces early heroin abstinence in an animal conflict model. *Pharmacology Biochemistry and Behavior, 138*, 20–25. https://doi.org/10.1016/j.pbb.2015.09.009.
Pooriamehr, A., Sabahi, P., & Miladi-Gorji, H. (2017). Effects of environmental enrichment during abstinence in morphine dependent parents on anxiety, depressive-like behaviors and voluntary morphine consumption in rat offspring. *Neuroscience Letters, 656*, 37–42. https://doi.org/10.1016/j.neulet.2017.07.024.
Raz, S., & Berger, B. D. (2010). Social isolation increases morphine intake: Behavioral and psychopharmacological aspects. *Behavioural Pharmacology, 21*(1), 39–46. https://doi.org/10.1097/FBP.0b013e32833470bd.
Rech, R. H., Mokler, D. J., & Briggs, S. L. (2012). Effects of combined opioids on pain and mood in mammals. *Pain Research and Treatment, 2012*, 145965. https://doi.org/10.1155/2012/145965.
Ribeiro Do Couto, B., Aguilar, M. A., Manzanedo, C., Rodríguez-Arias, M., Armario, A., & Miñarro, J. (2006). Social stress is as effective as physical stress in reinstating morphine-induced place preference in mice. *Psychopharmacology, 185*(4), 459–470. https://doi.org/10.1007/s00213-006-0345-z.
Robinson, T. E., & Berridge, K. C. (2000). The psychology and neurobiology of addiction: An incentive-sensitization view. *Addiction, 95*(2), 91–117. https://doi.org/10.1046/j.1360-0443.95.8s2.19.x.
Rudd, R. A. (2016). Increases in drug and opioid-involved overdose deaths—United States, 2010–2015. *MMWR. Morbidity and Mortality Weekly Report, 65*, 1445–1452. https://doi.org/10.15585/mmwr.mm655051e1.
Russell, B. S., Trudeau, J. J., & Leland, A. J. (2015). Social influence on adolescent polysubstance use: The escalation to opioid use. *Substance Use and Misuse, 50*(10), 1325–1331. https://doi.org/10.3109/10826084.2015.1013128.
Schenk, S., Hunt, T., Colle, L., & Amit, Z. (1982). Isolation versus grouped housing in rats: Differential effects of low doses of heroin in the place preference paradigm. *Life Sciences, 32*, 1129–1134. https://doi.org/10.1016/0024-3205(83)90118-2.
Scherbaum, N., & Specka, M. (2008). Factors influencing the course of opiate addiction. *International Journal of Methods in Psychiatric Research, 17*(S1), S39–S44. https://doi.org/10.1002/mpr.244.
Shaham, Y., Shalev, U., Lu, L., De Wit, H., & Stewart, J. (2003). The reinstatement model of drug relapse: History, methodology and major findings. *Psychopharmacology, 168*(1–2), 3–20. https://doi.org/10.1007/s00213-002-1224-x.

Sherman, S. G., Smith, L., Laney, G., & Strathdee, S. A. (2002). Social influences on the transition to injection drug use among young heroin sniffers: A qualitative analysis. *International Journal of Drug Policy*, *13*(2), 113–120. https://doi.org/10.1016/S0955-3959(02)00010-5.
Solinas, M., Chauvet, C., Thiriet, N., El Rawas, R., & Jaber, M. (2008). Reversal of cocaine addiction by environmental enrichment. *Proceedings of the National Academy of Sciences of the United States of America*, *105*(44), 17145–17150. https://doi.org/10.1073/pnas.0806889105.
Solinas, M., Thiriet, N., Chauvet, C., & Jaber, M. (2010). Prevention and treatment of drug addiction by environmental enrichment. *Progress in Neurobiology*, *92*(4), 572–592. https://doi.org/10.1016/j.pneurobio.2010.08.002.
Stanton, M. D., & Shadish, W. R. (1997). Outcome, attrition, and family-couples treatment for drug abuse: A meta-analysis and review of the controlled, comparative studies. *Psychological Bulletin*, *122*(2), 170–191. https://doi.org/10.1037/0033-2909.122.2.170.
Stein, C. (2016). Opioid receptors. *Annual Review of Medicine*, *67*(1), 433–451. https://doi.org/10.1146/annurev-med-062613-093100.
Steketee, J. D., & Kalivas, P. W. (2011). Drug wanting: Behavioral sensitization and relapse to drug-seeking behavior. *Pharmacological Reviews*, *63*(2), 348–365. https://doi.org/10.1124/pr.109.001933.
Tomek, S. E., Stegmann, G. M., & Olive, M. F. (2018). Effects of heroin on rat prosocial behavior. *Addiction Biology* (epub ahead of print). https://doi.org/10.1111/adb.12633.
Torabi, M., Pooriamehr, A., Bigdeli, I., & Miladi-Gorji, H. (2017). Maternal swimming exercise during pregnancy attenuates anxiety/depressive-like behaviors and voluntary morphine consumption in the pubertal male and female rat offspring born from morphine dependent mothers. *Neuroscience Letters*, *659*, 110–114. https://doi.org/10.1016/j.neulet.2017.08.074.
Tzschentke, T. M. (2007). Measuring reward with the conditioned place preference (CPP) paradigm: Update of the last decade. *Addiction Biology*, *12*(3–4), 227–462. https://doi.org/10.1111/j.1369-1600.2007.00070.x.
Varghese, S., Montalvo-Ortiz, J., Csernansky, J., Eiger, R., Herrold, A., Koola, M., et al. (2015). Early life stress as a risk factor for substance use disorders: Clinical and neurobiological substrates. *Indian Journal of Psychological Medicine*, *37*(1), 36. https://doi.org/10.4103/0253-7176.150816.
Vendruscolo, L. F., Schlosburg, J. E., Misra, K. K., Chen, S. A., Greenwell, T. N., & Koob, G. F. (2011). Escalation patterns of varying periods of heroin access. *Pharmacology Biochemistry and Behavior*, *98*(4), 570–574. https://doi.org/10.1016/j.pbb.2011.03.004.
Walsh, S. L., Comer, S. D., Lofwall, M. R., Vince, B., Levy-Cooperman, N., Kelsh, D., et al. (2017). Effect of buprenorphine weekly depot (CAM2038) and hydromorphone blockade in individuals with opioid use disorder: A randomized clinical trial. *JAMA Psychiatry*, *74*(9), 894–902. https://doi.org/10.1001/jamapsychiatry.2017.1874.
Wasserman, D. A., Stewart, A. L., & Delucchi, K. L. (2001). Social support and abstinence from opiates and cocaine during opioid maintenance treatment. *Drug and Alcohol Dependence*, *65*(1), 65–75. https://doi.org/10.1016/S0376-8716(01)00151-X.
Watterson, L. R., Kufahl, P. R., Nemirovsky, N. E., Sewalia, K., Grabenauer, M., Thomas, B. F., et al. (2014). Potent rewarding and reinforcing effects of the synthetic cathinone 3,4-methylenedioxypyrovalerone (MDPV). *Addiction Biology*, *19*(2), 165–174. https://doi.org/10.1111/j.1369-1600.2012.00474.x.
Weiss, F. (2005). Neurobiology of craving, conditioned reward and relapse. *Current Opinion in Pharmacology*, *5*(1), 9–19. https://doi.org/10.1016/j.coph.2004.11.001.
Williams, C. T., & Latkin, C. A. (2007). Neighborhood socioeconomic status, personal network attributes, and use of heroin and cocaine. *American Journal of Preventive Medicine*, *32*(6 Suppl), 203–210. https://doi.org/10.1016/j.amepre.2007.02.006.

Xigeng, Z., Yonghui, L., Xiaojing, L., Lin, X., Dongmei, W., Jie, L., et al. (2004). Social crowding sensitizes high-responding rats to psychomotor-stimulant effects of morphine. *Pharmacology, Biochemistry, and Behavior, 79*(2), 213–218. https://doi.org/10.1016/j.pbb.2004.07.012.

Xu, Z., Hou, B., Gao, Y., He, F., & Zhang, C. (2007). Effects of enriched environment on morphine-induced reward in mice. *Experimental Neurology, 204*(2), 714–719. https://doi.org/10.1016/j.expneurol.2006.12.027.

Zorrilla, E. P., Logrip, M. L., & Koob, G. F. (2014). Corticotropin releasing factor: A key role in the neurobiology of addiction. *Frontiers in Neuroendocrinology, 35*(2), 234–244. https://doi.org/10.1016/j.yfrne.2014.01.001.

CHAPTER FIVE

Social Modification of Amphetamine Reward

Shigeru Watanabe[1]
Keio University, Tokyo, Japan
[1]Corresponding author: e-mail address: swat@flet.keio.ac.jp

Contents

Abstract

Social modification of drug reward in general and amphetamine reward in particular is reviewed here. The simplest explanation for the social facilitation of drug reward in the conditioned place preference paradigm is the summation of drug and social reinforcing effects. However, experimental reports have shown that sharing a common pharmacological experience, simultaneously or successively, plays a crucial role in the social facilitation of methamphetamine reward. Therefore, social facilitation cannot be the simple summation of drug and social reinforcing effects. Other social factors, such as the dominant/subordinate relation, social defeat, and isolation, also affect drug reward. Because these factors induce stress, they may modulate drug effects through stress-induced mechanisms. However, it is unclear whether these mechanisms are social experience-specific or underlie all aversive experiences producing stress. Mutual modification of social and drug rewards is also discussed.

International Review of Neurobiology, Volume 140
ISSN 0074-7742
https://doi.org/10.1016/bs.irn.2018.07.023

Understanding drug reward is not only an important behavioral pharmacology topic but is critical to solving the drug abuse problem in society. Social influence is one of the most important factors in drug abuse (see Ellenbroek et al., 2005). Here, I review basic experimental research on social modification of amphetamine reward.

1. EXPERIMENTAL EVIDENCE OF SOCIAL MODIFICATION OF DRUG REWARD

Reinforcing properties of drugs are affected by social factors in humans. For example, individuals tend to drink more alcoholic beverages when they are with other drinkers and tend to start smoking if they have friends who smoke (Glynn, 1989). Initiating of methamphetamine injection is also influenced by social factors in humans (Marshall et al., 2011). Many different social influence on methamphetamine use in human adolescents has been reported (Embry, Hankins, Biglan, & Boles, 2009). However, the detailed mechanisms of stimulant reward modification by social factors remain unclear.

Methamphetamine reward has been examined in a social context using the conditioned place preference (CPP) paradigm in C57 male mice (Watanabe, 2011). I used conventional CPP chambers for mice. The apparatus had three compartments, one with a grid floor and black walls, another with a stainless steel mesh floor with a white walls and the other with a flat gray floor with gray walls. The gray center compartment was connected with the other two compartments through a guillotine door. The mice in a paired group received CPP training with a cage mate: the two mice were injected with methamphetamine (2mg/kg) and placed in one compartment together, or injected with saline and placed in a different compartment together. CPP-trained mice spent more time in the compartment associated with drug injection. Furthermore, the increment in the time spent in the methamphetamine-associated compartment was larger than that in mice that had undergone conventional CPP in a single-mouse design. Thus, the cage mate facilitated methamphetamine reward. Recently, I examined the effect of the number of cage mates receiving the same pharmacological treatment together on social facilitation. I compared 1-, 2-, and 3-mouse groups and found the facilitative social effect to depend on the number of cage mates. A dose–response analysis showed that the cage mates shifted the dose–response curves to the left without altering the maximum effect of methamphetamine suggesting social facilitation in potency not in efficacy (Watanabe, 2018).

Social modification of morphine effects is more complex. Kennedy, Panksepp, Runckel, and Lahvis (2012) used C57 and BALB male mice (representing social and asocial strains, respectively) to show that, in adolescent animals, morphine (0.25–5.0 mg/kg)-induced CPP was influenced by complex interactions between social context and genetic differences in social motivation, while social factors such as housing or group conditioning had little effect. Morphine suppresses or facilitates social behavior, such as play, in juvenile rodents depending on the context (Kennedy et al., 2012; Panksepp, Jalowiec, & Bishop, 1985; Vanderschuren, Niesink, Spruijt, & Van Ree, 1995).

I have examined social modification of morphine reward in C57 mice (Watanabe, 2013) using similar procedures to those used in the methamphetamine study (Watanabe, 2011). In a single-mouse group, CPP training resulted in conditioned preference when mice were injected with 1.0 or 3.0 mg/kg morphine. In contrast, a paired group showed reduced reward at 1.0 mg/kg, whereas no significant difference in CPP was observed with 3.0 mg/kg morphine. Because the results of these experiments differ from those of previous reports on methamphetamine CPP (Watanabe, 2011), social facilitation/suppression of drug reward may be drug-type-specific. In humans, recreational stimulant use tends to be more as a "party" drug (e.g., Nabben & Korf, 1999), whereas recreational morphine use may be more frequently enjoyed in private or in an opium den (e.g., Westermeyer, 1974). These different styles of consumption may be related to differential sensitivity to social factors.

2. IS SOCIAL FACILITATION THE SUM OF TWO REINFORCING EFFECTS?

The simplest explanation for social facilitation of drug reward is that it is the sum of the reinforcing effects of social stimuli and drugs. If the reinforcing properties of a drug and drug-injected conspecific animal are additive, social stimuli can enhance drug-induced CPP. Thiel, Sanabria, and Neisewander (2009) reported that, in adolescent rats, neither a low dose of nicotine (0.1–0.6 mg/kg) nor social partnering resulted in CPP when administered alone, whereas simultaneous application elicited CPP. Similarly, a low dose of cocaine or social pairing alone failed to produce CPP, while the combination of social reward and the drug (2 mg/kg) was effective (Thiel, Okun, & Neisewander, 2008). Grotewold, Wall, Goodell, et al. (2014)

also reported additive effects of social reward and cocaine on CPP in rats. These results indicate that the positive reinforcing properties of stimulants and social interaction can be summated.

2.1 Social Reward

Various physiological effects of the presence of a conspecific animal have been reported. Restraint causes so-called stress-induced hyperthermia (SIH) in mice. Non-shivering thermoregulation is maintained by metabolic activity of brown adipose tissue through the dorsomedial hypothalamic–raphe system (see DiMicco & Zaretsky, 2007; Kataoka, Hioki, Kaneko, & Nakamura, 2014). I used a thermograph to measure body surface temperature in ICR mice and found that SIH was reduced when other cage mates were also restrained (equality condition), while freely-moving cage mates enhanced SIH suggesting inequality aversion (Watanabe, 2015b). Hence, social relation among animals (equality or inequality) determines the vector of the social interaction effect (positive or negative).

Mice learned to press a lever to receive a visual display of a conspecific (Antonitis & Baron, 1964), and juvenile rats showed a preference for playing with cage mates in a T-maze (Humphreys & Einon, 1981). Calcagnetti and Schechter (1992) employed the CPP procedure to examine affective reward properties of rough-and-tumble play in rats. The rats demonstrated CPP in which playing was associated with a compartment. These CPP experiments demonstrated the positive affective properties of social stimuli, particularly play, and suggested the possibility of summation of social and drug reward effects.

Using a social choice paradigm in which two stimulus mice were placed in Plexiglas cylinders, Brodkin, Hagemann, Nemenski, and Silver (2004) demonstrated that C57 female mice (social strain) exhibited approach behavior to a conspecific, indicating a positive affective reaction, whereas female BALB mice (asocial strain) exhibited avoidance behavior, suggesting a negative affective reaction to the conspecific. Thus, preference for a cage mate is strain-specific in mice.

Social preferences (positive reward) in rodents are influenced by several other factors, such as developmental stage (Panksepp & Lahvis, 2007), environmental context (Pearson, Defensor, Blanchard, & Blanchard, 2010), and social status (Van Loo, de Groot, Van Zutphen, & Bauemans, 2001). Housing conditions also influenced social reward in a choice situation (Yates, Beckmanna, Meyer, & Bardo, 2013). Interaction with conspecifics induced

CPP only in individually housed adolescents, and amphetamine-induced CPP in both individually housed adolescents and adults; however, the effect of amphetamine was not statistically significant in pair-housed adults. When allowed to choose between compartments paired with either social interaction or amphetamine, individually housed adolescents preferred the compartment paired with social interaction, whereas pair-housed adolescents preferred the compartment paired with amphetamine. Thus, social isolation enhances the reinforcing property of social stimulus in adolescents. In contrast, adults showed similar preference for the amphetamine- and social-interaction-paired compartments regardless of the housing condition. Taken together, these results suggest that social reward was modulated by housing condition, sex and developmental age.

Interaction between social and drug reward suggests commonality of neural mechanisms. Cocaine-induced CPP and social CPP were found to activate Zif expression similar yet distinct brain regions in rats (El Rawas et al., 2012). The insular cortex was more strongly activated after cocaine CPP than social CPP, whereas the prelimbic cortex and core region of the nucleus accumbens were activated to a greater extent by social CPP. However, in concurrent CPP, in which one compartment was associated with cocaine (15 mg/kg) and the other with social interaction, lesions in the core of the nucleus accumbens resulted in CPP for only social interaction (Fritz et al., 2011), suggesting a role of the nucleus accumbens in cocaine reward. Morphine-, but not social-interaction-induced CPP resulted in augmented expression of the NR2B subunit of the NMDA receptor in the nucleus accumbens and hippocampus in rats (Ma et al., 2006). In addition, an NR2B selective antagonist (ifenprodil) blocked morphine CPP but not social-interaction-induced CPP. These results suggest that the effects of stimulants, morphine, and social interaction are mediated by overlapping but partially specialized brain regions to induce reward.

A conspecific in distress can have both positive and negative effects. Mice approached formalin-injected (displaying a pain response) cage mates, suggesting preference for suffering mates (Watanabe, 2014). However, when formalin-injected mates were used as the stimulus in CPP training, the test mice developed conditioned place aversion rather than conditioned preference, in an apparent contradiction with the social preference test results. Approach behavior or social stimulus preference, typically indicative of positive reward, may also result from curiosity for inherently aversive stimuli, such as pain response of conspecific.

2.2 Experience Sharing: Pharmacological States of Cage Mates

Another factor modifying the social effect on drug reward is the pharmacological state of the interacting partner. For example, adolescent rats consumed more ethanol following interactions with a social partner injected with ethanol than with rats injected with water (Hunt, Hollaway, & Scordalakes, 2009; Maldonado, Finkbeiner, & Kirstein, 2008). When attempting to elicit conditioned aversion to alcohol in rats, however, Gauvin, Briscoe, Goulden, and Hollowey (1994) found that the mere presence of a cage mate (with or without alcohol consumption) increased alcohol intake. Therefore, conditioned alcohol aversion is reduced by social interaction regardless of the pharmacological state of the cage mate.

In most studies of social facilitation, two animals receive similar pharmacological treatment, resulting in matched pharmacological states. If social facilitation is simple summation of social and drug reward, conspecific mates without pharmacological treatment should facilitate the reinforcing effect of the drug. To examine this possibility, I used a control-paired group in an amphetamine CPP experiment (Watanabe, 2011). Animals in the control-paired group received CPP training with their cage mates with reciprocal treatment: when one mouse was injected with the drug, the other was injected with saline. If the presence of a cage mate in any pharmacological state had an additive effect, the control-paired group would show CPP facilitation similar to that in a paired group (in which two mice received similar treatment together). However, the control-paired group did not show a social enhancement effect, while the paired group showed greater drug reward compared with a conventional single-mouse CPP group. These results suggest that experience sharing may be crucial for the social enhancement of methamphetamine reward. In addition to the presence of a cage mate, its pharmacological status modifies the reinforcing properties of a drug.

I have examined social modification of morphine reward in C57 mice using a similar experimental design (Watanabe, 2013). As described above, the paired group showed suppressed reward of the 1.0 mg/kg dose. In contrast, a control-paired group showed enhanced reward of the 3.0 mg/kg dose. These findings differ from previous results on methamphetamine reward (Watanabe, 2011). Probably stimulants, such as methamphetamine, and depressants, such as morphine, have different sensibility to social effects. We need more data of social effects on different drugs to make the reason of the difference clear.

The difference between paired and control-paired groups suggests that a common pharmacological experience is crucial for the facilitative effect of social interaction, at least in the case of methamphetamine. In a paired group, two mice experience a drug-induced state at the same time (simultaneous common experience). I hypothesized that prior exposure of a test animal to a drug may induce preference for a conspecific injected with the same drug (successive common experience). To test this hypothesis, the reinforcing effect of drug-injected cage mates was examined using a CPP paradigm in which one compartment was associated with the presence of a drug-injected cage mate, and the other with the presence of a saline-injected cage mate (Watanabe, 2015a). Mice were divided into two groups, one receiving methamphetamine and the other receiving morphine. The mice in each experiment were further divided into stimulus mice and subject mice. The subject mice were divided still further into common-experience and no-experience groups. The subject mice in the common-experience group had experienced drug injection (methamphetamine or morphine) before conditioning, and those in the no-experience group had experienced saline injection only.

Fig. 1 presents a flow diagram of the experiment. All subject mice received CPP training in which one compartment of a conventional CPP apparatus was associated with a drug-injected stimulus mouse, and the other with a saline-injected stimulus mouse. The subject mice did not receive any drug injection during the CPP training. Time spent in the compartment associated with a drug-injected cage mate was measured before and after training.

Subject mice that had previously experienced methamphetamine showed an increase in the time spent in the compartment associated with the methamphetamine-injected cage mate after CPP training (Fig. 2A). This effect was not observed in subject mice that had previously been injected with saline. In the morphine group (Fig. 2B), subject mice did not show an increase in the time spent in the compartment associated with the morphine-injected cage mate irrespective of whether they had previously received morphine or saline injections. Therefore, in agreement with previous reports (Watanabe, 2011, 2013), common experience with methamphetamine, but not with morphine, induced reward. Using similar design to Watanabe (2015a), Smith, Strickland, Bills, and Lacy (2015) reported preference for cocaine-treated mate after experience of the same drug in rats.

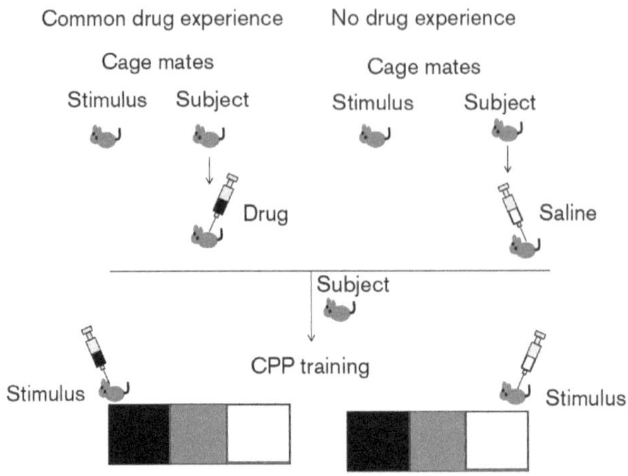

Fig. 1 Flow of experiment. The subject mice in the common drug-experience group were injected with drug, whereas those in the no drug-experience group were injected with saline. Then, the subject mice in each group received conditioned place preference (CPP) training during which one compartment apparatus was associated with a drug-injected stimulus mouse (cage mate) and the other compartment was associated with a stimulus mouse injected with saline. *From Watanabe, S. (2015). Common experience modifies the reinforcing properties of methamphetamine-injected cage mates but not morphine-injected cage mates in C57 mice.* Behavioral Pharmacology, *26, 636–641.*

Cole, Hofford, Evert, Wellman, and Eitan (2013) reported that pre-exposure to morphine (10–40 mg/kg) in mice in a group setting resulted in morphine-induced CPP; however, when the group was mixed (i.e., contained both morphine- and saline-injected mice housed together), CPP was not induced. Although the experiment did not directly examine the rewarding effect of morphine-injected mice on its cage mates, it suggests that morphine reward may be modified by common experience in some situations.

3. MODIFICATION OF DRUG EFFECTS BY SOCIAL STRESS

Drug reward has been shown to be affected by non-social stress, such as foot shock or restraint (Gawin, 1991; Gordon, 2002; Sinha, 2001; see reviews by Burke & Miczek, 2014; Logrip, Zorrilla, & Koob, 2012). Stressful experiences modify the activity of brain areas involved in stimulant reward (Belujon & Grace, 2011; Sinha, 2008). Stress exposure induced relapse of a drug effect. Foot shock reinstated nicotine self-administration (Buczek, Lê, Wang, & Stewart, 1999), whereas restraint stress reinstated nicotine (Leão, Cruz, & Planeta, 2009) and amphetamine

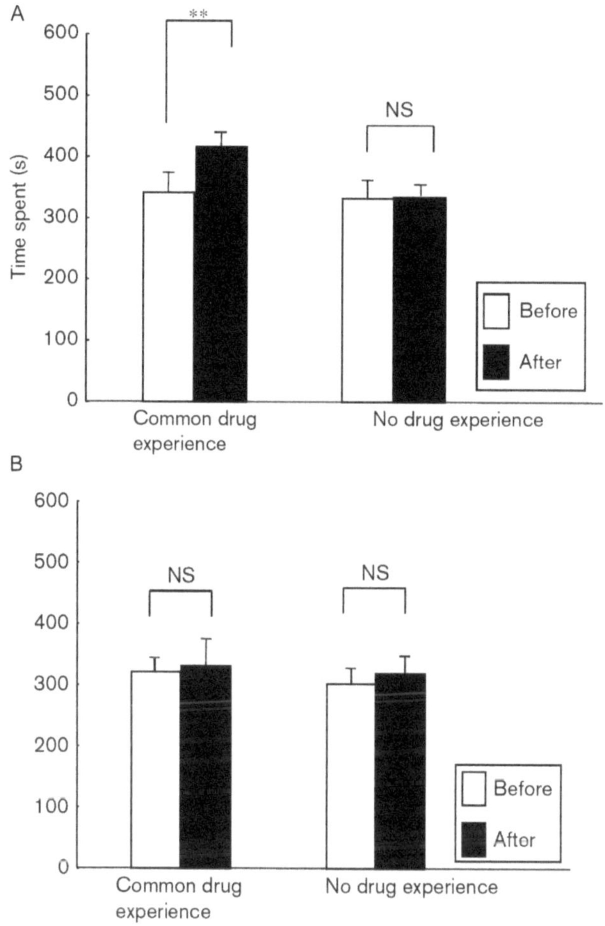

Fig. 2 (A) Mean time spent in the compartment associated with a methamphetamine-injected stimulus mouse before and after CPP. (B) Mean time spent in the compartment associated with a morphine-injected stimulus mouse before and after CPP. Error bars indicate the standard error of the mean. $^{**}P < 0.05$. *From Watanabe, S. (2015). Common experience modifies the reinforcing properties of methamphetamine-injected cage mates but not morphine-injected cage mates in C57 mice.* Behavioral Pharmacology, *26, 636–641.*

(Cruz, Leão, Marin, & Planeta, 2010) CPP in rats. However, Taylor et al. (2016) subjected rats to chronic variable stress for 14 days and found no effects on amphetamine CPP. Furthermore, Kabbaj et al. (2001) reported dissociation of the effects of social (isolation, defeat, rehousing, crowding, novelty, and litter shifting) and non-social (cold, forced swim, loud noise, ether, and restraint) stress on the sensitization of locomotor activity to amphetamine.

To differentiate the effects of social and non-social stress on stimulant reward, we must compare social and non-social stressors of similar intensity. In humans, the most common type of stress is social; thus, analysis of the modification of drug reward by social stress is important even though the difference between social and non-social stress is not clear. Here, I introduce examples of animal experiments on the modification of drug effects by social stress.

3.1 Presence of Dominant Mates

Social relationship, such as dominance or subordination, modifies the reinforcing effect of a conspecific animal. Virtually all human and animal societies have stratified social classes, or dominance hierarchies, which are fundamental properties of social groups. The adaptive basis of stratification is competition for food, water, territory, and reproductive opportunities. Social status, or the dominant/subordinate relation, plays a crucial role in social preference in mice (Fichett, Barnard, & Cassaday, 2006; Van Loo et al., 2001). Thus, subordinate mice spent more time with a dominant cage mate that had received a pain-inducing formalin injection than with a pain-free mid-rank cage mate. This preference was not evident when the dominant cage mate had not received a formalin injection (Watanabe, 2014), indicating that mice might prefer distress of dominant mice. Time spent at the formalin-injected dominant mate was negatively correlated with the "dominancy distance" between the two mice, and dominant mice did not display any preference for distressed subordinates (Watanabe, 2014). It is possible that, in mice, the pain response of a slightly more dominant mate signals a higher probability of overturning the relative hierarchical positions than does pain in a cage mate of a much higher dominance rank. The pain response of a dominant mate may have a rewarding effect on a subordinate, resulting in a Schadenfreude-like phenomenon. Schadenfreude occurs when misfortune of others induces pleasure in the observer. The term consists of "Schade" (sorry) and "Freude" (pleasure) in Germany. Consistent with pain display playing a negative role in social dominance, introduction of an unfamiliar male reduced the pain response in male mice, indicating an analgesic effect possibly serving to conceal the appearance of weakness (Langford et al., 2010).

Dominant rats showed morphine-induced CPP, whereas their submissive partners did not (Coventry, D'Aquila, Brain, & Willner, 1997). In a subordinate animal, the presence of a dominant mate induces social stress, which in turn elicits endocrine responses. The release of stress hormones has been demonstrated in subordinate monkeys (Coe, Franklin, Smith, & Levine, 1982; Steklis, Brummer, Raleigh, & McGuire, 1985) and hamsters (Huhman, Bunnell, Mougey, & Meyerhoff, 1990). These studies suggest that the presence of a dominant animal stimulates physiological changes in subordinate animals, resulting in reward value modification as well as other behavioral outcomes. Maxim (1977) reported that the bar-pressing rates for intracranial self-stimulation (ICSS) increased in subordinate monkeys in the presence of a dominant monkey. The question, however, arises whether the higher response rate indicates an increase or decrease in the reward associated with electrical stimulation. The animals may have responded more frequently because the same current intensity of stimulation became more rewarding, or they may have required more stimulation because each train of pulses was less rewarding. To answer this question, the ICSS thresholds of the lateral hypothalamic area were measured in hamsters (Kureta & Watanabe, 1996).

Fig. 3 shows that the presence of a dominant mate rendered ICSS less pleasurable, whereas the presence of a subordinate mate had no effect. This finding may represent another example of the modification of positive reward by the dominant/subordinate relation. We tried to mimic the effect of a dominant mate by injection of a benzodiazepine inverse agonist *N*-methyl-β-carboline-3-carboxamide (FG-7142), which produces a state of anxiety potentially similar to that induced in subordinate animals in the presence of a dominant mate. To test this possibility, the effect of FG-7142 on the ICSS threshold was measured in naive animals. FG-7142 caused a dose-dependent decrease in the rewarding effect of ICSS. These results suggest that the presence of a dominant animal may cause a state of anxiety similar to that caused by benzodiazepine receptor inverse agonists.

3.2 Social Defeat

Social defeat is experimentally initiated when a male rodent is introduced into the home cage of an older, aggressive, dominant male. Repeated exposures to social defeat stress in rodents, for example, cause a robust depression-like phenotype (Hollis & Kabbaj, 2014; Tornatzky & Miczek, 1993), increases vulnerability to cocaine self-administration (Covington & Miczek, 2001). Montagud-Romero et al. (2015) examined the influence of acute social defeat on the reinforcing effect of cocaine in adolescent

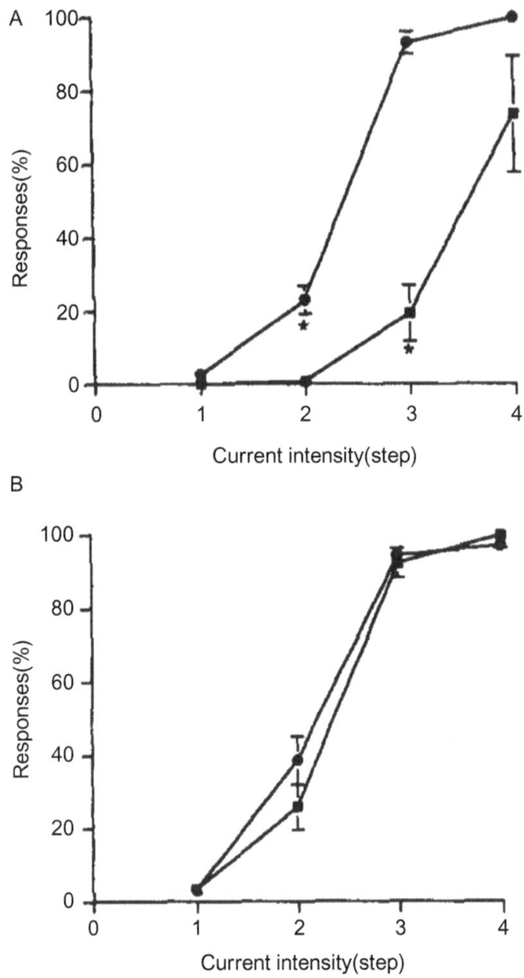

Fig. 3 (A) Effects of the presence of a dominant hamster on relative response rate for ICSS of the subordinate hamsters. *$P < 0.05$ (paired *t*-test). (B) Effects of the presence of subordinate animals on ICSS of the dominant animals. *From Kureta, Y., & Watanabe, S. (1996). Influence of social dominance on self-stimulation behavior in male golden hamsters.* Physiology & Behavior, *59, 621–624.*

and adult male mice using CPP. Mice were exposed to social defeat before each session of conditioning with cocaine. Adult mice exposed to acute social defeat showed an increase in conditioned cocaine preference, and needed a higher number of extinction sessions for the cocaine-induced CPP to be extinguished. In contrast, adolescent mice showed reduced cocaine CPP after acute social defeat. Adult rats that had experienced social defeat in adolescence showed facilitation of amphetamine CPP (Burke, Watt, & Forster, 2011), indicating that the timing of exposure to social defeat is important. Interestingly, the experience of non-social stress (foot shock) did not enhance amphetamine CPP (Burke et al., 2011), suggesting differential effects of social and non-social stress on amphetamine reward.

Effects of chronic defeat stress on drug-induced CPP have also been examined. Mathews et al. (2008) reported age-related differences in the effects of chronic social stress on amphetamine CPP. In a CPP experiment, adolescent female rats exposed to chronic social stress developed increased amphetamine CPP compared to that in unstressed rats. On the other hand, adults of both sexes showed a decrease in amphetamine CPP when exposed to chronic social stress. Thus, chronic stress facilitated amphetamine reward in adolescents while inhibiting it in adults. The age-dependent effects of social defeat have been observed in acute stress; however, their direction in chronic defeat is inconsistent with that in acute defeat.

García-Pardo et al. (2015) evaluated the long-term effects of repeated social defeat on CPP and young adult mice were exposed to repeated social defeat, followed by CPP training with (±)3,4-methylenedioxymethamphetamine (MDMA) 3 weeks later. Repeated social defeat during adolescence enhanced vulnerability to priming-induced reinstatement of MDMA CPP in conditioned animals. These results suggest that repeated social defeat exposed during adolescent causes a long-lasting facilitation of rewarding effects of MDMA.

The long-term effect of social defeat may be mediated by changes in the blood–brain barrier (BBB). Mice socially defeated during adolescence developed cocaine CPP, and exhibited reinstated preference for an ineffective dose of cocaine (Rodríguez-Arias et al., 2015) and social-defeat-associated changes in the BBB structure in adulthood. Reduction in expression of the tight junction protein and an increase in basal laminin degradation were observed in the nucleus accumbens and hippocampus.

3.3 Social Isolation

Social isolation reduced amphetamine CPP in adult rats (Schenk et al., 1986) and rats reared with social partners consumed less amphetamine than animals raised in social isolation (Bardo, Klebaur, Valone, & Deaton, 2001). Hofford et al. (2014) examined effects of social isolation on methamphetamine CPP. Rats were raised in social cohorts or in isolation.

Then they were then trained on methamphetamine CPP and methamphetamine self-administration. Neither CPP nor self-administration showed significant effects of social isolation. Similarly, Hensleigh and Pritchard (2014) demonstrated that neonatal separation did not affect methamphetamine CPP (while still affecting methamphetamine locomotor sensitivity). However, Hofford et al. (2014) showed that cue-induced reinstatement was reduced by social rearing. Thus, the effects of social isolation on the reinforcing properties of amphetamine remain to be clarified.

Contradictory results have also been reported on the effects of isolation on cocaine self-administration. Some studies found no effects (Bozarth, Murray, & Wise, 1989), but others reported a decrease in cocaine self-administration in rats (Hill & Powell, 1976; Phillips, Howes, Whitelaw, Robbins, & Everitt, 1994) and mice (Morse, Erwin, & Jones, 1993). Other researchers reported that social isolation enhanced cocaine CPP (Zakharova, Miller, Unterwald, Wade, & Izenwasser, 2009) and cocaine self-administration (Ding, Kang, & Li, 2005; Kosten, Miserendino, & Kehoe, 2000; Schenk, Lacelle, Gorman, & Amit, 1987). Thus, the effects of social isolation on cocaine reward remain controversial.

Rearing together with a conspecific has multiple aspects, with tactile stimulation and physical contact being particularly important. Artificial tactical stimulation may substitute for social stimulation: neonatal tactile stimulation reduced amphetamine CPP in socially isolated rats (Antoniazzi, Boufleur, Dolci, et al., 2014). Exercise also affects amphetamine CPP. Fontes-Ribeiro et al. (2011) forced rats to run on a treadmill for 8 weeks followed by amphetamine CPP training. The animals showed less conditioned preference than those without such exercise. Furthermore, swimming for 5 days after amphetamine CPP prevented drug relapse in rats (Segata et al., 2014). These results suggest that exercise may have a preventative effect on amphetamine addiction. Both social isolation and social defeat may result in reduced physical activity, which may underlie the enhanced amphetamine reward induced by these conditions.

4. PARENTAL EFFECTS AND MUTUAL MODIFICATION OF SOCIAL AND STIMULANT REWARDS

4.1 Parental Effects

Parental effects are another developmental factor modifying drug reward. Cocaine-sired male, but not female, rats displayed reduced cocaine self-administration compared to rats born to drug-free parents

(Vassoler, White, Schmidt, Sadri-Vakili, & Pierce, 2013). In contrast, no effects of cocaine reward in cocaine-sired mice was reported by another group of researchers (Killinger, Robinson, & Stanwood, 2012). Interestingly, Fischer et al. (2017) reported parental effects of male sires bred with drug-naïve females. Female cocaine-sired offspring showed lower cocaine CPP at low doses and consumed more sucrose (natural reward). The authors attributed the findings to lower sensitivity to reward in general, resulting in higher sucrose consumption. The paternal cocaine exposure could cause a shift in the threshold of rewarding effects non-specifically.

4.2 Mutual Modification

Social and drug reward effects interact mutually. Social interaction can reduce drug-induced CPP (Calcagnetti & Schechter, 1992; El Rawas et al., 2012; Fritz et al., 2011; Kummer, Klement, Eggart, et al., 2011; Thiel et al., 2008; Trezza, Damsteegt, & Vanderschuren, 2009). Salti et al. (2015) demonstrated that social interaction with a male early-adult conspecific rat as an alternative (non-drug-associated) stimulus reversed cocaine CPP. Conversely, exposure to drugs can modify social behavior of the prairie vole (*Microtus ochrogaster*), which forms long-term pair bonds after mating. Repeated amphetamine exposure impairs the formation of mating-induced partner preferences in male prairie voles and, conversely, pair bonding may impair amphetamine-induced CPP (Liu et al., 2010).

As described in Section 2, social and drug reward may share common neural mechanisms in the nucleus accumbens dopaminergic system. Amphetamine induces D1 type receptor activation in the nucleus accumbens, which may underlie pair bonding impairment (Liu et al., 2010). Liu, Young, Curtis, Aragona, and Wang (2011) demonstrated that amphetamine CPP in sexually naive, but not pair-bonded, males. In addition, amphetamine treatment increased D1 receptor autoradiographic binding in naive males, while decreasing it in pair-bonded males. Antagonizing D1 receptors in the nucleus accumbens blocked amphetamine CPP in sexually naïve males, and D1 receptor activation before conditioning resulted in amphetamine CPP in pair-bonded males. Thus, pair bonding experience decreases amphetamine reward through a D1-receptor-mediated mechanism.

Numerous reports have addressed the relationship between the neuropeptides oxytocin (OT) and vasopressin, and pair bonding (see Hammock & Young, 2006, for a review). The OT/vasopressin system can modulate some effects of MDMA on social behavior (Navarro, Rivera, Maldonado, Cavas, & de la Calle, 2003). In addition, Baracz and Cornish (2016) reported that OT modulated the reinforcing properties of methamphetamine. OT inhibited methamphetamine-induced hyperactivity, whereas injection of a selective OT inhibitor (atosiban) reduced this inhibitory effect (Qi, Yang, Song, et al., 2008). Ponzoni, Braida, Bondiolotti, and Sala (2017) examined MDMA induced CPP in zebrafish and observed that MDMA reward was blocked by a vasopressin antagonist (SR49059). These findings show that the OT/vasopressin system is involved not only in social reward but also in stimulant-induced reward in zebrafish.

5. FUTURE STUDY DIRECTIONS

Stimulant reward is not a simple pharmacological effect but a result of complicated interaction with other physiological and psychological systems, including genetics, development, individual history, and social relation. In social facilitation of drug reward, sharing a common pharmacological experience, simultaneously or successively, plays a crucial role. Other social factors, such as the dominant/subordinate relation, social defeat, and isolation, also affect drug reward but these factors induce stress, they may modulate drug effects through stress-induced mechanisms. Much remains to be elucidated to understand the interaction of social and drug effects. Social modification of drug effects has particular importance for understanding drug addiction and is widespread in the animal kingdom, including not only mammals but also fish. A comprehensive study of various species, particularly of non-social animals, is crucial for understanding the evolutionary origins of social modification of drug addiction. Another important area of future research is the study of neural mechanisms triggered by different types of reinforcing agents. In other words, is there only one common mechanism for any rewards or pleasure? Because reward is defined behaviorally and different neural mechanisms may produce the same behavioral effect, to understand reward-seeking behavior, it is imperative to characterize mechanisms common to different reinforcing agents as well as agent-specific mechanisms. Different drugs, for example methamphetamine and morphine, may have different sensitivity to social factors. To clarify drug-specific social modulation should improve our treatment of drug abuse.

ACKNOWLEDGMENTS

This research was supported by a Grant-in-Aid for Scientific Research on Innovative Areas, Japan Society for the Promotion of Science (25118001).

REFERENCES

Antoniazzi, C. T. D., Boufleur, N., Dolci, G., et al. (2014). Influence of neonatal tactile stimulation on amphetamine preference in young rats: Parameters of addiction and oxidative stress. *Pharmacology, Biochemistry, and Behavior, 124*, 341–349.

Antonitis, J., & Baron, A. (1964). Social reinforcement of bar pressing in mice. *The Journal of Genetic Psychology, 105*, 223–236.

Baracz, S. J., & Cornish, J. L. (2016). The neurocircuitry involved in oxytocin modulation of methamphetamine addiction. *Frontiers in Neuroendocrinology, 43*, 1–18.

Bardo, M. T., Klebaur, J. E., Valone, J. M., & Deaton, C. (2001). Environmental enrichment decreases intravenous sell-administration of amphetamine in female and male rats. *Psychopharmacology, 155*, 278–284.

Belujon, P., & Grace, A. A. (2011). Hippocampus, amygdala, and stress: Interacting systems that affect susceptibility to addiction. *Annals of the New York Academy of Sciences, 1216*, 114–121.

Bozarth, M. A., Murray, A., & Wise, R. A. (1989). Influence of housing conditions on the acquisition of intravenous heroin and cocaine self-administration in rats. *Pharmacology, Biochemistry, and Behavior, 33*, 903–907.

Brodkin, E. S., Hagemann, A., Nemenski, S. M., & Silver, L. M. (2004). Social approach avoidance behavior of inbred mouse strains to DBA/2 mice. *Brain Research, 1002*, 151–157.

Buczek, Y., Lê, A. D., Wang, A., & Stewart, J. (1999). Stress reinstates nicotine seeking but not sucrose solution seeking in rats. *Psychopharmacology, 144*, 183–188.

Burke, A. W., & Miczek, K. A. (2014). Stress in adolescence and drugs of abuse in rodent models: Role of dopamine, CRF, and HPA axis. *Psychopharmacology, 231*, 1557–1580.

Burke, A. R., Watt, M. J., & Forster, G. L. (2011). Adolescent social defeat increases adult amphetamine conditioned place preference and alters D2 dopamine receptor expression. *Neuroscience, 197*, 269–279.

Calcagnetti, D. J., & Schechter, M. D. (1992). Place conditioning reveals the rewarding aspect of social interaction in juvenile rats. *Physiology & Behavior, 51*, 667–672.

Coe, C. L., Franklin, D., Smith, E. R., & Levine, S. (1982). Hormonal responses accompanying fear and agitation in the squirrel monkey. *Physiology & Behavior, 29*, 1051–1057.

Cole, S. L., Hofford, R. S., Evert, D. J., Wellman, P. J., & Eitan, S. (2013). Social influences on morphine conditioned place preference in adolescent mice. *Addiction Biology, 18*, 274–285.

Coventry, T. L., D'Aquila, P. S., Brain, P., & Willner, P. (1997). Social influences on morphine conditioned place preference. *Behavioral Pharmacology, 8*, 575–584.

Covington, H. E., & Miczek, K. A. (2001). Repeated social-defeat stress, cocaine or morphine. Effects on behavioral sensitization and intravenous cocaine self-administration "binges". *Psychopharmacology, 158*, 388–398.

Cruz, F. C., Leão, R. M., Marin, M. T., & Planeta, C. L. (2010). Stress-induced reinstatement of amphetamine-conditioned place preference and changes in tyrosine hydroxylase in the nucleus accumbens in adolescent rats. *Pharmacology, Biochemistry, and Behavior, 96*, 160–165.

DiMicco, J. A., & Zaretsky, D. Z. (2007). The dorsomedial hypothalamus: A new player in thermoregulation. *American Journal of Physiology. Regulatory, Integrative and Comparative Physiology, 292*, R47–R63.

Ding, Y., Kang, L., & Li, M. L. (2005). Enhanced cocaine self-administration in adult rats with adolescent isolation experience. *Pharmacology, Biochemistry, and Behavior, 82*, 673–677.

Ellenbroek, B. A., van der Kam, E. L., van der Elst, M. C., & Cools, A. R. (2005). Individual differences in drug dependence in rats: The role of genetic factors and life events. *European Journal of Pharmacology, 526*, 251–258.

El Rawas, R., Klement, S., Kummer, K. K., Fritz, M., Dechant, G., Saria, A., et al. (2012). Brain regions associated with the acquisition of conditioned place preference for cocaine vs. social interaction. *Frontiers in Behavioral Neuroscience*, *6*, 63. https://doi.org/10.3389/fnbeh.2012.00063.

Embry, D., Hankins, M., Biglan, A., & Boles, S. (2009). Behavioral and social correlates of methamphetamine use in a population-based sample of early and later adolescents. *Addictive Behaviors*, *34*, 343–351.

Fichett, A. E., Barnard, C. J., & Cassaday, H. J. (2006). There's no place like home: Cage odours and place preference in subordinate CD-1 male mice. *Physiology & Behavior*, *87*, 955–962.

Fischer, D. K., Rice, R. C., Rivera, A. M., Donohoe, M., & Rajadhyaksha, A. M. (2017). Altered reward sensitivity in female offspring of cocaine-exposed fathers. *Behavioural Brain Research*, *332*, 23–31.

Fontes-Ribeiro, C. A., Marques, E., Pereira, F. C., Silva, A. P., & Macedo, T. R. A. (2011). May exercise prevent addiction? *Current Neuropharmacology*, *9*, 45–48.

Fritz, M., El Rawas, R., Klement, S., Kummer, K., Mayr, M. J., Eggart, V., et al. (2011). Differential effects of accumbens core vs. shell lesions in a rat concurrent conditioned place preference paradigm for cocaine vs. social interaction. *PLoS One*, *6*, e26761.

García-Pardo, M. P., Blanco-Gandía, M. C., Valiente-Lluch, M., Rodríguez-Arias, M., Miñarro, J., & Aguilar, M. A. (2015). Long-term effects of repeated social stress on the conditioned place preference induced by MDMA in mice. *Progress in Neuro-Psychopharmacology & Biological Psychiatry*, *63*, 98–109.

Gauvin, D. V., Briscoe, R. J., Goulden, K. L., & Hollowey, F. A. (1994). Aversive attribute of ethanol can be attenuated by dyadic social interaction. *Alcohol (Fayetteville, N.Y.)*, *11*, 247–251.

Gawin, F. H. (1991). Cocaine addiction: Psychology and neurophysiology. *Science*, *251*, 1580–1586.

Glynn, T. J. (1989). Essential elements of school-based smoking prevent programs. *Journal of School Health*, *59*, 181–188.

Gordon, H. W. (2002). Early environmental stress and biological vulnerability to drug abuse. *Psychoneuroendocrinology*, *27*, 115–126.

Grotewold, S. K., Wall, V. L., Goodell, D. J., et al. (2014). Effects of cocaine combined with a social cue on conditioned place preference and nucleus accumbens monoamines after isolation rearing in rats. *Psychopharmacology*, *231*, 3041–3053.

Hammock, E. A. D., & Young, L. J. (2006). Oxytocin, vasopressin and pair bonding: Implications for autism. *Philosophical Transactions of the Royal Society of London. Series B, Biological Sciences*, *361*, 2187–2198.

Hensleigh, E., & Pritchard, L. M. (2014). The effect of early environmental manipulation on locomotor sensitivity and methamphetamine conditioned place preference reward. *Behavioural Brain Research*, *268*, 66–71.

Hill, S. Y., & Powell, B. J. (1976). Cocaine and morphine self-administration: Effects of differential rearing. *Pharmacology, Biochemistry, and Behavior*, *5*, 701–704.

Hofford, R. C., Darna, M., Carrie, E., Wilmouth, C. E., Dwoskin, L. P., & Bardo, T. M. (2014). Environmental enrichment reduces methamphetamine cue-induced reinstatement but does not alter methamphetamine reward or VMAT2 function. *Behavioural Brain Research*, *270*, 151–158.

Hollis, F., & Kabbaj, M. (2014). Social defeat as an animal model for depression. *Institute for Laboratory Animal Research Journal*, *55*, 221–232.

Huhman, K. L., Bunnell, B. N., Mougey, E. H., & Meyerhoff, J. L. (1990). Effects of social conflict on POMC-derived peptides and glucocorticoids in male golden hamsters. *Physiology & Behavior*, *47*, 949–956.

Humphreys, A. P., & Einon, D. F. (1981). Play as a reinforcer for maze-learning in juvenile rats. *Animal Behavior*, *29*, 259–270.

Hunt, P. S., Hollaway, J. L., & Scordalakes, E. M. (2009). Social interaction with an intoxicated sibling can result in increased intake of ethanol by periadolescent rats. *Developmental Psychobiology*, *38*, 101–109.

Kabbaj, M., Nortonm, C. S., Kollack-Walker, S., Watson, S. J., Robinson, E., & Akil, H. (2001). Social defeat alters the acquisition of cocaine self-administration in rats: Role of individual differences in cocaine taking behavior. *Psychopharmacology*, *158*, 382–387.

Kataoka, N., Hioki, H., Kaneko, T., & Nakamura, K. (2014). Psychological stress activates a dorsomedial hypothalamus–medullary raphe circuit driving brown adipose tissue thermogenesis and hyperthermia. *Cell Metabolism*, *20*, 346–358.

Kennedy, B. C., Panksepp, J. B., Runckel, P. A., & Lahvis, G. P. (2012). Social influences on morphine-conditioned place preference in adolescent BALB/cJ and C57BL/6J mice. *Psychopharmacology*, *219*, 923–932.

Killinger, C. E., Robinson, S., & Stanwood, G. D. (2012). Subtle biobehavioral effects produced by paternal cocaine exposure. *Synapse*, *66*, 902–908.

Kosten, T. A., Miserendino, M. J. D., & Kehoe, P. (2000). Enhanced acquisition of cocaine self-administration in adult rats with neonatal isolation stress experience. *Brain Research*, *875*, 44–50.

Kummer, K., Klement, S., Eggart, V., et al. (2011). Conditioned place preference for social interaction in rats: Contribution of sensory components. *Frontiers in Behavioral Neuroscience*, *5*, 80. https://doi.org/10.3389/fnbeh.2011.00080.

Kureta, Y., & Watanabe, S. (1996). Influence of social dominance on self-stimulation behavior in male golden hamsters. *Physiology & Behavior*, *59*, 621–624.

Langford, D. J., Tuttle, A. H., Brown, K., Deschenes, S., Fischer, D. B., Mutso, A., et al. (2010). Social approach to pain in laboratory mice. *Social Neuroscience*, *5*, 163–170.

Leão, R. M., Cruz, F. C., & Planeta, C. S. (2009). Exposure to acute restraint stress reinstates nicotine-induced place preference in rats. *Behavioral Pharmacology*, *20*, 109–113.

Liu, Y., Aragona, B. J., Young, K. A., Dietz, D. M., Kabbaj, M., Mazei-Robison, M., et al. (2010). Nucleus accumbens dopamine mediates amphetamine-induced impairment of social bonding in a monogamous rodent species. *Proceedings of the National Academy of Sciences of the United States of America*, *107*, 1217–1222.

Liu, Y., Young, K. A., Curtis, J. T., Aragona, B. J., & Wang, Z. (2011). Social bonding decreases the rewarding properties of amphetamine through a dopamine D1 receptor-mediated mechanism. *The Journal of Neuroscience*, *31*, 7960–7966.

Logrip, M. L., Zorrilla, E. P., & Koob, G. F. (2012). Stress modulation of drug self-administration: Implications for addiction comorbidity with post-traumatic stress disorder. *Neuropharmacology*, *62*, 552–564.

Ma, Y. Y., Guo, C. Y., Yu, P., Lee, D. Y., Han, J. S., & Cui, C. L. (2006). The role of NR2B containing NMDA receptor in place preference conditioned with morphine and natural reinforcers in rats. *Experimental Neurology*, *200*, 343–355.

Maldonado, A. M., Finkbeiner, L. M., & Kirstein, C. L. (2008). Social interaction and partner familiarity differentially alter voluntary ethanol intake in adolescent male and female rats. *Alcohol (Fayetteville, N.Y.)*, *42*, 641–648.

Marshall, B. D. L., Wood, E., Shoveller, J. A., Buxton, J. A., Montaner, J. S. G., & Kerr, T. (2011). Individual, social, and environmental factors associated with initiating methamphetamine injection: Implications for drug use and HIV prevention strategies. *Prevention Science*, *12*, 173–180.

Mathews, I. Z., Mills, R. G., & McCormick, C. M. (2008). Chronic social stress in adolescence influenced both amphetamine conditioned place preference and locomotor sensitization. *Developmental Psychobiology*, *50*, 451–459.

Maxim, P. E. (1977). Self-stimulation of a hypothalamic site in response to tension or fear. *Physiology & Behavior*, *18*, 197–201.

Montagud-Romero, S., Aguilar, M. A., Maldonado, C., Manzanedo, C., Miñarro, J., & Rodríguez-Arias, M. (2015). Acute social defeat stress increases the conditioned rewarding effects of cocaine in adult but not in adolescent mice. *Pharmacology, Biochemistry, and Behavior, 135*, 1–12.

Morse, A. C., Erwin, V. G., & Jones, B. C. (1993). Strain and housing affect cocaine self-selection and open-field locomotor activity in mice. *Pharmacology, Biochemistry, and Behavior, 45*, 905–912.

Nabben, T., & Korf, D. J. (1999). Cocaine and crack in Amsterdam: Diverging subcultures. *Journal of Drug Issues, 29*, 627–652.

Navarro, J. F., Rivera, A., Maldonado, E., Cavas, M., & de la Calle, A. (2003). Anxiogenic-like activity of 3,4-methylenedioxy-methamphetamine (ecstasy) in the social interaction test is accompanied by an increase of c-fos expression in mice amygdala. *Progress in Neuro-Psychopharmacology & Biological Psychiatry, 28*, 249–254. https://doi.org/10.1016/j.pnpbp.

Panksepp, J. B., Jalowiec, J., DeEskinazi, F. G., & Bishop, P. (1985). Opiates and play dominance in juvenile rats. *Behavioral Neuroscience, 99*, 441–453.

Panksepp, J. B., & Lahvis, G. P. (2007). Social reward among juvenile mice. *Genes, Brain, and Behavior, 6*, 661–667.

Pearson, B. L., Defensor, E. B., Blanchard, D. C., & Blanchard, R. J. (2010). C57BL/6J mice fail to exhibit preference for social novelty in the three-chamber apparatus. *Behavioural Brain Research, 213*, 189–194.

Phillips, G. D., Howes, S. R., Whitelaw, R. B., Robbins, T. W., & Everitt, B. J. (1994). Isolation rearing impairs the reinforcing efficacy of intravenous cocaine or intra-accumbens d-amphetamine: Impaired response to intra-accumbens D1 and D2/D3 dopamine receptor antagonists. *Psychopharmacology, 115*, 419–429.

Ponzoni, L., Braida, D., Bondiolotti, G., & Sala, M. (2017). The non-peptide arginine-vasopressin selective receptor antagonist, sr49059, blocks the rewarding, prosocial, and anxiolytic effects of 3,4-methylenedioxymethamphetamine and its derivatives in zebra fish. *Frontiers in Psychiatry, 8*, 1–13. https://doi.org/10.3389/fpsyt.2017.00146.

Qi, J., Yang, J. Y., Song, M., et al. (2008). Inhibition by oxytocin of methamphetamine-induced hyperactivity related to dopamine turnover in the mesolimbic region in mice. *Naunyn-Schmiedeberg's Archives of Pharmacology, 376*, 441–448.

Rodríguez-Arias, M., Montagud-Romero, S., Rubio-Araiz, A., Aguilar, M. A., Martín-García, E., Cabrera, R., et al. (2015). Effects of repeated social defeat on adolescent mice on cocaine-induced CPP and self-administration in adulthood: Integrity of the blood–brain barrier. *Addiction Biology, 5*, 129–141. https://doi.org/10.1111/adb.12301.

Salti, A., Kummer, K. K., Sadangi, C., Dechant, G., Saria, A., & Rawas, R. E. (2015). Social interaction reward decreases p38 activation in the nucleus accumbens shell of rats. *Neuropharmacology, 99*, 510–516.

Schenk, S., Hunt, T., Malovechko, R., Robertson, A., Klukowski, G., & Amit, Z. (1986). Differential effects of isolation housing on the conditioned place preference produced by cocaine and amphetamine. *Pharmacology, Biochemistry, and Behavior, 24*, 1793–1796.

Schenk, S., Lacelle, G., Gorman, K., & Amit, Z. (1987). Cocaine self-administration in rats influenced by environmental conditions: Implications for the etiology of drug abuse. *Neuroscience Letters, 81*, 227–231.

Segata, H. J., Kronbauera, M., Roversic, K. R., Schusterc, A. J., Veyc, L. T., Roversic, K., et al. (2014). Exercise modifies amphetamine relapse: Behavioral and oxidative markers in rats. *Behavioural Brain Research, 262*, 94–100.

Sinha, R. (2001). How does stress increase risk of drug abuse and relapse? *Psychopharmacology, 158*, 343–359.

Sinha, R. (2008). Chronic stress, drug use, and vulnerability to addiction. *Annals of the New York Academy of Sciences, 1141*, 105–130.

Smith, M. A., Strickland, J. C., Bills, S. E., & Lacy, R. T. (2015). The effects of a shared history of drug exposure on social choice. *Behavioral Pharmacology, 26*, 631–635.

Steklis, H. D., Brummer, G. L., Raleigh, M. J., & McGuire, M. T. (1985). Serum testosterone, male dominance, and aggression in captive groups of vervet monkeys (*Cercopithecus aethiops sabaeus*). *Hormones and Behavior, 19*, 154–163.
Taylor, S. B., Watterson, L. R., Kufahl, P. R., Nemirovsky, N. E., Tomek, S. E., Conrad, C. D., et al. (2016). Chronic variable stress and intravenous methamphetamine self-administration—Role of individual differences in behavioral and physiological reactivity to novelty. *Neuropharmacology, 108*, 353–363.
Thiel, K. J., Okun, A., & Neisewander, J. L. (2008). Social reward-conditioned place preference: A model revealing an interaction between cocaine and social context rewards in rats. *Drug and Alcohol Dependence, 96*, 201–212.
Thiel, K. J., Sanabria, F., & Neisewander, J. L. (2009). Synergistic interaction between nicotine and social rewards in adolescent male rats. *Psychopharmacology (Berlin), 204*, 391–402.
Tornatzky, W., & Miczek, K. A. (1993). Long-term impairment of autonomic circadian rhythms after brief intermittent social stress. *Physiology & Behavior, 53*, 983–993.
Trezza, V., Damsteegt, R., & Vanderschuren, L. J. M. J. (2009). Conditioned place preference induced by social play behavior: Parametrics, extinction, reinstatement and disruption by methylphenidate. *European Neuropsychopharmacology, 19*, 659–669.
Vanderschuren, L. J., Niesink, R. J., Spruijt, B. M., & Van Ree, J. M. (1995). Effects of morphine on different aspects of social play in juvenile rats. *Psychopharmacology, 117*, 225–231.
Van Loo, P. L. P., de Groot, A. C., Van Zutphen, B. F. M., & Bauemans, V. (2001). Do male mice prefer or avoid each other's company? Influence of hierarchy, kinship and familiarity. *Journal of Applied Animal Welfare Science, 4*, 91–103.
Vassoler, F. M., White, S. L., Schmidt, H. D., Sadri-Vakili, G., & Pierce, R. C. (2013). Epigenetic inheritance of a cocaine-resistance phenotype. *Nature Neuroscience, 16*, 42–47.
Watanabe, S. (2011). Drug-social interactions in the reinforcing property of methamphetamine in mice. *Behavioral Pharmacology, 22*, 203–206.
Watanabe, S. (2013). Social factors in conditioned place preference with morphine in mice. *Pharmacology, Biochemistry, and Behavior, 103*, 440–443.
Watanabe, S. (2014). The dominant/subordinate relation between mice modifies the approach behavior toward a cage mate experiencing pain. *Behavioural Processes, 103*, 1–4.
Watanabe, S. (2015a). Common experience modifies the reinforcing properties of methamphetamine-injected cage mates but not morphine-injected cage mates in C57 mice. *Behavioral Pharmacology, 26*, 636–641.
Watanabe, S. (2015b). Social factors modulate restraint stress induced hyperthermia in mice. *Brain Research, 162*, 134–139.
Watanabe, S. (2018). Social facilitation of methamphetamine-induced conditioned place preference in mice: Analysis of dose–response relationship. *Behavioral Pharmacology*. (submitted for publication).
Westermeyer, J. (1974). Opium dens. A social resource for addicts in Laos. *Archives of General Psychiatry, 31*, 237–240.
Yates, J. R., Beckmanna, J. S., Meyer, A. C., & Bardo, M. T. (2013). Concurrent choice for social interaction and amphetamine using conditioned place preference in rats: Effects of age and housing condition. *Drug and Alcohol Dependence, 129*, 240–246.
Zakharova, E., Miller, J., Unterwald, E., Wade, D., & Izenwasser, S. (2009). Social and physical environment alter cocaine conditioned place preference and dopaminergic markers in adolescent male rats. *Neuroscience, 163*, 890–897.

FURTHER READING

Watanabe, S. (2017). Social inequality aversion in mice: Analysis with stress-induced hyperthermia and behavioral preference. *Learning and Motivation, 59*, 38–46.

CHAPTER SIX

Animal Models of the Impact of Social Stress on Cocaine Use Disorders

Annika Vannan, Gregory L. Powell, Samantha N. Scott, Broc A. Pagni, Janet L. Neisewander[1]
School of Life Sciences, Arizona State University, Tempe, AZ, United States
[1]Corresponding author: e-mail address: janet.neisewander@asu.edu

Contents

International Review of Neurobiology, Volume 140
ISSN 0074-7742
https://doi.org/10.1016/bs.irn.2018.07.005

Abstract

Cocaine use disorders are strongly influenced by the social conditions prior, during, and after exposure to cocaine. In this chapter, we discuss how social factors such as early life stress, social rank stress, and environmental stress impact vulnerability and resilience to cocaine. The discussion of each animal model begins with a brief review of examples from the human literature, which provide the psychosocial background these models attempt to capture. We then discuss preclinical findings from use of each model, with emphasis on how social factors influence cocaine-related behaviors and how sex and age influence the behaviors and neurobiology. Models discussed include (1) early life social stress, such as maternal separation and neonatal isolation, (2) social defeat stress, (3) social hierarchies, and (4) social isolation and environmental enrichment. The cocaine-related behaviors reviewed for each of these animal models include cocaine-induced conditioned place preference, behavioral sensitization, and self-administration. Together, our review suggests that the degree of psychosocial stress experienced yields robust effects on cocaine-related behaviors and neurobiology, and these preclinical findings have translational impact for the future of cocaine use disorder treatment.

1. INTRODUCTION

Cocaine use disorders have been a costly and prevalent problem in the United States for several decades. Even though there has been a decline in use over the past 15 years, approximately 1.6% of the population in North America (5.1 million people) uses cocaine (United Nations Office on Drugs and Crime, 2016). In the United States, 4.2% of 12th graders surveyed from 2014 to 2017 reported cocaine use (Gorelick, 2017; National Institute on Drug Abuse, 2017), and in 2011 cocaine was associated with the most emergency room visits among all illegal drugs (Gorelick, 2017; Substance Abuse and Mental Health Services Administration, 2014). Unlike other addictive drugs such as opiates, alcohol, and tobacco, safe and effective pharmacotherapies are not available for treating cocaine use disorders. Understanding the mechanisms that underlie cocaine effects on the brain and risk factors for developing cocaine dependence is essential for developing novel pharmacological approaches for treating cocaine use disorders.

Much of the current knowledge of the neurobiology of cocaine use disorders comes from preclinical research using animal models and a recent thrust of research in this field is on the impact of social experiences at various phases of cocaine use and dependence. Social factors play a major role in initiation of cocaine use (Degenhardt, Stockings, Patton, Hall, & Lynskey, 2016; Kirst, Mecredy, Borland, & Chaiton, 2014) and psychosocial stress plays a major role in relapse to cocaine use even after prolonged period of abstinence (Sinha, 2001). Animal models used to examine phases of cocaine use disorders include (1) acquisition of cocaine self-administration to examine initiation of use, (2) long access and intermittent access self-administration sessions to examine escalation and binge patterns of use, and (3) extinction and reinstatement of cocaine-seeking behavior, defined as operant responding previously reinforced by cocaine infusions in the absence of cocaine availability, to examine incentive motivational effects of reinstating stimuli. Stimuli capable of reinstating extinguished cocaine-seeking behavior include re-exposure to drug-associated contextual and discrete cues, drug priming, and stress, all of which elicit craving in human cocaine abusers (Childress, McLellan, Ehrman, & O'Brien, 1988; Jaffe, Cascella, Kumor, & Sherer, 1989; Sinha, 2001). Another relevant behavior model is cocaine-conditioned place preference (CPP), which measures the incentive motivational effects of a cocaine-associated environment. A shift in preference for the cocaine-associated environment over an initially preferred environment when rodents are given free-access to both simultaneously is also thought to reflect the rewarding effects of cocaine (Bardo & Bevins, 2000). Cocaine effects on behavior often become enhanced as a result of intermittent exposure in rodents, referred to as sensitization (Huang, Liang, Lee, Wu, & Hsu, 2009; Post, Weiss, Fontana, & Pert, 1992). This behavioral phenomenon is relevant because there is cross-sensitization between social stress and cocaine whereby prior exposure to the stress renders animals more sensitive to cocaine (Tidey & Miczek, 1997).

In this chapter, we review some of the most prominent models that are used in preclinical research investigating the role of social factors in cocaine use disorders. We begin with maternal separation and neonatal isolation (NI) in rodents, which model psychosocial stress that occurs early in life and contributes to the likelihood of initiation of cocaine use and development of dependence later in life. Next, we review social defeat stress in rodents, which models psychosocial stress that occurs more proximal to opportunities for cocaine use. Finally, we review how social hierarchies and environmental living conditions impact various phases of cocaine use. As we review these models, we discuss their parallels to human cocaine use and we highlight some of the research on the underlying neural mechanisms. Two important biological variables, sex and age—both of which have only recently gained prominence in animal research—are discussed throughout the chapter.

2. MODELS OF EARLY LIFE, SOCIAL STRESS

2.1 Early Life, Social Stress in Humans and Cocaine Use

Human development involves a complex biological–environmental interplay that can influence stress reactivity, decision-making and mental health later in life. Physical and psychological development proceed in orderly patterns that operate with critical windows of time during which distinct experiences leave long-lasting impact. For instance, adverse childhood experiences (ACEs) can lead to lasting neurological, social, and emotional deficits (Behen et al., 2009; Ege, Messias, Thapa, & Krain, 2015; Kajeepeta, Gelaye, Jackson, & Williams, 2015; Koss et al., 2014), which may in turn heighten vulnerability to substance use disorders (Choi, Dinitto, Marti, & Choi, 2017; Dube et al., 2003). In England, an alarming 58.7% of heroin/crack cocaine usage is associated with ACEs (Bellis, Hughes, Leckenby, Perkins, & Lowey, 2014). Even among adult addicts who are abstinent from cocaine, childhood maltreatment can increase stress reactivity and perceived stress levels, making relapse more likely (Hyman, Paliwal, & Sinha, 2007). To understand how early life social stress impacts cocaine addiction in adulthood, animal models are used that capture psychosocial aspects of ACEs.

2.2 The Maternal Separation and Neonatal Isolation Models

The bond between infant and mother is the earliest and strongest social attachment observed in most mammals (Bowlby, 1982), and the amount and quality of maternal care can influence future behavior and physiology of offspring during social encounters (Francis & Meaney, 1999; Gunnar & Donzella, 2002). The presence of a caregiver may support proper neural development of brain regions involved in emotional regulation (Gee et al., 2014), while neglect is associated with behavioral, social, emotional, and cognitive difficulties later in life (Delker, Noll, Kim, & Fisher, 2014; Madigan, Vaillancourt, McKibbon, & Benoit, 2012). Overall, in both human and nonhuman species, mother–offspring social interactions play a crucial role in development.

Maternal separation (MS) in rodents was developed as a model to study the consequences of inconsistent maternal caregiving, including effects of this early life stress on substance use and dependence later in life. For the rodent model of MS, the litter is removed from the dam for a variable period of time, typically on a daily basis (Aisa, Tordera, Lasheras, Del Río, & Ramírez, 2007; Daniels, Pietersen, Carstens, & Stein, 2004; Moffett et al., 2006). The behavioral and neurobiological adaptations that result from MS vary depending on the timing, duration, and circumstances of the separation, as well as sex and age at which offspring are examined.

Duration of MS shows a hyperbolic function in terms of impact on drug self-administration later in life, with no separation or lengthy separation producing deleterious effects and brief separation producing protective effects. For instance, Moffett and colleagues compared cocaine self-administration in adulthood of rats that, as pups, were separated from dams for 0 (brief handling), 15, or 180 min/day for 2 weeks or that were not handled or removed from the dam during the 2-week period. Rats separated for 180 min/day acquired self-administration at a lower cocaine dose than the other groups, suggesting an increase in sensitivity to cocaine reinforcement. Surprisingly, non-separated controls had the highest daily intake of cocaine at high doses (Moffett et al., 2006, 2007). The brief handling and 15 min separation groups overall showed less vulnerability to cocaine self-administration than the 180 min separation and no separation/handling groups. One possible reason for these findings is that the pups learn coping strategies as a result of brief handling/MS that protects them against cocaine use later in life. Because handling by the experimenter allows rats to acclimate to handling stress, the no handling/MS group may have experienced greater stress or exhibited greater stress reactivity to handling during the cocaine self-administration training. Interestingly, the opposite relationship between separation and cocaine self-administration is observed in dams. Specifically, dams of both separated and handled pups had higher cocaine intake than those of non-handled pups when later given access to cocaine. Moffett and colleagues posited that for the dams, both separation from pups and experimenter handling of pups are stressors that increase the mothers' vulnerability to cocaine use.

There are sex differences in vulnerability to early life stress, including MS stress or NI stress. The NI model is more severe than the MS model because NI separates pups not only from the biological dam, but from other littermates as well. Sex differences were first reported after rat pups were exposed to MS 6 h/day for 10 days spaced randomly over postnatal days 5–20

(Matthews, Robbins, Everitt, & Caine, 1999). This study found that MS facilitates the acquisition and maintenance of cocaine self-administration at a high dose in adult females, but not in adult males. Even on day 1 of cocaine self-administration, MS females took more infusions of low-dose cocaine than non-separated controls. With NI, seven daily exposures for 1 h/day enhances acquisition and maintenance of cocaine self-administration in both males and females (Kosten, Lee, & Kim, 2006; Kosten, Miserendino, & Kehoe, 2000; Kosten, Sanchez, Zhang, & Kehoe, 2004), however the effects of NI on responding for food vary by sex. NI fails to alter food reinforcement in males (Kosten et al., 2000; Kosten, Sanchez, et al., 2004; Zhang, Sanchez, Kehoe, & Kosten, 2005), but decreases food reinforcement rates in females on both fixed (Kosten, Sanchez, et al., 2004) and progressive ratio (PR) schedules of reinforcement (Kosten et al., 2006). Thus, while NI enhances cocaine self-administration, this effect may be due to a general enhancement of reinforcement or motivation, at least in females, rather than a cocaine-specific effect. In support of a sex-dependent effect on motivation for cocaine, an 11-day MS exposure increases cocaine-seeking behavior during cue-induced reinstatement in male rats compared to female rats (Lynch, Mangini, & Taylor, 2005). Overall, these findings indicate that MS and NI early in life can alter cocaine reinforcement and/or motivation later in life in a sex-dependent manner.

Early life stress can cross sensitize animals to a cocaine challenge later in life, suggesting overlap in the mechanisms of early life stress and sensitivity to cocaine. For instance, Kikusui, Faccidomo, and Miczek (2005) demonstrated that 2 weeks of repeated, 1 h/day MS sensitized locomotor activity in response to low doses of cocaine later in life in both female and male mice, and a test for expression of sensitization showed a greater effect in males than females. Although similar sensitization results have been found in rats (Brake, Zhang, Diorio, Meaney, & Gratton, 2004), other studies showed contrasting findings, in which 2–3 weeks of MS *decreased* locomotor sensitization to a high dose of cocaine i.p. (Moffett et al., 2006) and to a moderate dose i.v. (Li, Robinson, & Bhatnagar, 2003). These results suggest MS can increase or decrease sensitivity to cocaine-induced locomotion depending on the paradigm, species, and sex.

2.3 Early Life Stress and the HPA Axis

Exposure to a stressor activates the hypothalamic–pituitary–adrenal (HPA) axis, initially causing release of corticotropin-releasing hormone (CRH; also commonly referred to as corticotropin-releasing factor, CRF) from the hypothalamus, which in turn causes secretion of adrenocorticotropic hormone (ACTH) from the pituitary gland. ACTH travels through the bloodstream to the adrenal gland where it stimulates production of glucocorticoids, i.e., cortisol in humans and corticosterone in rodents (CORT). Importantly, the HPA axis is regulated by feedback inhibition, whereby CORT and ACTH act at receptors in the hypothalamus and pituitary gland to inhibit further CRH and ACTH release. This cascade results in short-term increases in HPA axis hormones, which initiates physiological responses that supply energy to the brain and body to cope with the stress, and the negative feedback then serves to re-establish homeostasis. With chronic stress, however, the HPA axis becomes dysregulated and this dysregulation is thought to be a risk factor for the development of addiction (Koob, 2008; Sinha, Garcia, Paliwal, Kreek, & Rounsaville, 2006). Cocaine is a stressor that activates the HPA axis in humans (Baumann et al., 1995; Mendelson, Teoh, Mello, Ellingboe, & Rhoades, 1992) and rodents (Rivier & Vale, 1987; Sarnyai, Bíró, Penke, & Telegdy, 1992), and this effect of cocaine is involved in the neural mechanisms underlying cocaine seeking and consumption. For instance, both exogenous administration of CORT and increased levels of endogenous CORT modulate acquisition and maintenance of cocaine self-administration in rats (Goeders & Guerin, 1996; Mantsch, Saphier, & Goeders, 1998), and CORT also plays a role in reinstatement of extinguished cocaine-seeking behavior in rodents (Graf et al., 2013; McReynolds et al., 2016).

Childhood trauma and ACEs lead to deficits in emotional regulation (Douglas et al., 2010), resulting in increased circulating cortisol (Flanagan, Baker, McRae-Clark, Brady, & Moran-Santa Maria, 2015; Moran-Santa Maria et al., 2010). Furthermore, Moran-Santa Maria et al. (2010) found that childhood trauma affects the HPA axis differently in cocaine-dependent and healthy adults. In subjects with a history of childhood trauma, controls show blunted cortisol levels in response to CRH administered intravenously, whereas cocaine-dependent subjects failed to show an increase in ACTH levels (Moran-Santa Maria et al., 2010). Similarly, rodent MS induces long-lasting dysregulation of the HPA axis (Aisa et al., 2007; Faure et al., 2006; Kikusui et al., 2005). For example, female rats exposed to MS have higher plasma CORT compared to non-separated female rats (Aisa et al., 2007) and a significant increase in ACTH in response to stress (Daniels et al., 2004). Overall data from both rodents and humans suggest that MS-induced stress and ACEs alter stress reactivity via dysregulation of the HPA axis, and that this in turn increases vulnerability to cocaine use disorders.

2.4 Early Life Stress and Monoaminergic Neurotransmitter Systems

In addition to the HPA axis, MS also dysregulates the mesocorticolimbic dopamine (DA) system, which originates in the ventral tegmental area (VTA) and projects to the nucleus accumbens (NAc), amygdala, hippocampus, and prefrontal cortex (PFC). This dopaminergic system is essential for processing rewarding experiences, such as those derived from cocaine. Furthermore, low DA D2 receptor (D2R) concentrations in the striatum is associated with vulnerability to cocaine use disorders (Volkow et al., 1993). In mice, MS impairs neuroplasticity in the striatal dopaminergic pathways and decreases D2Rs in the NAc early in life (Gracia-Rubio et al., 2016). However, when MS mice reach adolescence they show increases in DA turnover, DA transporter (DAT) protein expression, and D2R levels when exposed to cocaine (Gracia-Rubio et al., 2016). These findings differ partly with those of rats in the NI model, where isolated rats show unaltered D2R expression, but decreased DAT levels in the NAc core and increased DA D3 receptors in the NAc shell (Brake et al., 2004).

In addition to inhibiting DATs, cocaine inhibits serotonin (5-HT) transporters (SERTs) resulting in dysregulation of 5-HT systems as well. Addiction research indicates a role of 5-HT in modulating cocaine reinforcement and cocaine-seeking behaviors (Wiers et al., 2016) (Cunningham, Paris, & Goeders, 1992; Filip, 2005; Neisewander, Cheung, & Pentkowski, 2014; Parsons, Weiss, & Koob, 1998). Moreover, 5-HT systems are widely known to play a role in mood, appetite, sleep, libido, anxiety, and depression and several of these processes are implicated in early life stress and substance use disorders (Clark & Neighbors, 1996). For instance, MS reduces basal levels of SERT and 5-HT in the hippocampus and raphe nucleus, which are brain regions associated with anxiety and depressive-like behaviors (Lee et al., 2007). In response to restraint stress, however, MS increases 5-HT turnover in the hypothalamus (Daniels et al., 2004). Alterations of 5-HT systems have also been found using NI models. After 8 days of NI, 5-HT levels in the ventral striatum decrease at baseline and in response to cocaine administration in infant rats (Kosten, Zhang, & Kehoe, 2004). Together these data suggest that early life stress dysregulates serotonin systems, and further research is needed to investigate the role of 5-HT in MS-induced changes in cocaine abuse-related behaviors.

2.5 Gaps in the Literature: Paternal Deprivation, Cross-Fostering, and Cocaine Effects on Maternal Care

Although preclinical addiction research generally utilizes rodent species that have limited paternal involvement in offspring development, the mandarin vole is a biparental rodent used to study effects of paternal deprivation (PD). In this model, vole pups are separated from the father, but are cared for by the mother. Like MS, PD has effects on DA systems in juvenile rats observed as decreases in immunolabeling of mesocorticolimbic DA cells, with the exception of the orbitofrontal cortex, whereas adult rats exhibit a selective decrease in the NAc (Braun, Seidel, Holetschka, Groeger, & Poeggel, 2013). Yu et al. (2012) found sex differences in DA receptor changes in response to PD. Specifically during adulthood, PD decreases dopamine 1 (D1) and D2 receptor mRNA expression in females while increasing these measures in males (Yu et al., 2012). However, a recent study found that PD had no effects on cocaine-induced anxiety or locomotor sensitization in offspring tested later in life (Wang, Fang, & Yang, 2017). The lack of a PD effect on cocaine-induced behavior warrants additional research on this issue.

Cross-fostering is another model for studying the role of parental care on behavior in adulthood. In this model, pups are separated from their biological dam and then cared for by another female simultaneously rearing her own pups of similar age (Denenberg, Grota, & Zarrow, 1963). As with MS, duration is a critical component of the cross-fostering model. Short-term (180 min/day for 13 days) cross-fostering in rats does not dysregulate HPA axis activity (Huot, Gonzalez, Ladd, Thrivikraman, & Plotsky, 2004). However, long-term cross-fostering for 3 weeks in mice results in emotional abnormalities and social interaction deficits in adulthood (Lu et al., 2009). Research with nicotine suggests that cross-fostering may increase drug self-administration in males (Roguski, Chen, Sharp, & Matta, 2013), but blunts drug-induced DA increases in the NAc in females (Kane, Fu, Matta, & Sharp, 2004). Despite links to emotionality and stress, however, cross-fostering remains relatively unexplored in models of cocaine abuse. This is an important question for future research given that in humans, children of drug-dependent parents are often placed in foster homes or otherwise cared for by a non-biological parent.

Another understudied area of research is the effect of drug use on maternal care. In humans, substance abuse by a parent is considered an ACE and is a risk factor for drug dependence in adulthood (Dube et al., 2003). In particular maternal cocaine use leads to deficits in mother–child play interactions (Johnson et al., 2002). In rats, cocaine exposure in-utero dysregulates the monoamine systems (Mayes, 1994) and enhances drug-seeking later in life (Keller, Lefevre, Raucci, Carlson, & Glick, 1996). Exposure of dams to cocaine after pregnancy reduces maternal care, as measured by licking and sniffing of pups (Zimmerberg & Gray, 1992), as well as nest-building and assumption of the nursing "crouch" position (Johns et al., 1998; Johns, Noonan, Zimmerman, Li, & Pedersen, 1994). More work is needed in this area to determine whether these alterations in maternal care by cocaine lead to increased propensity to self-administration cocaine in adulthood.

2.6 Conclusions

Models for understanding the role of early life social stress in vulnerability to cocaine use disorders later in life suggest that inconsistent maternal/parental caregiving has long-term impacts on behavior and the neurochemical pathways involved in addiction. In general, the findings from these models suggest that both early life stress and cocaine dysregulate DA and serotonin systems, as well as the HPA axis. The specific effects of early life stress on these systems and their relation to cocaine abuse-related behaviors are complicated by several factors that vary across studies, such as length and mode of early life stress and age, sex, and behavioral measure in adulthood. A general pattern emerging from studies of MS and NI is a hyperbolic, inverted U-shaped relationship akin to the Yerkes–Dodson function, whereby a history of transient exposure to early life stress may build resilience whereas either no stress, severe, or chronic stress may enhance vulnerability. In the future, models of early life stress will be helpful in identifying circuitry changes, as well as genetic and epigenetic mechanisms, that underlie resilience and vulnerability to cocaine use disorders.

3. MODELS OF AVERSIVE SOCIAL STRESS IN COCAINE ADDICTION

3.1 Aversive Social Experiences and Drug Use in Humans

People begin taking drugs as early as childhood, with the majority beginning during adolescence and young adulthood as they are developing a desire for independence. Both positive and negative social experiences at this development stage can influence initiation and maintenance of cocaine use (Björkqvist, 2001). Some common negative experiences are bullying and child abuse, both of which are correlated with psychiatric outcomes in early adulthood (Copeland, Wolke, Angold, & Costello, 2013).

Victims of bullying are known to suffer from increased rates of depression, anxiety, social phobias, self-esteem issues, and drug use (Björkqvist, 2001; Radliff, Wheaton, Robinson, & Morris, 2012; Tharp-Taylor, Haviland, & D'Amico, 2009). Thus, it is important to understand how aversive social experiences alter the brain to produce vulnerability to drug use.

3.2 The Social Defeat Stress Model

A useful animal model of physical abuse is the rodent social defeat stress (SDS) model (Koolhaas et al., 2013). In the SDS model, a smaller "intruder" rodent is placed into the home cage of an aggressive "resident," who will then threaten and physically assault the intruder until there are clear signs of submission. Animals exposed to SDS during adolescence exhibit anxiety and depression-like behaviors in adulthood (Vidal et al., 2007; Weder et al., 2014), as well as changes in sensitivity to cocaine (Burke & Miczek, 2015). The effects of SDS on the rewarding and motivational properties of psychostimulants vary in a manner dependent on the timing, severity, and frequency at which defeat episodes occur. SDS models may employ a single defeat episode (acute), daily episodes (chronic), or episodes every few days (intermittent).

3.3 The Effects of SDS on Cocaine-Related Behaviors

Frequency of defeat episodes contribute to SDS effects on cocaine self-administration. Chronic social defeat decreases motivation to obtain a high dose of cocaine on a PR schedule, whereas this same study found no effect with intermittent social defeat (Miczek, Nikulina, Shimamoto, & Covington, 2011). Compared to controls, chronic social defeat in rodents also reduces body weight, sucrose consumption, motility and exploratory behavior, and operant responding for saccharin on a PR schedule (Rygula et al., 2005), all of which are indicative of a depression-like phenotype. Collectively, the findings suggest bidirectional SDS effects with chronic SDS producing anhedonia-related behaviors, which have been associated with attenuated cocaine self-administration, and intermittent SDS producing cross-sensitization to cocaine and enhancing cocaine self-administration.

SDS effects on cocaine self-administration depend on the schedule of reinforcement and dose of cocaine available. For example, SDS increases response rates for low doses of cocaine (Miczek & Mutschler, 1996), which may reflect an increase in sensitivity to the reinforcing effects of cocaine; however, because this effect was observed primarily during a time-out period immediately after an infusion when drug was temporarily unavailable, the effect may be due to cocaine-induced hyperactivity or increased motivation to seek cocaine. Consistent with the latter idea, under a high-effort, PR schedule of reinforcement, SDS-experienced animals work harder for cocaine at moderate (Quadros & Miczek, 2009), but not high doses (Covington & Miczek, 2001; Covington, Tropea, Rajadhyaksha, Kosofsky, & Miczek, 2008). However, SDS does not affect drug consumption during short or long access to cocaine (Quadros & Miczek, 2009), nor acquisition of cocaine self-administration on low effort, fixed ratio (FR) schedules (Covington & Miczek, 2001). SDS effects on cocaine self-administration vary by age as well. Enhancement of cocaine self-administration in adolescents can last up to 3 weeks after the last defeat; on the other hand, enhanced acquisition and intake in adult SDS animals only lasts for a few days after the defeat (Haney, Maccari, Le Moal, Simon, & Piazza, 1995; Kabbaj et al., 2001; Tidey & Miczek, 1997).

SDS affects cocaine intake observed under "binge" self-administration conditions are thought to model cocaine use disorders in humans. A binge refers to consumption of high amounts of cocaine over several hours to days, with periods of voluntary abstinence between binges. To model binge self-administration in rodents, cocaine is available continuously over a 24-h session. Quadros and Miczek (2009) observed that intermittent SDS rats self-administer more cocaine and continue to lever press for a longer duration compared to controls during a 24-h, multi-dose binge session. Additionally, intermittent SDS animals show an increase in cocaine intake on a PR schedule, while chronic SDS animals show a decrease in intake (Miczek et al., 2011).

Housing conditions influence SDS effects on cocaine binge behavior. Burke and Miczek (2015) found that pair-housed adolescent rats take more cocaine infusions during a 24-h binge or on a PR schedule after SDS than single-housed SDS rats. Pair-housed SDS rats also show increased freezing during the defeat episodes, which is thought to be adaptive for reducing the probability of attack. Single-housed SDS rats, on the other hand, show decreased freezing and increased rates of following and crawling under the resident, but no change in binge cocaine intake compared to non-stressed, single-housed controls. It is possible that the enhanced cocaine intake in the social-housed rats occurs because social defeat is more stressful to social-housed rats than isolate-housed rats, perhaps because of competing desire for social interaction in the isolate-housed rats.

SDS alters other behavioral effects of cocaine as well, including sensitization of cocaine-induced locomotion and cocaine-CPP. Previous studies have shown that stimulant-induced sensitization correlates with compulsive drug-seeking and drug-taking (Robinson & Berridge, 1993), suggesting that sensitization predicts vulnerability to drug abuse. SDS accentuates cocaine-induced increases in locomotion (Covington et al., 2005; Covington & Miczek, 2001, 2005), and these effects are persistent after the last defeat episode (Covington & Miczek, 2001, 2005; Quadros & Miczek, 2009) and also interact with housing conditions. For instance, Burke and Miczek (2015) found that adolescent pair-housed, but not single-housed, SDS animals exhibited sensitized locomotor activity following a cocaine challenge. The lack of sensitization in SDS single-housed animals is counterintuitive because both SDS and isolate housing typically accelerate acquisition of cocaine self-administration (Boyson, Miguel, Quadros, DeBold, & Miczek, 2011; Schenk, Lacelle, Gorman, & Amit, 1987). Perhaps the severity of stress from the combination of isolation and SDS induces anhedonia which counters sensitization effects.

Studies on cocaine-CPP offer more evidence that SDS effects vary with age. Adult mice subjected to acute SDS show potentiated cocaine-CPP and corticosterone levels compared to adolescent mice (Montagud-Romero et al., 2015). After acute SDS, adolescent mice require a higher dose of cocaine to establish CPP and show weaker CPP reflected by a faster rate of extinction compared to adults. Although these age-dependent effects could be due to a difference in sensitivity to cocaine, it seems more likely that the adolescents have more difficulty learning drug–environment associations and therefore show weaker CPP. Intermittent SDS, on the other hand, enhances cocaine effects in defeated adolescent mice evident from establishment of cocaine-CPP using a subthreshold dose (Rodríguez-Arias et al., 2017). These animals also require protracted trials to extinguish the CPP and exhibit reinstatement of CPP with a low-dose priming injection compared to non-stressed animals. Together these findings suggest that age and severity of social stress interact such that in adults acute SDS renders mice more vulnerable to cocaine conditioning effects, whereas in adolescents intermittent SDS renders them more vulnerable to cocaine conditioning effects. These findings are consistent with humans, who are also more susceptible to the reward conditioning effects after social stress (Björkqvist, 2001).

Despite increased behavioral sensitivity to CPP, adolescent mice acquire stable cocaine self-administration more slowly compared to adults after intermittent SDS (Rodríguez-Arias et al., 2017). Again, it is possible that adolescent SDS impedes cocaine reinforcement and/or learning and memory. Consistent with the latter explanation, deficits in spatial navigation are observed in adolescents after SDS (McCormick et al., 2012). Overall, the research suggests that SDS effects on cocaine intake are influenced by age, housing conditions, cocaine dose, and duration of social stress. Careful consideration of each parameter is necessary to determine how SDS shapes vulnerability to cocaine use disorders.

3.4 The Role of Individual Differences in SDS Effects on Cocaine Self-Administration

Individual differences are important contributors to SDS influences on cocaine abuse-related behaviors. The intruder and resident animals show marked differences in cocaine self-administration even though both show comparable augmentation of plasma corticosterone in response to defeat episodes and similar corticosterone levels after the first and last SDS episodes. The latter findings suggest that stress reactivity does not habituate in response to intermittent SDS. Despite the hormonal similarity, intruder animals have higher breakpoints on a PR schedule of cocaine reinforcement and they self-administer more cocaine than resident and control animals during a 24-h binge of cocaine (Covington et al., 2005; Covington & Miczek, 2005). Resident animals may be buffered from cocaine susceptibility because, unlike intruders, residents exert control over the social encounter.

Cocaine abuse-related behaviors in rodents are related to individual differences in risk-taking and novelty-seeking (Bardo, Neisewander, & Kelly, 2013). Rat screened for high (HR) and low (LR) responsivity to a novel environment provide a model to study these differences. Under non-stress conditions, HR rats acquire cocaine self-administration more rapidly compared to LR rats (Kabbaj et al., 2001). In contrast, following four days of social defeat, HR and LR rats display blunted and enhanced cocaine acquisition, respectively. The authors suggest that SDS may be the "equalizer" of individual differences in cocaine administration. High sensation seekers are the human equivalent to HR rodents and both populations may be at greater risk for drug abuse (Ball, Carroll, & Rounsaville, 1994; Bardo, Donohew, & Harrington, 1996; Kelly et al., 2006). Future research is needed to examine whether aversive social stress differentially affects high and low sensation seekers, including their propensity for cocaine use. If such findings are similar to the animal research, social stress may have different effects on cocaine use depending on one's disposition toward sensation seeking.

3.5 Social Defeat Stress Effects on Glutamate and Dopamine Systems

Cocaine reward and reward learning, which underlie the development of addiction, are mediated in part by mesolimbic dopamine neurons that originate in the VTA and project to the NAc. SDS is thought to modulate the mesolimbic neurons via glutamatergic innervations from the PFC. Indeed, this pathway is involved in stress-induced cross-sensitization to cocaine (Covington et al., 2008; Wang et al., 2016). The glutamate α-amino-3-hydroxy-5-methyl-4-isoxazolepropionic acid (AMPA) and *N*-methyl-D-aspartate (NMDA) receptor subtypes play a key role in SDS effects. Cross-sensitized cocaine-induced locomotor activity is reduced in defeated animals, but not controls, by systemic administration of an AMPA GluA1 subunit antagonist and by systemic or intra-VTA administration of a NMDA GluN1 subtype antagonist. Downstream, AMPA and NMDA receptor signaling activates extracellular signal-regulated kinase (ERK) and blocking the phosphorylation of ERK in the VTA before each defeat episode prevents cross-sensitization to cocaine (Yap et al., 2015). Other brain regions implicated in SDS effects include the dorsal raphe, periaqueductal gray, and locus coeruleus, which exhibit peak increases in neuronal activity one week after acute SDS, corresponding to the time course of cocaine cross-sensitization (Miczek, Nikulina, Kream, Carter, & Espejo, 1999).

Modulation of the dopaminergic VTA projections to NAc shell via the HPA axis is also critically involved in SDS effects on cocaine-related behaviors. Antagonizing CRH receptors 1 and 2 (CRH-R1 and CRH-R2) in the VTA prior to social defeat episodes prevents sensitized DA release in the NAc shell, and escalates cocaine intake during a 24-h binge (Boyson et al., 2014). Additionally, systemic and intra-VTA injection of a CRH-R1 antagonist is sufficient to attenuate SDS-induced locomotor sensitization to a cocaine injection and cocaine self-administration in a 24-h binge session (Boyson et al., 2011). It is likely that these effects involve dopamine D1 receptor-expressing neurons based on the electrophysiological responses of these neurons. Mice given non-contingent cocaine injections and chronic SDS show diminished amplitude of excitatory post-synaptic currents (EPSCs) in D1-expressing medium spiny neurons (MSNs) after terminating the cocaine injections, while there was no such change in D2-expressing MSNs (Khibnik et al., 2016). Conversely, mice that were resilient to chronic SDS displayed augmented EPSCs in D1-expressing MSNs and attenuated EPSCs in D2-expressing MSNs (Khibnik et al., 2016). These results indicate that while cocaine and stress both act on D1 MSNs of the NAc shell, they do so in a manner dependent on sensitivity to SDS.

3.6 Gaps in the Literature: Sex and Age Differences

The SDS model is less reliable when tested in females because of sex differences in aggressive behavior; however, phenotypic differences have been observed after chronic SDS in females (Shimamoto, Holly, Boyson, Debold, & Miczek, 2015). After chronic SDS, the majority of females display anhedonia-like behavior as measured by decreased saccharine consumption, whereas the others appeared resistant to SDS. The stress-sensitive females self-administered less cocaine during a 24-h binge, however, a small number of stress-resistant females show control-level sucrose consumption along with high rates of cocaine intake in a binge session. Relative to stress-sensitive rats, stress-resistant females show less increase in DA levels in the NAc following a cocaine challenge, suggesting a potential role of the mesolimbic circuit in these individual differences. Future research aimed at examining cocaine effects after defeat stress in females could employ a more reliable model of social defeat in females whereby reliable aggression is achieved by pairing intruders with same-sex lactating resident dams (Holly, Shimamoto, Debold, & Miczek, 2012).

Most investigations of age-related SDS effects have focused on stress during adolescence and subsequent cocaine use either immediately after the SDS or in adulthood. There is a gap in the literature specifically geared toward SDS effects in adolescents versus adults. Though CPP studies have shown that acute SDS potentiates conditioned reward in adults compared to adolescents, further research is needed to delineate age-related sensitivity to SDS-induced cocaine effects, especially using self-administration models (Montagud-Romero et al., 2015).

3.7 Conclusions

The SDS model has been useful for examining how aversive social experiences impact vulnerability to cocaine use disorders. SDS-induced cocaine effects depend on parameters such as housing conditions, duration of SDS paradigm, age of defeat, sex, and individual differences. The studies that have been conducted in this area suggest that exposure to social defeat enhances cross-sensitization, cocaine-CPP, binge-like cocaine intake, and motivation for cocaine. Glutamatergic, dopaminergic and CRF signaling in brain stem nuclei are integral to some of these effects, although much is left to explore in terms of neurobiological mechanisms.

4. SOCIAL HIERARCHY MODELS

4.1 Comparative Social Status and Cocaine Use

Though people of all social strata abuse cocaine, those of lower socioeconomic status (SES) are more susceptible both to the negative social consequences of drug abuse, particularly in terms of jail time and arrests, and cocaine dependence in general (Palamar, Davies, Ompad, Cleland, & Weitzman, 2015). Friends and family provide pressure to quit drug use (Cheney, Booth, Borders, & Curran, 2016) and access to larger social support networks correlates with higher SES (Campbell, Marsden, & Hurlbert, 1986). Although social support and employment are partial safeguards against cocaine use, neither can combat the role of neighborhood poverty, which is associated with higher drug use (Williams & Latkin, 2007). Notably, while adolescents with higher family incomes are more likely to report lifetime use of cocaine (Humensky, 2010; Palamar & Ompad, 2014), higher parental education and family income are negatively associated with crack cocaine usage (Palamar et al., 2015). This has important implications as the addictive potential of crack is greater compared to other routes of administration. Smoked, free-base cocaine (i.e., crack) produces the fastest and strongest subjective high compared to intranasal and intravenous cocaine (Volkow et al., 2000). Among adults, cocaine use of any kind is decreased in those with higher education (Chatterji, 2006; Harder & Chilcoat, 2007). Together, these findings suggest that the social stress experienced at differing levels of SES and social mobility influence vulnerability to drug abuse.

4.2 Models of Social Hierarchy in Cocaine Addiction

As direct testing of the effects of SES on human cocaine is extremely difficult, social hierarchies in nonhuman primates offer a useful model with translational potential. In this model, adult cynomolgus monkeys (*Macaca fascicularis*) are socially-housed in groups of four and social rankings are determined by repeated observations of dyadic interactions. The "winner" of an interaction is the animal showing aggression against the other. After a clear, linear hierarchy is formed, the two animals that ranked first and second are considered dominant, while the remaining two animals are considered subordinate (e.g., Czoty & Nader, 2013, 2015; Gould, Czoty, Porrino, & Nader, 2017). Using this paradigm, Czoty, McCabe, and Nader (2005) placed male monkeys on a concurrent FR food, PR cocaine schedule of reinforcement. For the PR schedule, the response requirement remains the same within-session, but is doubled between sessions until obtaining cocaine is so effortful that monkeys show a stronger preference for food than cocaine. On average, the PR response requirement that causes the monkeys to switch from cocaine to food is higher in subordinates than dominants, suggesting that cocaine is more reinforcing to subordinate monkeys.

The effects of stressors other than ranking itself affect cocaine self-administration in monkeys differently depending on social status. Following a social confrontation, cocaine self-administration increases in subordinate monkeys whereas no change is observed in dominant monkeys (Gould et al., 2017). On the other hand, placing a rubber snake in the home cage, which elicits a fear response and is representative of a biologically relevant stressor, shifts the dose–response curve to the left in dominants, but not subordinates (Czoty & Nader, 2012). These contrasting results suggest that the influence of stress on motivation for cocaine depends on social rank, but varies depending on the type of stressor.

4.3 Behavioral and Neurobiological Consequences of Social Rank

Neurobiological studies in both humans and animals have demonstrated that dopaminergic systems, particularly D2R and D3Rs in the basal ganglia, contribute to the effects of social hierarchy on motivation for cocaine. Using repeated positron emission tomography (PET), Morgan et al. (2002) founds that monkeys that became subordinate during social housing show no change in D2R availability and display the typical, inverted U-shaped dose–response curve during cocaine self-administration. However, dominant monkeys consume less cocaine, with similar rates of cocaine and saline infusions. Furthermore, these monkeys show increases D2Rs in the basal ganglia after social housing. The results suggest that subordinate monkeys are more vulnerable to the reinforcing effects of cocaine due to decreased levels of D2Rs. Compared to dominants, subordinates also have lower D2Rs in the caudate nucleus after prolonged abstinence and increased novelty-seeking in a novel object test (Czoty, Gage, & Nader, 2010). Together, these results support previous research indicating a correlation between low striatal D2Rs and psychostimulant intake observed previously in humans (Volkow et al., 1999) and rats (Dalley et al., 2007).

Human PET studies also show a relationship between social status, cocaine, and D2/3R availability in the striatum (Wiers et al., 2016). In male subjects, SES was positively correlated with D2/3Rs in healthy volunteers, as found previously in both men and women (Martinez et al., 2010). Thus, Wiers et al. (2016) conclude that chronic cocaine abuse, which reduces D2 receptor density (Volkow et al., 1993), may negate the benefits of higher SES in addicts. In support of this view, there are no social rank differences in striatal D2Rs in male monkeys after long-term cocaine use (Czoty, Morgan, Shannon, Gage, & Nader 2004; Nader & Czoty, 2005). Although female monkeys also show a positive relationship between social status and D2/3R availability in the striatum, with higher D2/3R availability in dominant females (Nader, Nader, et al., 2012; Shively, Grant, Ehrenkaufer, Mach, & Nader, 1997), the dominant females acquire cocaine self-administration at lower doses than subordinates (Nader, Nader, et al., 2012). These findings suggest that while dominance is correlated with higher D2R availability in reward regions of the brain, cocaine self-administration in dominant, high D2R-expressing monkeys varies sex-dependently, with the males showing less propensity to acquire cocaine self-administration and the females showing enhanced propensity. Whether this result mirrors human females has not yet been established. Adding to this complexity, D2/3R ligands have mixed success in reducing motivation for cocaine (Neisewander et al., 2014) and the efficacy of several D2/3R ligands varies depending on the particular compound and social rank in male monkeys (Czoty & Nader, 2013, 2015).

Despite evidence that mice form stable, linear dominance hierarchies in large groups (Williamson, Lee, & Curley, 2016), rodent research on social dominance and addiction has largely been limited to dyadic interactions outside the home cage, such as the SDS model mentioned earlier, perhaps because of the ease of determining rank in the model. Jupp et al. (2016) used competition for access to a drinking spout to examine social ranking in male rats housed in groups of four. The two rats that drank the most were considered dominant, and the other two were ranked as subordinate. Subsequently, they found no social rank differences in acquisition of self-administration, but did find that dominant animals self-administered cocaine at higher rates than subordinates. This finding may be explained in part by increased novelty-seeking in dominant animals, which has been suggested to occur in experiments using the five-choice serial-reaction time task (5-CSRTT). Subsequent autoradiography showed that dominant rats have increased D2/3Rs in the dorsal striatum along with increased D2/3Rs and DAT in the NAc shell. In contrast, subordinates displayed higher DA levels in the NAc shell. Notably, these results for male rats are more similar to those of female monkeys (Nader, Nader, et al., 2012) than the male monkeys (Morgan et al., 2002).

Social rankings in mice have been assessed using the dominant–submissive relationship (DSR) test (Feder et al., 2010). In this test, pairs of animals compete for access to a food target on the far side of an arena comprised of two chambers connected by a small central tunnel. A recent study assessed cocaine-CPP using socially-dominant and socially-submissive mice selectively bred based on the DSR test (Yanovich, Kirby, Michaelevski, Yadid, & Pinhasov, 2018). In the initial tests, dominant mice displayed cocaine-CPP, while at the same cocaine dose (15 mg/kg) submissive mice showed cocaine-conditioned place aversion. Next, all animals were subjected to a chronic mild stress (CMS) regimen, which has been shown to induce anhedonia in submissive, but not dominant, mice (Gross & Pinhasov, 2016). After CMS, submissive mice displayed higher cocaine-CPP, which was correlated with stress-induced decreases in *Drd1* and *Drd2* mRNA expression in the hippocampus (Yanovich et al., 2018). Submissive mice also had markedly higher CRH expression in the hippocampus following CMS. Thus, lower social status appears to be a stressor (Czoty, Gould, & Nader, 2009; Lupien, King, Meaney, & Mcewen, 2001) that enhances cocaine-CPP (Feder et al., 2010).

4.4 Conclusions and Gaps in the Literature

From research in humans, rodents, and nonhuman primates, it is clear that social ranking and societal status influence dopaminergic systems in a manner that varies by sex and species. Some of this complexity may be due to current limited knowledge of hierarchies both in rodents and overall. For example, while ranking in the social dominance hierarchy influences patterns and severity of stress, lower social rank does not necessarily indicate higher stress. Lower social status can be associated with lower stress and cortisol levels depending on social context. A study of army recruits found that socially-dominant men had increased reactivity to psychological and physical stress, as assessed by cortisol levels (Hellhammer, Buchtal, Gutberlet, & Kirschbaum, 1997). There are several societal characteristics, such as stability of the hierarchy (Creel, Creel, & Monfort, 1996; Sapolsky, 1992, 2005) and the coping strategies available to subordinates (Abbott et al., 2003; Creel, 2001), which can alter the social stratum that experiences the most stress. Furthermore, the substantial complexity of human societies limits the translational relevance of these animal models. In the monkey and rat studies, animals were housed in groups of four, allowing for simple linear hierarchies to form. By contrast, humans take on both low- and high-ranking roles in different social arenas (e.g., at work, school, church, etc.) (Sapolsky, 2005), leading to more intricate hierarchies that are non-linear and overlapping, and which are not fully captured in animal models.

Another complication is that social rank and the dopaminergic circuitry underlying motivation for cocaine are dynamic and adaptable. Recent work in monkeys suggests that altering the social hierarchy changes D2/3R availability and cocaine self-administration (Czoty, Gould, Gage, & Nader, 2017). For animals that moved up in rank after social reorganization, D2/3R availability increased in the caudate nucleus and putamen. Regardless of initial or subsequent rank, the majority of animals also had decreased cocaine self-administration after the hierarchy shifted. This suggests that social mobility and increased social capital may combat motivation for drug, similar to human research which shows that access to social support increases desire to quit drug use (Cheney et al., 2016). Therefore, reducing the stress of low social rank, either through improving the perception of social status or by increasing social mobility, may be helpful in combating cocaine addiction.

Research is needed using more diverse animal models with more intricate hierarchy designs. While nonhuman primate models have high validity due to strong neurophysiological similarities between monkeys and humans, ethical and financial considerations limit the neurobiological information that can be obtained from monkey models. Therefore, concurrent rodent research is necessary to illuminate the brain mechanisms underpinning addiction and motivation for cocaine. Finally, the dearth of research on sex differences and how mixed sex social contexts affect cocaine self-administration needs to be addressed.

5. SOCIAL ISOLATION AND ENVIRONMENTAL ENRICHMENT MODELS

5.1 Social Isolation and Cocaine Use in Humans

Lack of social support and loneliness contribute to cocaine use disorders. The ability of cocaine to induce euphoria while simultaneously reducing intense feelings of loneliness is a strong attractor for recreational use (Haddad, 1979). Conversely, providing social support networks during treatment reduces cocaine use and facilitates quitting entirely (Cheney et al., 2016). Furthermore, the level of social support available predicts the likelihood of a patient remaining in treatment (Dobkin, De Civita, Paraherakis, & Gill, 2002). According to a study using the Cocaine Relapse

Interview, the strongest unpleasant feeling experienced by cocaine-dependent patients prior to relapse was extreme loneliness, with other relapse cues such as being alone and lacking social and self-help support also ranking at the top of responses (McKay, Rutherford, Alterman, Cacciola, & Kaplan, 1995). Additionally, male drug users in detox programs experience greater levels of loneliness variables including emotional distress, social inadequacy and alienation, interpersonal isolation, and self-alienation compared to the general population (Orzeck & Rokach, 2004). In the same study when methadone treatment was used in conjunction with group therapy, loneliness subscale measures were comparable to the general population. McMahon (2001) determined that increased levels of detached personality predict relapse to cocaine, whereas components of antisocial personality do not, suggesting that those with detached personalities are less likely to benefit from socially-focused therapies. These examples support the idea that loneliness makes one more vulnerable to cocaine use disorders, whereas strong social support improves treatment outcomes.

5.2 Social Isolation Versus Environmental Enrichment Model

Isolation housing was first used to study phenomena such as sensory deprivation, isolation-induced aggression, early life enrichment versus isolation, and "crowding" versus isolation (Brain & Benton, 1979). Now, it is well-known that isolation has negative effects on brain weight, neuronal size, dendritic spine number, dendritic branching, and glial cell population. These effects can be counteracted by environmental enrichment (EE) (van Praag, Kempermann, & Gage, 2000), which typically includes larger group housing environments to provide social stimulation, toys changed periodically to provide novelty, and access to a running wheel or climbing structures to provide opportunity for exercise (Crofton, Zhang, & Green, 2015). In contrast to isolation, EE enhances learning and its associated neurochemistry, regardless of drug history (Bardo et al., 1995; Pham et al., 1999; Rosenzweig & Bennett, 1996; van Praag et al., 2000). Isolation and EE housing in rodents provides an animal analog of living in social isolation in an impoverished environment versus living in an enriched environment.

5.3 Isolation Versus EE and Cocaine-Related Behaviors

Preclinical models of social experiences on cocaine use disorders often vary by when the differential housing occurs, such as those that occur prior to drug use and those that occur during drug use or abstinence (Bardo et al., 2013). When applied prior to drug use, the housing manipulation is useful for identifying psychosocial influences on initiation of drug use, whereas introducing EE during abstinence is useful for examining potential therapeutic effects of enrichment for maintaining abstinence.

Rearing rodents in social isolation alters effects of cocaine on locomotion and CPP. Mixed results have been reported for cocaine-induced locomotion, with studies suggesting that rearing in isolation decreases (Boyle, Gill, Smith, & Amit, 1991; Morse, Erwin, & Jones, 1993), increases (Phillips, Howes, Whitelaw, Robbins, & Everitt, 1994), or has no effect (Sahakian, Robbins, Morgan, & Iversen, 1975). Discrepancies among these studies may be due to differences in species and sex of the rodents, numbers of conspecifics present in the control condition, and/or dose of cocaine investigated. Specifically, the only study of mice among the group demonstrated increases in cocaine-induced locomotor activity and also tested for the shortest duration (15 min) with only one dose of cocaine (Morse et al., 1993), potentially limiting time and dose effects of cocaine that are observed with other studies. Furthermore, in the study where isolation rearing had no effect on cocaine-induced locomotor activity, the socially-deprived rats displayed higher baseline locomotor activity in a novel environment compared to group-housed controls (Sahakian et al., 1975). Among all studies, cocaine dose administered ranged from 2.5 to 20 mg/kg, with two of the studies administering only a single dose. Together, these differences increase the difficulty of interpretation for differential housing and behavioral sensitization.

EE decreases effects of cocaine compared to social housing, suggesting that other elements of EE contribute to its effects besides the social housing. Compared to social-housed controls, mice reared in EE show a decrease in cocaine-induced locomotion (Bezard et al., 2003; Solinas, Thiriet, Rawas, Lardeux, & Jaber, 2009) and cocaine-CPP (Chauvet, Lardeux, Jaber, & Solinas, 2011; Solinas et al., 2009), suggesting that EE reduces sensitivity to cocaine. Removal of mice from EE during conditioning increases cocaine-CPP, suggesting that the stress associated with a loss of EE increases sensitivity to the rewarding effects of cocaine (Nader, Chauvet, et al., 2012). Further support for stress caused by the loss of EE is that animals exhibited anxiety-like and depression-like behaviors, including reduced time spent in the center of an open field, increased immobility time in a forced swim test, and decreased grooming compared to mice that remained in enrichment.

Social housing and EE implemented during adolescence or adulthood blunt the effects of cocaine on locomotion and CPP. Socially-housed adolescent rats fail to acquire cocaine-CPP across multiple doses, including 5 and 10 mg/kg doses, IP which produce CPP in isolated rats (Zakharova, Miller, Unterwald, Wade, & Izenwasser, 2009). Adolescent mice housed under different social conditions, including isolation, group housing, crowded housing, and mixed sex pairs, showed no differences in acquisition of cocaine-CPP across doses, but only the isolated and group-housed mice exhibited cocaine-primed reinstatement of extinguished CPP (Do Couto, Aguilar, Lluch, Rodríguez-Arias, & Miñarro, 2009). Similarly, Solinas, Chauvet, Thiriet, El Rawas, and Jaber (2008) found that adult, social-housed mice exhibit cocaine-primed reinstatement of extinguished cocaine-CPP, whereas EE-housed mice fail to reinstate. These researchers also showed that EE housing during a 30-day period after establishing cocaine-CPP prevents CPP expression in a subsequent test. Finally, this same study found that mice repeatedly injected with cocaine to induce behavioral sensitization show loss of sensitization after 30 days of EE (Solinas et al., 2008). These results suggest that isolation enhances, while enrichment reduces, stimulant and motivational effects of cocaine and cocaine-associated stimuli.

Housing conditions during rearing also affect the reinforcing and motivational effects of self-administered cocaine. Rearing in isolation produces a leftward shift in the dose–response curve for cocaine self-administration compared to social rearing (Howes, Dalley, Morrison, Robbins, & Everitt, 2000). The leftward shift in this study suggests that isolation rearing enhances sensitivity to cocaine reinforcement. Puhl, Blum, Acosta-Torres, and Grigson (2012) show a clear dichotomy of low and high cocaine consumers among rats reared in isolation whereas nearly all rats reared in EE were low cocaine consumers. This dichotomy persisted under PR self-administration testing, suggesting that enriched animals are less motivated to work for cocaine. Gipson, Beckmann, El-Maraghi, Marusich, and Bardo (2011) investigated the effect of enrichment rearing on escalation of drug use during extended access session of 6 h/day as this measure is thought to provide a preclinical phenotype of addiction (Edwards & Koob, 2013). Housing effects are observed for intake of low-dose cocaine, but not high-dose cocaine, where the social component of EE alone attenuates escalation of low-dose intake of cocaine, while access to novel toys alone fails to be beneficial (Gipson et al., 2011). In aggregate, these studies suggest that enrichment protects against the reinforcing effects of cocaine.

Introducing EE during an abstinence period from self-administration appears to protect against cue-elicited cocaine seeking. Over the first weeks to months of abstinence, cue reactivity becomes stronger (Grimm, Hope, Wise, & Shaham, 2001; Neisewander et al., 2000; Tran-Nguyen et al., 1998) and EE introduced during abstinence blunts this effect (Chauvet, Lardeux, Goldberg, Jaber, & Solinas, 2009; Thiel, Sanabria, Pentkowski, & Neisewander, 2009). These findings suggest that providing enrichment during abstinence may help to prevent relapse. Furthermore, the effects of EE may be further enhanced when combined with extinction sessions aimed at reducing incentive motivational impact of drug-associated cues (Thiel, Engelhardt, Hood, Peartree, & Neisewander, 2011). The beneficial effects of EE during abstinence do not persist if animals return to isolation (Chauvet et al., 2011; Thiel et al., 2011). Together, these results suggest that enrichment during abstinence may be useful as an intervention to prevent relapse if implemented consistently over a long period.

5.4 The Neurobiology of Differential Housing and Cocaine

The effects of EE on cocaine-related behaviors are accompanied by changes in DA systems at the cellular and network levels. Among saline-exposed animals, enrichment reduces DA within the NAc and striatum when measured post-mortem, but does not change DA synthesis or metabolism (Bowling, Rowlett, & Bardo, 1993). DAT, a primary target of cocaine in the brain, demonstrates reduced binding affinity and mRNA expression following enrichment (Bezard et al., 2003), reducing the ability of cocaine to induce a rewarding effect. Additionally, enrichment reduces DA release in the NAc in response to cocaine (Howes et al., 2000). Similarly in the medial prefrontal cortex, enrichment in cocaine-naïve rats reduces DA uptake (Zhu, Apparsundaram, Bardo, & Dwoskin, 2005; Zhu, Green, Bardo, & Dwoskin, 2004), DAT function, and DA metabolism (Zhu et al., 2005) compared to isolated control rats. At post-synaptic sites within the PFC, enrichment decreases expression of DA D1 receptors (Del Arco et al., 2007). Collectively, the literature suggests that enrichment reduces the impact of DA in the brain.

As mentioned previously, isolation is a severe stressor that dysregulates physiological responses to stress. Additionally, switching from EE to isolation increases expression of CRF and CREB phosphorylation within the bed nucleus of the stria terminalis (Nader, Chauvet, et al., 2012), consistent with the idea that removal from EE is stressful. Antagonizing CRF-1 receptors blocks the increase in cocaine-induced CPP that occurs with a change in environment, reinforcing the hypothesis that isolation is a more stressful environment and leads to increased cocaine consumption and seeking-behavior.

The Fos family of transcription factors within the striatum are commonly regulated by cocaine use and withdrawal (Larson et al., 2010), and are differentially regulated by housing environment. Following CPP testing, enriched rats shift *c-fos* expression from the medioventral to the dorsolateral striatum (Bezard et al., 2003). Striatum expression levels of *ΔFosB*, a splice variant of the *FosB* gene linked to the development of addiction behaviors, increase with cocaine sensitization in isolated mice, but decrease due to cocaine if mice were instead raised in an enriched environment (Solinas et al., 2009). Similarly, Chauvet et al. (2011) observed enriched animals eliminate their cocaine-CPP and Fos protein expression was at control levels, whereas animals housed in a standard environment increase Fos expression in multiple brain regions involved with relapse processes. As with previous studies, recent investigation indicated that the transcription factor *ΔFosB* is elevated in both the NAc core and shell in rats reared in enrichment (Zhang et al., 2014). Furthermore, *ΔFosB* is commonly elevated due to cocaine exposure, but because enrichment increased baseline levels of expression, no additional increase was seen in animals reared in enrichment and then self-administered cocaine (Zhang et al., 2014). However, viral overexpression of *ΔFosB* decreased cocaine intake and reduced cocaine-seeking behavior in extinction and reinstatement in patterns similar to that observed with EE (Zhang et al., 2014). In contrast, overexpression of *ΔFosB* did not affect anxiety- or depression-related behaviors, suggesting that the effects on cocaine-related behaviors were not related to these behaviors.

The onset of next generation sequencing techniques allows for thorough investigation of the cellular and molecular changes brought about by housing conditions. In particular, recent studies of the differential effects caused by cocaine exposure as well as housing type during rearing have begun to illuminate changes in the transcriptome (Zhang, Crofton, et al., 2016; Zhang, Kong, et al., 2016) and proteome (Lichti et al., 2014; Zhang, Crofton, et al., 2016) while highlighting potential signaling pathways to target and thus reduce cocaine dependence (Zhang, Crofton, et al., 2016; Zhang, Kong, et al., 2016). In particular, isolation prior to acquisition of cocaine self-administration modulated a broad range of proteins, including those related to neuronal functions such as energy production and mRNA splicing, molecular cascade signaling through protein kinases, as well as proteins involved neurodegenerative diseases such as Huntington's disease and Alzheimer's disease (Lichti et al., 2014). At the transcriptome level, Zhang and colleagues noted the retinoic acid signaling pathway as a potential candidate for isolation's deleterious effects, confirming their post hoc RNA sequencing data by knocking down a retinoic acid degradation enzyme only to see increases in cocaine self-administration and seeking-behaviors (Zhang, Kong, et al., 2016). Continued emphasis on "-omic" studies has the potential to unlock testable pathways and individual targets to reduce cocaine-related behaviors, and is therefore of utmost importance.

5.5 Conclusions and Gaps in the Literature

Considerable progress has been made toward understanding the effects of social isolation and enrichment on cocaine-related behaviors, yet some gaps remain. Much of the literature currently available uses only male rats. Sex differences in cocaine use in rats and humans have been documented (Anker, Zlebnik, Navin, & Carroll, 2011; Becker & Hu, 2008; Zilberman, Tavares, & El-Guebaly, 2003), however research investigating sex differences in housing effects on cocaine-related behaviors is needed. Another area of investigation ripe for discovery of novel molecular mechanisms involved in social impact on cocaine-related behavior is the use of high throughput transcriptomics and proteomics. These powerful techniques will likely lead to innovation in the development of treatment strategies.

6. CONCLUDING REMARKS

In general, the literature reviewed in this chapter suggests that severe forms of psychosocial stress, such as NI and chronic social defeat, produce symptoms of anxiety, depression, and anhedonia, and that the latter reduces motivation for cocaine. Less severe forms of psychosocial stress, on the other hand, increase vulnerability to initiate cocaine use, escalate intake, and to relapse after periods of abstinence. In contrast, EE and social networks/companions are protective against these same behaviors. There are quantitative, and in some cases qualitative, differences in social stress effects across age groups and sex indicating that these are important variables to consider moving forward. Although there are limitations with the models reviewed in terms of capturing the complexity of human social interactions and environments that create challenges for translation, the models have been useful for studying neurobiological underpinnings of elements of human social experiences. It is important to note that this chapter focuses on social stress and does not include models that examine the influence of positive social encounters during drug-taking. These models capture social learning and drug:social reward interactions, which are also important for understanding social influences on cocaine use disorders (Smith, Lacy, & Strickland, 2014; Thiel, Okun, & Neisewander, 2008). The neurobiological mechanisms discussed in this chapter focused primarily on DA systems and the HPA axis involvement in the social stress models, but other systems, such as opioids, oxytocin, acetylcholine, serotonin, and endocannabinoids also play a role as well (Koob & Volkow, 2016). Future research in this area is essential for understanding mechanisms of cocaine use disorders and is promising for identifying novel neural and behavioral treatment strategies.

REFERENCES

Abbott, D. H., Keverne, E. B., Bercovitch, F. B., Shively, C. A., Mendoza, S. P., Saltzman, W., et al. (2003). Are subordinates always stressed? A comparative analysis of rank differences in cortisol levels among primates. *Hormones and Behavior*, *43*(1), 67–82.

Aisa, B., Tordera, R., Lasheras, B., Del Río, J., & Ramírez, M. J. (2007). Cognitive impairment associated to HPA axis hyperactivity after maternal separation in rats. *Psychoneuroendocrinology*, *32*(3), 256–266.

Anker, J. J., Zlebnik, N. E., Navin, S. F., & Carroll, M. E. (2011). Responding during signaled availability and nonavailability of iv cocaine and food in rats: Age and sex differences. *Psychopharmacology*, *215*(4), 785–799.

Ball, S. A., Carroll, K. M., & Rounsaville, B. J. (1994). Sensation seeking, substance abuse, and psychopathology in treatment-seeking and community cocaine abusers. *Journal of Consulting and Clinical Psychology*, *62*(5), 1053.

Bardo, M. T., & Bevins, R. A. (2000). Conditioned place preference: What does it add to our preclinical understanding of drug reward? *Psychopharmacology*, *153*(1), 31–43.

Bardo, M. T., Bowling, S. L., Rowlett, J. K., Manderscheid, P., Buxton, S. T., & Dwoskin, L. P. (1995). Environmental enrichment attenuates locomotor sensitization, but not in vitro dopamine release, induced by amphetamine. *Pharmacology, Biochemistry, and Behavior*, *51*(2–3), 397–405.

Bardo, M. T., Donohew, R. L., & Harrington, N. G. (1996). Psychobiology of novelty seeking and drug seeking behavior. *Behavioural Brain Research*, *77*(1–2), 23–43.

Bardo, M. T., Neisewander, J. L., & Kelly, T. H. (2013). Individual differences and social influences on the neurobehavioral pharmacology of abused drugs. *Pharmacological Reviews*, *65*(1), 255–290.

Baumann, M. H., Gendron, T. M., Becketts, K. M., Henningfield, J. E., Gorelick, D. A., & Rothman, R. B. (1995). Effects of intravenous cocaine on plasma cortisol and prolactin in human cocaine abusers. *Biological Psychiatry*, *38*(11), 751–755.

Becker, J. B., & Hu, M. (2008). Sex differences in drug abuse. *Frontiers in Neuroendocrinology*, *29*(1), 36–47.

Behen, M. E., Muzik, O., Saporta, A. S. D., Wilson, B. J., Pai, D., Hua, J., et al. (2009). Abnormal fronto-striatal connectivity in children with histories of early deprivation: A diffusion tensor imaging study. *Brain Imaging and Behavior*, *3*(3), 292–297.

Bellis, M. A., Hughes, K., Leckenby, N., Perkins, C., & Lowey, H. (2014). National household survey of adverse childhood experiences and their relationship with resilience to health-harming behaviors in England. *BMC Medicine*, *12*(1), 1–10.

Bezard, E., Dovero, S., Belin, D., Duconger, S., Jackson-Lewis, V., Przedborski, S., et al. (2003). Enriched environment confers resistance to 1-methyl-4-phenyl-1,2,3,6-tetrahydropyridine and cocaine: Involvement of dopamine transporter and trophic factors. *The Journal of Neuroscience*, *23*(35), 10999–11007.

Björkqvist, K. (2001). Social defeat as a stressor in humans. *Physiology and Behavior*, *73*(3), 435–442.

Bowlby, J. (1982). Attachment and loss: Retrospect and prospect. *American Journal of Orthopsychiatry*, *52*(4), 664–678.

Bowling, S. L., Rowlett, J. K., & Bardo, M. T. (1993). The effect of environmental enrichment on amphetamine-stimulated locomotor activity, dopamine synthesis and dopamine release. *Neuropharmacology*, *32*(9), 885–893.

Boyle, A. E., Gill, K., Smith, B. R., & Amit, Z. (1991). Differential effects of an early housing manipulation on cocaine-induced activity and self-administration in laboratory rats. *Pharmacology, Biochemistry, and Behavior*, *39*(2), 269–274.

Boyson, C. O., Holly, E. N., Shimamoto, A., Albrechet-Souza, L., Weiner, L. A., DeBold, J. F., et al. (2014). Social stress and CRF-dopamine interactions in the VTA: Role in long-term escalation of cocaine self-administration. *The Journal of Neuroscience*, *34*(19), 6659–6667.

Boyson, C. O., Miguel, T. T., Quadros, I. M., DeBold, J. F., & Miczek, K. A. (2011). Prevention of social stress-escalated cocaine self-administration by CRF-R1 antagonist in the rat VTA. *Psychopharmacology*, *218*(1), 257–269.

Brain, P., & Benton, D. (1979). The interpretation of physiological correlates of differential housing in laboratory rats. *Life Sciences*, *24*(2), 99–115.

Brake, W. G., Zhang, T. Y., Diorio, J., Meaney, M. J., & Gratton, A. (2004). Influence of early postnatal rearing conditions on mesocorticolimbic dopamine and behavioural responses to psychostimulants and stressors in adult rats. *European Journal of Neuroscience*, *19*(7), 1863–1874.

Braun, K., Seidel, K., Holetschka, R., Groeger, N., & Poeggel, G. (2013). Paternal deprivation alters the development of catecholaminergic innervation in the prefrontal cortex and related limbic brain regions. *Brain Structure and Function*, *218*(4), 859–872.

Burke, A. R., & Miczek, K. A. (2015). Escalation of cocaine self-administration in adulthood after social defeat of adolescent rats: Role of social experience and adaptive coping behavior. *Psychopharmacology*, *232*(16), 3067–3079.

Campbell, K. E., Marsden, P. V., & Hurlbert, J. S. (1986). Social resources and socioeconomic status. *Social Networks*, *8*, 97–117.

Chatterji, P. (2006). Illicit drug use and educational attainment. *Health Economics*, *15*(5), 489–511.

Chauvet, C., Lardeux, V., Goldberg, S. R., Jaber, M., & Solinas, M. (2009). Environmental enrichment reduces cocaine seeking and reinstatement induced by cues and stress but not by cocaine. *Neuropsychopharmacology*, *34*(10), 2767–2778.

Chauvet, C., Lardeux, V., Jaber, M., & Solinas, M. (2011). Brain regions associated with the reversal of cocaine conditioned place preference by environmental enrichment. *Neuroscience*, *184*, 88–96.

Cheney, A. M., Booth, B. M., Borders, T. F., & Curran, G. M. (2016). The role of social capital in African Americans' attempts to reduce and quit cocaine use. *Substance Use and Misuse*, *51*(6), 777–787.

Childress, A. R., McLellan, A. T., Ehrman, R., & O'Brien, C. P. (1988). Classically conditioned responses in opioid and cocaine dependence: A role in relapse. *NIDA Research Monograph*, *84*, 25–43.

Choi, N. G., Dinitto, D. M., Marti, C. N., & Choi, B. Y. (2017). Association of adverse childhood experiences with lifetime mental and substance use disorders among men and women aged 50+ years. *International Psychogeriatrics*, *29*(3), 359–372.

Clark, D. B., & Neighbors, B. (1996). Adolescent substance abuse and internalizing disorders. *Child and Adolescent Psychiatric Clinics of North America*, *5*(1), 45–57.

Copeland, W. E., Wolke, D., Angold, A., & Costello, E. J. (2013). Adult psychiatric outcomes of bullying and being bullied by peers in childhood and adolescence. *JAMA Psychiatry*, *70*(4), 419–426.

Covington, H. E., Kikusui, T., Goodhue, J., Nikulina, E. M., Hammer, R. P., & Miczek, K. A. (2005). Brief social defeat stress: Long lasting effects on cocaine taking during a binge and *zif268* mRNA expression in the amygdala and prefrontal cortex. *Neuropsychopharmacology*, *30*(2), 310–321.

Covington, H. E., & Miczek, K. A. (2001). Repeated social-defeat stress, cocaine or morphine: Effects on behavioral sensitization and intravenous cocaine self-administration "binges" *Psychopharmacology*, *158*(4), 388–398.

Covington, H. E., & Miczek, K. A. (2005). Intense cocaine self-administration after episodic social defeat stress, but not after aggressive behavior: Dissociation from corticosterone activation. *Psychopharmacology*, *183*(3), 331–340.

Covington, H. E., Tropea, T. F., Rajadhyaksha, A. M., Kosofsky, B. E., & Miczek, K. A. (2008). NMDA receptors in the rat VTA: A critical site for social stress to intensify cocaine taking. *Psychopharmacology*, *197*(2), 203–216.

Creel, S. (2001). Social dominance and stress hormones. *Trends in Ecology and Evolution*, *16*(9), 491–497.

Creel, S., Creel, N. M., & Monfort, S. L. (1996). Social stress and dominance. *Nature*, *379*(6562), 212.

Crofton, E. J., Zhang, Y., & Green, T. A. (2015). Inoculation stress hypothesis of environmental enrichment. *Neuroscience and Biobehavioral Reviews*, *49*, 19–31.

Cunningham, K. A., Paris, J. M., & Goeders, N. E. (1992). Serotonin neurotransmission in cocaine sensitization. *Annals of the New York Academy of Sciences*, *654*(1), 117–127.

Czoty, P. W., Gage, H. D., & Nader, M. A. (2010). Differences in D2 dopamine receptor availability and reaction to novelty in socially housed male monkeys during abstinence from cocaine. *Psychopharmacology*, *208*(4), 585–592.

Czoty, P. W., Gould, R. W., Gage, H. D., & Nader, M. A. (2017). Effects of social reorganization on dopamine D2/D3 receptor availability and cocaine self-administration in male cynomolgus monkeys. *Psychopharmacology*, *234*(18), 2673–2682.

Czoty, P. W., Gould, R. W., & Nader, M. A. (2009). Relationship between social rank and cortisol and testosterone concentrations in male cynomolgus monkeys (*Macaca fascicularis*). *Journal of Neuroendocrinology*, *21*(1), 68–76.

Czoty, P. W., McCabe, C., & Nader, M. A. (2005). Assessment of the relative reinforcing strength of cocaine in socially housed monkeys using a choice procedure. *Journal of Pharmacology and Experimental Therapeutics*, *312*(1), 96–102.

Czoty, P. W., Morgan, D., Shannon, E. E., Gage, H. D., & Nader, M. A. (2004). Characterization of dopamine D1 and D2 receptor function in socially housed cynomolgus monkeys self-administering cocaine. *Psychopharmacology*, *174*(3), 381–388.

Czoty, P. W., & Nader, M. A. (2012). Individual differences in the effects of environmental stimuli on cocaine choice in socially housed male cynomolgus monkeys. *Psychopharmacology*, *224*(1), 69–79.

Czoty, P. W., & Nader, M. A. (2013). Effects of dopamine D2/D3 receptor ligands on food-cocaine choice in socially housed male cynomolgus monkeys. *Journal of Pharmacology and Experimental Therapeutics*, *344*(2), 329–338.

Czoty, P. W., & Nader, M. A. (2015). Effects of oral and intravenous administration of buspirone on food-cocaine choice in socially housed male cynomolgus monkeys. *Neuropsychopharmacology*, *40*(10), 1072–1083.

Dalley, J. W., Fryer, T. D., Brichard, L., Robinson, E. S. J., Theobald, D. E. H., Lääne, K., et al. (2007). Nucleus accumbens D2/3 receptors predict trait impulsivity and cocaine reinforcement. *Science, 315*(5816), 1267–1270.

Daniels, W. M. U., Pietersen, C. Y., Carstens, M. E., & Stein, D. J. (2004). Maternal separation in rats leads to anxiety-like behavior and a blunted ACTH response and altered neurotransmitter levels in response to a subsequent stressor. *Metabolic Brain Disease, 19*(1–2), 3–14.

Degenhardt, L., Stockings, E., Patton, G., Hall, W. D., & Lynskey, M. (2016). The increasing global health priority of substance use in young people. *The Lancet. Psychiatry, 3*(3), 251–264.

Del Arco, A., Segovia, G., Canales, J. J., Garrido, P., De Blas, M., García-Verdugo, J. M., et al. (2007). Environmental enrichment reduces the function of D1 dopamine receptors in the prefrontal cortex of the rat. *Journal of Neural Transmission, 114*(1), 43–48.

Delker, B. C., Noll, L. K., Kim, H. K., & Fisher, P. A. (2014). Maternal abuse history and self-regulation difficulties in preadolescence. *Child Abuse & Neglect, 38*, 2033–2043.

Denenberg, V., Grota, L., & Zarrow, M. (1963). Maternal behaviour in the rat: Analysis of cross-fostering. *Journal of Reproduction and Fertility, 5*, 133–141.

Dobkin, P. L., De Civita, M., Paraherakis, A., & Gill, K. (2002). The role of functional social support in treatment retention and outcomes among outpatient adult substance abusers. *Addiction, 97*(3), 347–356.

Do Couto, B. R., Aguilar, M. A., Lluch, J., Rodríguez-Arias, M., & Miñarro, J. (2009). Social experiences affect reinstatement of cocaine-induced place preference in mice. *Psychopharmacology, 207*(3), 485–498.

Douglas, K. R., Chan, G., Gelernter, J., Arias, A. J., Anton, R. F., Weiss, R. D., et al. (2010). Adverse childhood events as risk factors for substance dependence: Partial mediation by mood and anxiety disorders. *Addictive Behaviors, 35*(1), 7–13.

Dube, S. R., Dong, M., Chapman, D. P., Giles, W. H., Anda, R. F., & Felitti, V. J. (2003). Childhood abuse, neglect, and household dysfunction and the risk of illicit drug use: The Adverse Childhood Experiences Study. *Pediatrics, 111*(3), 564.

Edwards, S., & Koob, G. F. (2013). Escalation of drug self-administration as a hallmark of persistent addiction liability. *Behavioural Pharmacology, 24*(5–6), 356–362.

Ege, M. A., Messias, E., Thapa, P. B., & Krain, L. P. (2015). Adverse childhood experiences and geriatric depression: Results from the 2010 BRFSS. *American Journal of Geriatric Psychiatry, 23*(1), 110–114.

Faure, J., Uys, J. D. K., Marais, L., Stein, D. J., Daniels, W. M. U., Uys, J. D. K., et al. (2006). Early maternal separation followed by later stressors leads to dysregulation of the HPA-axis and increases in hippocampal NGF and NT-3 levels in a rat model. *Metabolic Brain Disease, 21*(21), 181–74181.

Feder, Y., Nesher, E., Ogran, A., Kreinin, A., Malatynska, E., Yadid, G., et al. (2010). Selective breeding for dominant and submissive behavior in Sabra mice. *Journal of Affective Disorders, 126*(1–2), 214–222.

Filip, M. (2005). Role of serotonin (5-HT)2 receptors in cocaine self-administration and seeking behavior in rats. *Pharmacological Reports, 57*(1), 35–46.

Flanagan, J. C., Baker, N. L., McRae-Clark, A. L., Brady, K. T., & Moran-Santa Maria, M. M. (2015). Effects of adverse childhood experiences on the association between intranasal oxytocin and social stress reactivity among individuals with cocaine dependence. *Psychiatry Research, 229*(1–2), 94–100.

Francis, D. D., & Meaney, M. J. (1999). Maternal care and the development of stress responses. *Current Opinion in Neurobiology, 9*(1), 128–134.

Gee, D. G., Gabard-Durnam, L., Telzer, E. H., Humphreys, K. L., Goff, B., Shapiro, M., et al. (2014). Maternal buffering of human amygdala-prefrontal circuitry during childhood but not during adolescence. *Psychological Science, 25*(11), 2067–2078.

Gipson, C. D., Beckmann, J. S., El-Maraghi, S., Marusich, J. A., & Bardo, M. T. (2011). Effect of environmental enrichment on escalation of cocaine self-administration in rats. *Psychopharmacology*, *214*(2), 557–566.

Goeders, N. E., & Guerin, G. F. (1996). Role of corticosterone in intravenous cocaine self-administration in rats. *Neuroendocrinology*, *64*(5), 337–348.

Gorelick, D. A. (2017). Cocaine use disorder in adults: Epidemiology, pharmacology, clinical manifestations, medical consequences, and diagnosis. Retrieved June 1, 2018, from http://www.uptodate.com.

Gould, R. W., Czoty, P. W., Porrino, L. J., & Nader, M. A. (2017). Social status in monkeys: Effects of social confrontation on brain function and cocaine self-administration. *Neuropsychopharmacology*, *42*(5), 1093–1102.

Gracia-Rubio, I., Moscoso-Castro, M., Pozo, O. J., Marcos, J., Nadal, R., & Valverde, O. (2016). Maternal separation induces neuroinflammation and long-lasting emotional alterations in mice. *Progress in Neuropsychopharmacology & Biological Psychiatry*, *65*, 104–117.

Graf, E. N., Wheeler, R. A., Baker, D. A., Ebben, A. L., Hill, J. E., McReynolds, J. R., et al. (2013). Corticosterone acts in the nucleus accumbens to enhance dopamine signaling and potentiate reinstatement of cocaine seeking. *Journal of Neuroscience*, *33*(29), 11800–11810.

Grimm, J. W., Hope, B. T., Wise, R. A., & Shaham, Y. (2001). Incubation of cocaine craving after withdrawal. *Nature*, *412*(6843), 141–142.

Gross, M., & Pinhasov, A. (2016). Chronic mild stress in submissive mice: Marked polydipsia and social avoidance without hedonic deficit in the sucrose preference test. *Behavioural Brain Research*, *298*, 25–34.

Gunnar, M. R., & Donzella, B. (2002). Social regulation of the cortisol levels in early human development. *Psychoneuroendocrinology*, *27*, 199–220.

Haddad, L. M. (1979). 1978: Cocaine in perspective. *Journal of the American College of Emergency Physicians*, *8*(9), 374–376.

Haney, M., Maccari, S., Le Moal, M., Simon, H., & Piazza, P. (1995). Social stress increases the acquisition of cocaine self-administration in male and female rats. *Brain Research*, *698*(1–2), 46–52.

Harder, V. S., & Chilcoat, H. D. (2007). Cocaine use and educational achievement: Understanding a changing association over the past 2 decades. *American Journal of Public Health*, *97*(10), 1790–1793.

Hellhammer, D. H., Buchtal, J., Gutberlet, I., & Kirschbaum, C. (1997). Social hierarchy and adrenocortical stress reactivity in men. *Psychoneuroendocrinology*, *22*(8), 643–650.

Holly, E. N., Shimamoto, A., Debold, J. F., & Miczek, K. A. (2012). Sex differences in behavioral and neural cross-sensitization and escalated cocaine taking as a result of episodic social defeat stress in rats. *Psychopharmacology*, *224*(1), 179–188.

Howes, S. R., Dalley, J. W., Morrison, C. H., Robbins, T. W., & Everitt, B. J. (2000). Leftward shift in the acquisition of cocaine self-administration in isolation-reared rats: Relationship to extracellular levels of dopamine, serotonin and glutamate in the nucleus accumbens and amygdala-striatal FOS expression. *Psychopharmacology*, *151*(1), 55–63.

Huang, C.-C., Liang, Y.-C., Lee, C.-C., Wu, M.-Y., & Hsu, K.-S. (2009). Repeated cocaine administration decreases 5-HT2A receptor-mediated serotonergic enhancement of synaptic activity in rat medial prefrontal cortex. *Neuropsychopharmacology*, *34*(8), 1979–1992.

Humensky, J. L. (2010). Are adolescents with high socioeconomic status more likely to engage in alcohol and illicit drug use in early adulthood? *Substance Abuse: Treatment, Prevention, and Policy*, *5*(1), 19.

Huot, R. L., Gonzalez, M. E., Ladd, C. O., Thrivikraman, K. V., & Plotsky, P. M. (2004). Foster litters prevent hypothalamic-pituitary-adrenal axis sensitization mediated by neonatal maternal separation. *Psychoneuroendocrinology*, *29*(2), 279–289.

Hyman, S. M., Paliwal, P., & Sinha, R. (2007). Childhood maltreatment, perceived stress, and stress-related coping in recently abstinent cocaine dependent adults. *Psychology of Addictive Behaviors*, *21*(2), 233–238.
Jaffe, J. H., Cascella, N. G., Kumor, K. M., & Sherer, M. A. (1989). Cocaine-induced cocaine craving. *Psychopharmacology*, *97*(1), 59–64.
Johns, J. M., Nelson, C. J., Meter, K. E., Lubin, D. A., Couch, C. D., Ayers, A., et al. (1998). Dose-dependent effects of multiple acute cocaine injections on maternal behavior and aggression in Sprague-Dawley rats. *Developmental Neuroscience*, *20*(6), 525–532.
Johns, J. M., Noonan, L. R., Zimmerman, L. I., Li, L., & Pedersen, C. A. (1994). Effects of chronic and acute cocaine treatment on the onset of maternal behavior and aggression in Sprague-Dawley rats. *Behavioral Neuroscience*, *108*(1), 107–112.
Johnson, A. L., Morrow, C. E., Accornero, V. H., Xue, L., Anthony, J. C., & Bandstra, E. S. (2002). Maternal cocaine use: Estimated effects on mother-child play interactions in the preschool period. *Journal of Developmental and Behavioral Pediatrics*, *23*(4), 191–202.
Jupp, B., Murray, J. E., Jordan, E. R., Xia, J., Fluharty, M., Shrestha, S., et al. (2016). Social dominance in rats: Effects on cocaine self-administration, novelty reactivity and dopamine receptor binding and content in the striatum. *Psychopharmacology*, *233*(4), 579–589.
Kabbaj, M., Norton, C. S., Kollack-Walker, S., Watson, S. J., Robinson, T., & Akil, H. (2001). Social defeat alters the acquisition of cocaine self-administration in rats: Role of individual differences in cocaine-taking behavior. *Psychopharmacology*, *158*(4), 382–387.
Kajeepeta, S., Gelaye, B., Jackson, C. L., & Williams, M. A. (2015). Adverse childhood experiences are associated with adult sleep disorders: A systematic review. *Sleep Medicine*, *16*(3), 320–330.
Kane, V. B., Fu, Y., Matta, S. G., & Sharp, B. M. (2004). Gestational nicotine exposure attenuates nicotine-stimulated dopamine release in the nucleus accumbens shell of adolescent Lewis rats. *Journal of Pharmacology and Experimental Therapeutics*, *308*(2), 521–528.
Keller, R. W., Lefevre, R., Raucci, J., Carlson, J. N., & Glick, S. D. (1996). Enhanced cocaine self-administration in adult rats prenatally exposed to cocaine. *Neuroscience Letters*, *205*, 153–156.
Kelly, T. H., Robbins, G., Martin, C. A., Fillmore, M. T., Lane, S. D., Harrington, N. G., et al. (2006). Individual differences in drug abuse vulnerability: d-Amphetamine and sensation-seeking status. *Psychopharmacology*, *189*(1), 17–25.
Khibnik, L. A., Beaumont, M., Doyle, M., Heshmati, M., Slesinger, P. A., Nestler, E. J., et al. (2016). Stress and cocaine trigger divergent and cell type-specific regulation of synaptic transmission at single spines in nucleus accumbens. *Biological Psychiatry*, *79*(11), 898–905.
Kikusui, T., Faccidomo, S., & Miczek, K. A. (2005). Repeated maternal separation: Differences in cocaine-induced behavioral sensitization in adult male and female mice. *Psychopharmacology*, *178*(2–3), 202–210.
Kirst, M., Mecredy, G., Borland, T., & Chaiton, M. (2014). Predictors of substance use among young adults transitioning away from high school: A narrative review. *Substance Use & Misuse*, *49*(13), 1795–1807.
Koob, G. F. (2008). A role for brain stress systems in addiction. *Neuron*, *59*(1), 11–34.
Koob, G. F., & Volkow, N. D. (2016). Neurobiology of addiction: A neurcircuitry analysis. *The Lancet. Psychiatry*, *3*(8), 760–773.
Koolhaas, J. M., Coppens, C. M., de Boer, S. F., Buwalda, B., Meerlo, P., & Timmermans, P. J. A. (2013). The resident-intruder paradigm: A standardized test for aggression, violence and social stress. *Journal of Visualized Experiments*, (77), 1–7.
Koss, K. J., George, M. R. W., Cummings, E. M., Davies, P. T., El-Sheikh, M., & Cicchetti, D. (2014). Asymmetry in children's salivary cortisol and alpha-amylase in the context of marital conflict: Links to children's emotional security and adjustment. *Developmental Psychobiology*, *56*(4), 836–849.

Kosten, T. A., Lee, H. J., & Kim, J. J. (2006). Early life stress impairs fear conditioning in adult male and female rats. *Brain Research, 1087*(1), 142–150.

Kosten, T. A., Miserendino, M. J. D., & Kehoe, P. (2000). Enhanced acquisition of cocaine self-administration in adult rats with neonatal isolation stress experience. *Brain Research, 875*(1–2), 44–50.

Kosten, T. A., Sanchez, H., Zhang, X. Y., & Kehoe, P. (2004). Neonatal isolation enhances acquisition of cocaine self-administration and food responding in female rats. *Behavioural Brain Research, 151*(1–2), 137–149.

Kosten, T. A., Zhang, X. Y., & Kehoe, P. (2004). Infant rats with chronic neonatal isolation experience show decreased extracellular serotonin levels in ventral striatum at baseline and in response to cocaine. *Developmental Brain Research, 152*(1), 19–24.

Larson, E. B., Akkentli, F., Edwards, S., Graham, D. L., Simmons, D. L., Alibhai, I. N., et al. (2010). Striatal regulation of delta-fosB, FosB, and cFos during cocaine self-administration and withdrawal. *Journal of Neurochemistry, 115*(1), 112–122.

Lee, J.-H., Kim, H. J., Kim, J. G., Ryu, V., Kim, B.-T., Kang, D.-W., et al. (2007). Depressive behaviors and decreased expression of serotonin reuptake transporter in rats that experienced neonatal maternal separation. *Neuroscience Research, 58*(1), 32–39.

Li, Y., Robinson, T. E., & Bhatnagar, S. (2003). Effects of maternal separation on behavioural sensitization produced by repeated cocaine administration in adulthood. *Brain Research, 960*(1–2), 42–47.

Lichti, C. F., Fan, X., English, R. D., Zhang, Y., Li, D., Kong, F., et al. (2014). Environmental enrichment alters protein expression as well as the proteomic response to cocaine in rat nucleus accumbens. *Frontiers in Behavioral Neuroscience, 8*, 246.

Lu, L., Mamiya, T., Lu, P., Niwa, M., Mouri, A., Zou, L. B., et al. (2009). The long-lasting effects of cross-fostering on the emotional behavior in ICR mice. *Behavioural Brain Research, 198*(1), 172–178.

Lupien, S. J., King, S., Meaney, M. J., & Mcewen, B. S. (2001). Can poverty get under your skin? Basal cortisol levels and cognitive function in children from low and high socioeconomic status. *Development and Psychopathology, 13*(3), 653–676.

Lynch, W. J., Mangini, L. D., & Taylor, J. R. (2005). Neonatal isolation stress potentiates cocaine seeking behavior in adult male and female rats. *Neuropsychopharmacology, 30*(2), 322–329.

Madigan, S., Vaillancourt, K., McKibbon, A., & Benoit, D. (2012). The reporting of maltreatment experiences during the adult attachment interview in a sample of pregnant adolescents. *Attachment & Human Development, 14*(2), 119–143.

Mantsch, J., Saphier, D., & Goeders, N. E. (1998). Corticosterone facilitates the acquisition of cocaine self-administration in rats: Opposite effects of the type II glucocorticoid receptor agonist dexamethasone. *Journal of Pharmacology and Experimental Therapeutics, 287*(1), 72–80.

Martinez, D., Orlowska, D., Narendran, R., Slifstein, M., Liu, F., Kumar, D., et al. (2010). Dopamine type 2/3 receptor availability in the striatum and social status in human volunteers. *Biological Psychiatry, 67*(3), 275–278.

Matthews, K., Robbins, T. W., Everitt, B. J., & Caine, S. B. (1999). Repeated neonatal maternal separation alters intravenous cocaine self-administration in adult rats. *Psychopharmacology, 141*(2), 123–134.

Mayes, L. C. (1994). Neurobiology of prenatal cocaine exposure effect on developing monoamine systems. *Infant Mental Health Journal, 15*(2), 121–133.

McCormick, C. M., Thomas, C. M., Sheridan, C. S., Nixon, F., Flynn, J. A., & Mathews, I. Z. (2012). Social instability stress in adolescent male rats alters hippocampal neurogenesis and produces deficits in spatial location memory in adulthood. *Hippocampus, 22*(6), 1300–1312.

McKay, J. R., Rutherford, M. J., Alterman, A. I., Cacciola, J. S., & Kaplan, M. R. (1995). An examination of the cocaine relapse process. *Drug and Alcohol Dependence, 38*(1), 35–43.

McMahon, R. C. (2001). Personality, stress, and social support in cocaine relapse prediction. *Journal of Substance Abuse Treatment, 21*(2), 77–87.

McReynolds, J. R., Doncheck, E. M., Vranjkovic, O., Ganzman, G. S., Baker, D. A., Hillard, C. J., et al. (2016). CB1 receptor antagonism blocks stress-potentiated reinstatement of cocaine seeking in rats. *Psychopharmacology, 233*(1), 99–109.

Mendelson, J. H., Teoh, S. K., Mello, N. K., Ellingboe, J., & Rhoades, E. (1992). Acute effects of cocaine on plasma adrenocorticotropic hormone, luteinizing hormone and prolactin levels in cocaine-dependent men. *The Journal of Pharmacology and Experimental Therapeutics, 263*(2), 505–509.

Miczek, K. A., & Mutschler, N. H. (1996). Activational effects of social stress on IV cocaine self-administration in rats. *Psychopharmacology, 128*, 256–264.

Miczek, K. A., Nikulina, E., Kream, R. M., Carter, G., & Espejo, E. F. (1999). Behavioral sensitization to cocaine after a brief social defeat stress: c-fos expression in the PAG. *Psychopharmacology, 141*, 225–234.

Miczek, K. A., Nikulina, E. M., Shimamoto, A., & Covington, H. E. (2011). Escalated or suppressed cocaine reward, tegmental BDNF, and accumbal dopamine caused by episodic versus continuous social stress in rats. *Journal of Neuroscience, 31*(27), 9848–9857.

Moffett, M. C., Harley, J., Francis, D., Sanghani, S. P., Davis, W. I., & Kuhar, M. J. (2006). Maternal separation and handling affects cocaine self-administration in both the treated pups as adults and the dams. *The Journal of Pharmacology and Experimental Therapeutics, 317*(3), 1210–1218.

Moffett, M. C., Vicentic, A., Kozel, M., Plotsky, P., Francis, D. D., & Kuhar, M. J. (2007). Maternal separation alters drug intake patterns in adulthood in rats. *Biochemical Pharmacology, 73*(3), 321–330.

Montagud-Romero, S., Aguilar, M. A., Maldonado, C., Manzanedo, C., Miñarro, J., & Rodríguez-Arias, M. (2015). Acute social defeat stress increases the conditioned rewarding effects of cocaine in adult but not in adolescent mice. *Pharmacology Biochemistry and Behavior, 135*, 1–12.

Moran-Santa Maria, M. M., McRae-Clark, A. L., Back, S. E., DeSantis, S. M., Baker, N. L., Spratt, E. G., et al. (2010). Influence of cocaine dependence and early life stress on pituitary-adrenal axis responses to CRH and the Trier social stressor. *Psychoneuroendocrinology, 35*(10), 1492–1500.

Morgan, D., Grant, K. A., Gage, H. D., Mach, R. H., Kaplan, J. R., Prioleau, O., et al. (2002). Social dominance in monkeys: Dopamine D2 receptors and cocaine self-administration. *Nature Neuroscience, 5*(2), 169–174.

Morse, A. C., Erwin, V. G., & Jones, B. C. (1993). Strain and housing affect cocaine self-selection and open-field locomotor activity in mice. *Pharmacology Biochemistry and Behavior, 45*(4), 905–912.

Nader, J., Chauvet, C., Rawas, R. E., Lavot, L., Jaber, M., Thiriet, N., et al. (2012). Loss of environmental enrichment increases vulnerability to cocaine addiction. *Neuropsychopharmacology, 37*(7), 1579–1587.

Nader, M. A., & Czoty, P. W. (2005). PET imaging of dopamine D2 receptors in monkey models of cocaine abuse: Genetic predisposition versus environmental modulation. *American Journal of Psychiatry, 162*(8), 1473–1482.

Nader, M. A., Nader, S. H., Czoty, P. W., Riddick, N. V., Gage, H. D., Gould, R. W., et al. (2012). Social dominance in female monkeys: Dopamine receptor function and cocaine reinforcement. *Biological Psychiatry, 72*(5), 414–421.

National Institute on Drug Abuse. (2017, April). Trends & Statistics | National Institute on Drug Abuse. Retrieved June 1, 2018, from https://www.drugabuse.gov/related-topics/trends-statistics

Neisewander, J. L., Baker, D. A., Fuchs, R. A., Tran-Nguyen, L. T., Palmer, A., & Marshall, J. F. (2000). Fos protein expression and cocaine-seeking behavior in rats after exposure to a cocaine self-administration environment. *The Journal of Neuroscience*, *20*(2), 798–805.

Neisewander, J. L., Cheung, T. H. C., & Pentkowski, N. S. (2014). Dopamine D3 and 5-HT1B receptor dysregulation as a result of psychostimulant intake and forced abstinence: Implications for medications development. *Neuropharmacology*, *76*(Pt B), 301–319.

Orzeck, T., & Rokach, A. (2004). Men who abuse drugs and their experience of loneliness. *European Psychologist*, *9*(3), 163–169.

Palamar, J. J., Davies, S., Ompad, D. C., Cleland, C. M., & Weitzman, M. (2015). Powder cocaine and crack use in the United States: An examination of risk for arrest and socioeconomic disparities in use. *Drug and Alcohol Dependence*, *149*, 108–116.

Palamar, J. J., & Ompad, D. C. (2014). Demographic and socioeconomic correlates of powder cocaine and crack use among high school seniors in the United States. *American Journal of Drug and Alcohol Abuse*, *40*(1), 37–43.

Parsons, L. H., Weiss, F., & Koob, G. (1998). Serotonin 1B receptor stimulation enhances cocaine reinforcement. *The Journal of Neuroscience*, *18*(23), 10078–10089.

Pham, T. M., Ickes, B., Albeck, D., Soderstrom, S., Granholm, A. C., & Mohammed, A. H. (1999). Changes in brain nerve growth factor levels and nerve growth factor receptors in rats exposed to environmental enrichment for one year. *Neuroscience*, *94*(1), 279–286.

Phillips, G. D., Howes, S. R., Whitelaw, R. B., Robbins, T. W., & Everitt, B. J. (1994). Isolation rearing impairs the reinforcing efficacy of intravenous cocaine or intra-accumbens d-amphetamine: Impaired response to intra-accumbens D1 and D2/D3 dopamine receptor antagonists. *Psychopharmacology*, *115*(3), 419–429.

Post, R. M., Weiss, S. R. B., Fontana, D., & Pert, A. (1992). Conditioned sensitization to the psychomotor stimulant cocaine. *Annals of the New York Academy of Sciences*, *654*(1), 386–399.

Puhl, M. D., Blum, J. S., Acosta-Torres, S., & Grigson, P. S. (2012). Environmental enrichment protects against the acquisition of cocaine self-administration in adult male rats, but does not eliminate avoidance of a drug-associated saccharin cue. *Behavioural Pharmacology*, *23*(1), 43–53.

Quadros, I. M. H., & Miczek, K. A. (2009). Two modes of intense cocaine bingeing: Increased persistence after social defeat stress and increased rate of intake due to extended access conditions in rats. *Psychopharmacology*, *206*(1), 109–120.

Radliff, K. M., Wheaton, J. E., Robinson, K., & Morris, J. (2012). Illuminating the relationship between bullying and substance use among middle and high school youth. *Addictive Behaviors*, *37*(4), 569–572.

Rivier, C., & Vale, W. (1987). Cocaine stimulates adrenocorticotropin (ACTH) secretion through a corticotropin-releasing factor (CRF)-mediated mechanism. *Brain Research*, *422*(2), 403–406.

Robinson, T. E., & Berridge, K. C. (1993). The neural basis of drug craving: An incentive-sensitization theory of addiction. *Brain Research Reviews*, *18*(3), 247–291.

Rodríguez-Arias, M., Montagud-Romero, S., Rubio-Araiz, A., Aguilar, M. A., Martín-García, E., Cabrera, R., et al. (2017). Effects of repeated social defeat on adolescent mice on cocaine-induced CPP and self-administration in adulthood: Integrity of the blood–brain barrier. *Addiction Biology*, *22*(1), 129–141.

Roguski, E. E., Chen, H., Sharp, B. M., & Matta, S. G. (2013). Fostering itself increases nicotine self-administration in young adult male rats. *Psychopharmacology*, *229*(2), 227–234.

Rosenzweig, M. R., & Bennett, E. L. (1996). Psychobiology of plasticity: Effects of training and experience on brain and behavior. *Behavioural Brain Research, 78*(1), 57–65.

Rygula, R., Abumaria, N., Flügge, G., Fuchs, E., Rüther, E., & Havemann-Reinecke, U. (2005). Anhedonia and motivational deficits in rats: Impact of chronic social stress. *Behavioural Brain Research, 162*(1), 127–134.

Sahakian, B. J., Robbins, T. W., Morgan, M. J., & Iversen, S. D. (1975). The effects of psychomotor stimulants on stereotypy and locomotor activity in socially-deprived and control rats. *Brain Research, 84*(2), 195–205.

Sapolsky, R. M. (1992). Cortisol concentrations and the social significance of rank instability among wild baboons. *Psychoneuroendocrinology, 17*(6), 701–709.

Sapolsky, R. M. (2005). The influence of social hierarchy on primate health. *Science, 308*(5722), 648–652.

Sarnyai, Z., Bíró, É., Penke, B., & Telegdy, G. (1992). The cocaine-induced elevation of plasma corticosterone is mediated by endogenous corticotropin-releasing factor (CRF) in rats. *Brain Research, 589*(1), 154–156.

Schenk, S., Lacelle, G., Gorman, K., & Amit, Z. (1987). Cocaine self-administration in rats influenced by environmental conditions: Implications for the etiology of drug abuse. *Neuroscience Letters, 81*(1–2), 227–231.

Shimamoto, A., Holly, E. N., Boyson, C. O., Debold, J. F., & Miczek, K. A. (2015). Individual differences in anhedonic and accumbal dopamine responses to chronic social stress and their link to cocaine self-administration in female rats. *Psychopharmacology, 232*(4), 825–834.

Shively, C. A., Grant, K. A., Ehrenkaufer, R. L., Mach, R. H., & Nader, M. A. (1997). Social stress, depression, and brain dopamine in female cynomolgus monkeys. *Annals of the New York Academy of Sciences, 807*(1), 574–577.

Sinha, R. (2001). How does stress increase risk of drug abuse and relapse? *Psychopharmacology, 158*(4), 343–359.

Sinha, R., Garcia, M., Paliwal, P., Kreek, M. J., & Rounsaville, B. J. (2006). Stress-induced cocaine craving and hypothalamic-pituitary-adrenal responses are predictive of cocaine relapse outcomes. *Archives of General Psychiatry, 63*(3), 324–331.

Smith, M. A., Lacy, R. T., & Strickland, J. C. (2014). The effects of social learning on the acquisition of cocaine self-administration. *Drug & Alcohol Dependence, 141*, 1–8.

Solinas, M., Chauvet, C., Thiriet, N., El Rawas, R., & Jaber, M. (2008). Reversal of cocaine addiction by environmental enrichment. *Proceedings of the National Academy of Sciences of the United States of America, 105*(44), 17145–17150.

Solinas, M., Thiriet, N., Rawas, R. E., Lardeux, V., & Jaber, M. (2009). Environmental enrichment during early stages of life reduces the behavioral, neurochemical, and molecular effects of cocaine. *Neuropsychopharmacology, 34*(5), 1102–1111.

Substance Abuse and Mental Health Services Administration. (2014). *Results from the 2013 National Survey on Drug Use and Health: Summary of National Findings*, NSDUH Series H-48, HHS Publication No. (SMA) 14-4863. Rockville, MD: Substance Abuse and Mental Health Services Administration.

Tharp-Taylor, S., Haviland, A., & D'Amico, E. J. (2009). Victimization from mental and physical bullying and substance use in early adolescence. *Addictive Behaviors, 34*(6–7), 561–567.

Thiel, K. J., Engelhardt, B., Hood, L. E., Peartree, N. A., & Neisewander, J. L. (2011). The interactive effects of environmental enrichment and extinction interventions in attenuating cue-elicited cocaine-seeking behavior in rats. *Pharmacology Biochemistry and Behavior, 97*(3), 595–602.

Thiel, K. J., Okun, A. C., & Neisewander, J. L. (2008). Social reward-conditioned place preference: A model revealing an interaction between cocaine and social context rewards in rats. *Drug and Alcohol Dependence, 96*(3), 202–212.

Thiel, K. J., Sanabria, F., Pentkowski, N. S., & Neisewander, J. L. (2009). Anti-craving effects of environmental enrichment.

Tidey, J. W., & Miczek, K. A. (1997). Acquisition of cocaine self-administration after social stress: Role of accumbens dopamine. *Psychopharmacology*, *130*(3), 203–212.

Tran-Nguyen, L. T. L., Fuchs, R. A., Coffey, G. P., Baker, D. A., O'Dell, L. E., & Neisewander, J. L. (1998). Time-dependent changes in cocaine-seeking behavior and extracellular dopamine levels in the amygdala during cocaine withdrawal. *Neuropsychopharmacology*, *19*(1), 48–59.

United Nations Office on Drugs and Crime. (2016). World Drug Report 2016 (United Nations publication, Sales No. E.16.XI.7).

van Praag, H., Kempermann, G., & Gage, F. H. (2000). Neural consequences of enviromental enrichment. *Nature Reviews Neuroscience*, *1*(3), 191–198.

Vidal, J., de Bie, J., Granneman, R. A., Wallinga, A. E., Koolhaas, J. M., & Buwalda, B. (2007). Social stress during adolescence in Wistar rats induces social anxiety in adulthood without affecting brain monoaminergic content and activity. *Physiology and Behavior*, *92*(5), 824–830.

Volkow, N. D., Fowler, J. S., Wang, G., Hitzemann, R., Logan, J., Schlyer, D. J., et al. (1993). Decreased dopamine D2 receptor availability is associated with reduced frontal metabolism in cocaine abusers. *Synapse*, *14*(2), 169–177.

Volkow, N. D., Wang, G. J., Fischman, M. W., Foltin, R., Fowler, J. S., Franceschi, D., et al. (2000). Effects of route of administration on cocaine induced dopamine transporter blockade in the human brain. *Life Sciences*, *67*(12), 1507–1515.

Volkow, N. D., Wang, G. J., Fowler, J. S., Logan, J., Gatley, S. J., Wong, C., et al. (1999). Reinforcing effects of psychostimulants in humans are associated with increases in brain dopamine and occupancy of D2 receptors. *The Journal of Pharmacology and Experimental Therapeutics*, *291*(1), 409–415.

Wang, J., Bastle, R. M., Bass, C. E., Hammer, R. P., Neisewander, J. L., & Nikulina, E. M. (2016). Overexpression of BDNF in the ventral tegmental area enhances binge cocaine self-administration in rats exposed to repeated social defeat. *Neuropharmacology*, *109*, 121–130.

Wang, J., Fang, Q., & Yang, C. (2017). Effects of paternal deprivation on cocaine-induced behavioral response and hypothalamic oxytocin immunoreactivity and serum oxytocin level in female mandarin voles. *Behavioural Brain Research*, *334*, 135–141.

Weder, N., Zhang, H., Jensen, K., Yang, B. Z., Simen, A., Jackowski, A., et al. (2014). Child abuse, depression, and methylation in genes involved with stress, neural plasticity, and brain circuitry. *Journal of the American Academy of Child and Adolescent Psychiatry*, *53*(4), 417–424. e5.

Wiers, C. E., Shokri-Kojori, E., Cabrera, E., Cunningham, S., Wong, C., Tomasi, D., et al. (2016). Socioeconomic status is associated with striatal dopamine D2/D3 receptors in healthy volunteers but not in cocaine abusers. *Neuroscience Letters*, *617*, 27–31.

Williams, C. T., & Latkin, C. A. (2007). Neighborhood socioeconomic status, personal network attributes, and use of heroin and cocaine. *American Journal of Preventive Medicine*, *32*(6 Suppl), S203–S210.

Williamson, C. M., Lee, W., & Curley, J. P. (2016). Temporal dynamics of social hierarchy formation and maintenance in male mice. *Animal Behaviour*, *115*, 259–272.

Yanovich, C., Kirby, M. L., Michaelevski, I., Yadid, G., & Pinhasov, A. (2018). Social rank-associated stress vulnerability predisposes individuals to cocaine attraction. *Scientific Reports*, *8*(1), 1–9.

Yap, J. J., Chartoff, E. H., Holly, E. N., Potter, D. N., Carlezon, W. A., & Miczek, K. A. (2015). Social defeat stress-induced sensitization and escalated cocaine self-administration: The role of ERK signaling in the rat ventral tegmental area. *Psychopharmacology*, *232*(9), 1555–1569.

Yu, P., An, S., Tai, F., Zhang, X., He, F., Wang, J., et al. (2012). The effects of neonatal paternal deprivation on pair bonding, NAcc dopamine receptor mRNA expression and serum corticosterone in mandarin voles. *Hormones and Behavior, 61*(5), 669–677.

Zakharova, E., Miller, J., Unterwald, E., Wade, D., & Izenwasser, S. (2009). Social and physical environment alter cocaine conditioned place preference and dopaminergic markers in adolescent male rats. *Neuroscience, 163*(3), 890–897.

Zhang, Y., Crofton, E. J., Fan, X., Li, D., Kong, F., Sinha, M., et al. (2016). Convergent transcriptomics and proteomics of environmental enrichment and cocaine identifies novel therapeutic strategies for addiction. *Neuroscience, 339*, 254–266.

Zhang, Y., Crofton, E. J., Li, D., Lobo, M. K., Fan, X., Nestler, E. J., et al. (2014). Overexpression of DeltaFosB in nucleus accumbens mimics the protective addiction phenotype, but not the protective depression phenotype of environmental enrichment. *Frontiers in Behavioral Neuroscience, 8*(August), 1–10.

Zhang, Y., Kong, F., Crofton, E. J., Dragosljvich, S. N., Sinha, M., Li, D., et al. (2016). Transcriptomics of environmental enrichment reveals a role for retinoic acid signaling in addiction. *Frontiers in Molecular Neuroscience, 9*, 119.

Zhang, X. Y., Sanchez, H., Kehoe, P., & Kosten, T. A. (2005). Neonatal isolation enhances maintenance but not reinstatement of cocaine self-administration in adult male rats. *Psychopharmacology, 177*(4), 1–9.

Zhu, J., Apparsundaram, S., Bardo, M. T., & Dwoskin, L. P. (2005). Environmental enrichment decreases cell surface expression of the dopamine transporter in rat medial prefrontal cortex. *Journal of Neurochemistry, 93*(6), 1434–1443.

Zhu, J., Green, T., Bardo, M. T., & Dwoskin, L. P. (2004). Environmental enrichment enhances sensitization to GBR 12935-induced activity and decreases dopamine transporter function in the medial prefrontal cortex. *Behavioural Brain Research, 148*(1–2), 107–117.

Zilberman, M., Tavares, H., & El-Guebaly, N. (2003). Gender similarities and differences: The prevalence and course of alcohol- and other substance-related disorders. *Journal of Addictive Diseases, 22*(4), 61–74.

Zimmerberg, B., & Gray, M. S. (1992). The effects of cocaine on maternal behaviors in the rat. *Physiology and Behavior, 52*(2), 379–384.

CHAPTER SEVEN

Social Factors and Animal Models of Cannabis Use

Lauren Hood[1]
University of Washington, Neuroscience Graduate Student, Seattle, WA, United States
[1]Corresponding author: e-mail address: hoodl@uw.edu

Contents

Abstract

Cannabis, or the dried leaves, stems, and seeds of the hemp plant *Cannabis sativa*, is the most widely used illicit drug in America. Typically smoked, vaporized or ingested orally, cannabis is used primarily for recreational purposes, though a few synthetic cannabinoids have been approved for medicinal treatments. Psychoactive cannabinoids, or the pharmacologically active compounds within cannabis, are responsible for producing the infamous "high" sensation, characterized by feelings of euphoria and relaxation, though can also provoke hallucinations, paranoia and anxiety. Cannabinoids act on G-protein coupled receptors in the brain, primarilyCB1 receptors, that typically decrease neural activity and modulate transmitter release. Compared to other drugs of abuse, cannabis use has minimal health risks and almost no potential for fatal overdose, though the trademark method of administration (smoking) has detrimental consequences. Chronic heavy use can also lead to changes in memory, cognitive deficits, psychosis and dependence. Up to 9% of users can develop a cannabis dependence, characterized by a characteristic withdrawal syndrome. The growing prevalence of cannabis use has spurred the development of animals models to research the neurobehavioral basis of cannabis use. Traditional animal models of drug abuse (i.e., conditioned place preference (CPP) and self-administration) have historically struggled to establish rewarding or reinforcing effects of individual cannabinoid molecules. Decades of research have been needed to reveal the appropriate dosage and conditions to promote reward and reinforcement in animal models. While the field has made great strides in

International Review of Neurobiology, Volume 140
ISSN 0074-7742
https://doi.org/10.1016/bs.irn.2018.07.006

elucidating the mechanisms involved in behavioral pharmacology cannabinoids, the social aspects of cannabis use remains underrepresented in animal models. Social interactions are vital to the initiation and continuation of cannabis use in humans, and this component has yet to be accurately captured in current animal models.

1. OVERVIEW OF CANNABIS USE

The dried leaves, stems, and seeds of the hemp plant *Cannabis sativa* are most frequently ingested orally, smoked or vaporized for recreational and medicinal purposes. Cannabis contains over 60 pharmacologically active compounds called cannabinoids, the most well-studied of which are delta-9-tetrahydrocannabinol (THC) and cannabidiol (World Health Organization, 2016). THC was identified as the primary psychoactive component in marijuana in 1964 and has been a topic of research ever since. It is responsible for producing the infamous "high" sensation, characterized by feelings of euphoria and relaxation, though can also provoke hallucinations, paranoia and anxiety (Trezza, Cuomo, & Vanderschuren, 2008). Cannabidiol is not psychoactive and is thought to temper the psychoactive effects of THC. Therefore, the psychoactive properties of a given strain of marijuana plant is not only dependent on the concentration of THC, but also on the ratio of THC to cannabidiol content. Cannabis strains yielding different ratios can be bred to produce different effects when smoked. THC can also be extracted in fats or ethanol and ingested as a food additive. Marijuana is the most widely used federally illicit drug in the United States (Center for Behavioral Health Statistics and Quality, 2015). Worldwide, its high usage is surpassed only by nicotine and alcohol (Bonnet & Preuss, 2017).

In the United States, the laws governing marijuana use are both confusing and slowly changing. The federal government classifies marijuana as a Schedule I controlled substance, meaning it has a high potential for abuse and no accepted medical use (https://www.dea.gov/druginfo/ds.shtml).It is therefore federally illegal for doctors to prescribe marijuana for medical purposes, but state drug regulations regarding marijuana use vary widely. As of 2017, medical marijuana was legal in 29 states, with recreational use legal in 9 states, including the nation's capital Washington DC (Bonnet & Preuss, 2017). The discrepancies in the federal and state laws make it difficult for people to maintain access to marijuana for effective treatment. It also is difficult for suppliers to regulate the quality of the product across states, limiting its usefulness as a medicine (Gostin, Hodge, & Wetter, 2018).

There is a growing body of evidence to support the medicinal benefits of marijuana for treating a range of disorders. Managing chronic pain and nausea, particularly nausea induced by chemotherapy, are commonly reported applications for cannabis treatment (Hill & Palastro, 2017; https://www.fda.gov/NewsEvents/PublicHealthFocus/ucm421168.htm). Many studies also report antiepileptic properties of cannabis in children (Hill & Palastro, 2017). While anything botanically-derived from the cannabis plant is prohibited at the federal level, there are two synthetic cannabinoids available in pill form that are approved by the FDA (Hill & Palastro, 2017; Tsang & Giudice, 2016). Dronabinol is a synthetic form of THC accepted for use as an appetite stimulant for wasting symptoms of AIDS and an antiemetic for people undergoing chemotherapy. Nabilone is another synthetic cannabinoid that is accepted for the same uses, and is gaining traction for the treatment of chronic pain in diseases like fibromyalgia and multiple sclerosis (Hill & Palastro, 2017; Tsang & Giudice, 2016).

Hundreds of different synthetic cannabinoids have been produced since the 1980s (Trecki, Gerona, & Schwartz, 2015). These designer drugs, like the popularly known Spice, K2, and Kronic, are classified as cannabinoids because they act on the same receptors in the brain as THC (the CB1 cannabinoid receptor, discussed below) (Zawilska & Wojcieszak, 2014). The formulation of these chemicals is not regulated, as many are illegal, resulting in highly variable compositions, quality and dosages among batches. More aversive symptoms are reported from consuming these drugs compared to marijuana, including anxiety, nausea, seizures, and kidney problems (Zawilska & Wojcieszak, 2014). Also unlike marijuana, synthetic cannabinoids present a risk of lethality, as their use has been linked to deaths across the United States (Trecki et al., 2015).

Perhaps fueled by the fact that few, if any, reported fatalities have been reported as a direct result of toxicity from natural cannabinoids, even at very high doses, there is a common misconception that recreational marijuana use poses minimal personal risks (World Health Organization, 2016). In reality, natural cannabinoids in marijuana do have the potential for abuse and adverse consequences for overall wellbeing. High usage is associated with some cognitive deficits (i.e., memory loss), dependence and psychosis (Bonnet & Preuss, 2017). Chronic marijuana use is also correlated with higher rates of unemployment, lower income, diminished life satisfaction, and criminal behavior (Han, Compton, Blanco, & Jones, 2018). Additionally, heavy marijuana smoking is known to have detrimental health repercussions, including increased risk for cancer and heart disease, as well as respiratory maladies.

Humans also develop a tolerance to various effects of THC, which can lead to escalated drug usage over time (Lichtman & Martin, 2005).

The National Institute on Drug Abuse (NIDA) reports that contrary to popular belief, marijuana is addictive for about 9% of users; this number increases among those who initiate use at younger ages (to about 17%, or 1 in 6) and among people who use marijuana daily (to 25–50%). NIDA defines addiction as a chronic relapsing disease characterized in part by compulsive seeking of the substance and use despite harmful consequences. Chronic marijuana use is known to desensitize and downregulate of the primary cannabinoid receptor in the brain (the CB1 receptor), producing a decrease that takes a month of abstinence from marijuana use to reverse (Bonnet & Preuss, 2017). NIDA also highlights their distinction of "addiction" from research performed by Lopez-Quintero et al. (2011) and an epidemiological study by Anthony, Warner, and Kessler (1997). These studies posit that dependence becomes addiction when the individuals continue to use marijuana even though they experience problems in their lives because of the marijuana use. While the DSM-V (American Psychiatric Association, 2013) does not utilize the word "addiction," NIDA's definition roughly correlates to the DSM's use of the word "dependence."

One criterion for diagnosing cannabis dependency is the development of cannabis withdrawal syndrome after the abrupt cessation of consistent use. While the symptoms are relatively mild compared to those of other drug dependencies (i.e., opiates), withdrawal symptoms do occur after long term cannabis use. The symptoms typically come in waves and are predominantly behavioral and mood-based, including dysphoria, insomnia, anxiety, depression, agitation and drug-craving. Some more physical symptoms are also reported, like weakness, sweating and chills. The longer and heavier the cannabis use, the worse the withdrawal symptoms experienced upon cessation (Bonnet & Preuss, 2017). Over 4 million people were estimated to satisfy the DSM-IV requirements for cannabis dependency in the United States (Crean, Crane, & Mason, 2011). There is also a high prevalence of people seeking treatment for cannabis dependency, which even surpassed the number of people seeking treatment for cocaine dependency in 2007 (Panlilio, Justinova, & Goldberg, 2010; World Health Organization, 2016).

Neither these risks nor federal laws have stopped Americans from using marijuana; the large user base has been steadily increasing since 2007. In 2012, there were 18.9 million current (past-month) users—about 7.3% of people aged 12 or older—up from 14.4 million (5.8%) in 2007.

The prevalence of worldwide cannabis use as of December 2013 is shown in Fig. 1. Previous research indicates that individuals who are Caucasian, male, or single are more likely to use marijuana than people who are non-Caucasian, female, or married (Bell, Weschler, & Johnston, 1997; Kaplan, Martin, Johnson, & Robbins, 1986). A large percentage of users (approximately 23%) are individuals in their late teens or early 20s (NIDA, 2014).

Therefore, a portion of our young population is at risk for marijuana abuse and developing an addiction. Adolescence is considered an important developmental window because active brain maturation is still occurring, including synapse reformation and refinement (Ellgren, Spano, & Hurd, 2007). These processes are susceptible to fluctuations in brain chemistry, thus altered cannabinoid signaling has the potential to negatively impact adolescent brain development. The peak age range of marijuana use also encompasses the prime reproductive ages; not surprisingly, marijuana is the most widely used illicit drug during pregnancy (Trezza et al., 2008). Cannabinoids are known to cross the placenta and are secreted into the milk of lactating mothers, where they could impact both embryonic, neonatal and postnatal development (Trezza et al., 2008).

Prenatal exposure to cannabis has been implicated in abnormal neural development. Longitudinal studies on the children of mothers who smoked marijuana while pregnant show high instances of behavioral problems, as well as impairments in attention, learning and memory (Trezza et al., 2008; World Health Organization, 2016). The behavioral and cognitive consequences of exposure to cannabis in utero often do not manifest until later on during maturation and adolescence (World Health Organization, 2016). Prenatal exposure is also a risk factor for later drug use, as affected offspring are also more likely to smoke and use marijuana in adolescence (Porath & Fried, 2005). Taken altogether, this information raises concerns about the abuse of marijuana during critical developmental periods.

While no gross physical abnormalities have been reported in human studies of prenatal cannabis exposure, chronic marijuana usage in adulthood has been correlated to regional structural changes in the brain (Moreno-Alcázar et al., 2018; Trezza et al., 2008). However, there are many conflicting reports as to how the adult brain changes in response to chronic cannabis. Neuroimaging techniques have reported reductions in the gray matter of the hippocampus, amygdala and prefrontal cortex compared to cannabis-naïve controls (Lorenzetti, Lubman, Whittle, Solowij, & Yücel, 2010). These are all regions involved in the cognitive processes that are altered while high (i.e., memory and attention). Similar techniques have

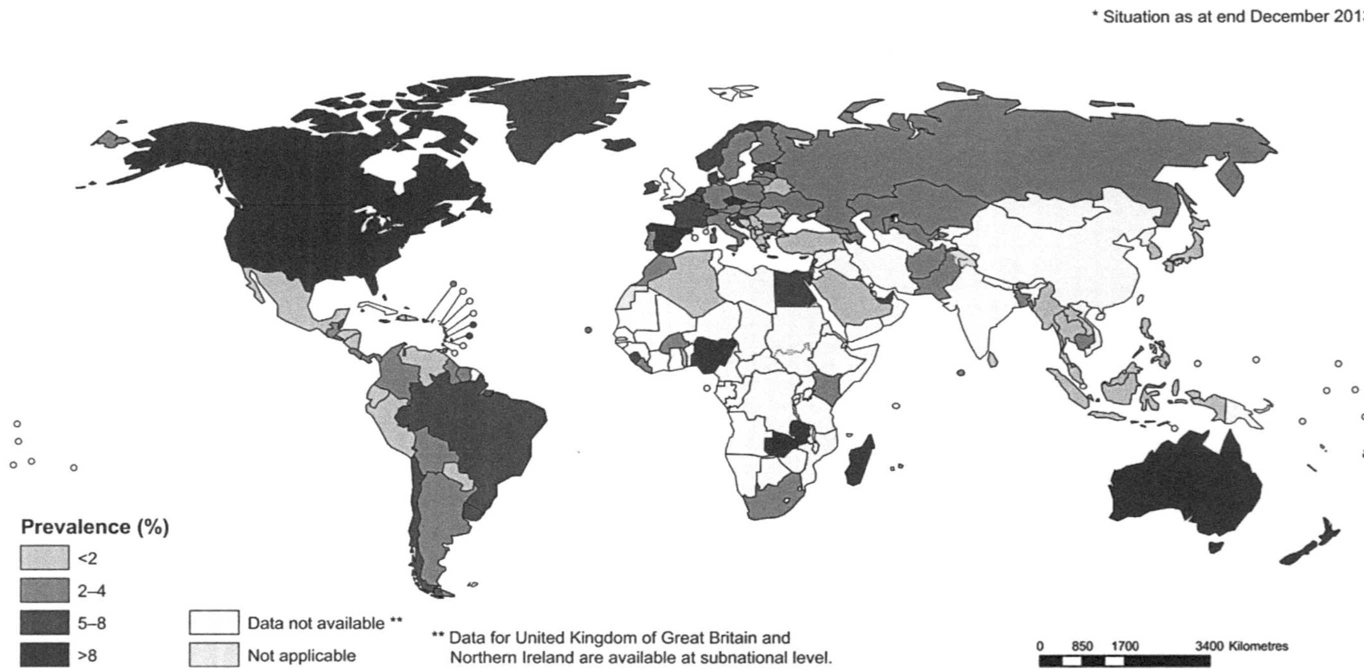

Fig. 1 Annual prevalence of cannabis use for population aged 15–64 years as of December 2013 (World Health Organization, 2016).

been used to find no changes in the gray matter volume of these structures, while some groups report regional volume increases (Lorenzetti et al., 2010; Moreno-Alcázar et al., 2018). Relatively low sample sizes, confounds due to poly-drug use and inconsistent criteria for defining chronic cannabis use contribute to this variability (Moreno-Alcázar et al., 2018).

2. GENERAL OVERVIEW OF ENDOCANNABINOID SYSTEM

Many mammalian species utilize endogenous cannabinoid signaling in both the development and maintenance of a functional, balanced nervous system. The two main endocannabinoid ligands are 2-arachidonoyl-glycerol (2-AG) and N-arachidonoyl-ethanolamine (anandamide or AEA). These lipid-derived neurotransmitters each have unique enzymes for synthesis and degradation that are depicted in Fig. 2 (Trezza et al., 2008; Zou & Kumar, 2018). The distribution of these enzymes dictates the mechanism of action for each endocannabinoid (Di Marzo, Stella, & Zimmer, 2015).

2-AG is a low affinity but high efficacy agonist for CB1 and CB2 receptors. It is synthesized in postsynaptic cells by diacylglycerol lipase-α (DAGLα) and acts as a retrograde transmitter, being released into the synaptic cleft and binding to CB receptors on the presynaptic terminal to modulate further transmitter release. 2-AG is degraded by monoacylglycerol lipase (MAGL). AEA is synthesized in the both the presynaptic and postsynaptic neuron by N-acylphosphatidylethanolamine (NAPE)-specific phospholipase D (NAPE-PLD), where it can be transported intracellularly, in anterograde or in retrograde to ultimately be degraded by fatty acid amide hydrolase 1 (FAAH). AEA is a high-affinity, low efficacy agonist with a primary affinity for CB1 receptors (Di Marzo et al., 2015). Due to their lipophilic nature, endocannabinoids are not stored in classical synaptic vesicles. Rather, they undergo de novo synthesis in response to postsynaptic membrane depolarizations and increased levels of intracellular calcium (Trezza et al., 2008).

CB receptors are G protein coupled receptors (GPCRs), with CB1 widely expressed throughout the brain and CB2 expressed mainly in immune cells and in the peripheral nervous system (Trezza et al., 2008). CB1 and CB2 are the only CB receptors that have been characterized thus far, though mounting evidence argues for the inclusion of other GPCRs that can bind endogenous and exogenous cannabinoids such as the orphan GPCR GPR55 (Ryberg et al., 2007). The persistence of endocannabinoid

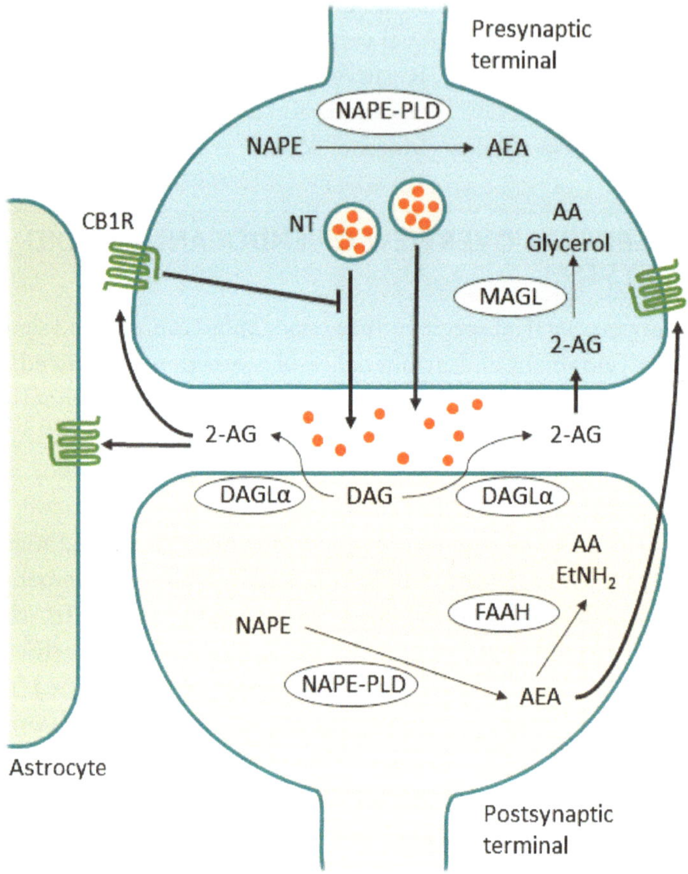

Fig. 2 The endocannabinoid synapse. 2-AG acts in retrograde on CB1 receptors of the presynaptic neuron to modulate transmitter release. Diacylglycerol lipase-α (DAGLα) and monoacylglycerol lipase (MAGL) are the enzymes that produce and break down 2-AG. AEA is synthesized in the both the presynaptic and postsynaptic neuron by N-acylphosphatidylethanolamine (NAPE)-specific phospholipase D (NAPE-PLD), where it can act intracellularly, in retrograde, or in anterograde (not pictured). *Reprinted from Zou, S., & Kumar, U. (2018). Cannabinoid receptors and the endocannabinoid system: Signaling and function in the central nervous system.* International Journal of Molecular Sciences, 19*(3). https://doi.org/10.3390/ijms19030833. With permission.*

signaling in CB-receptor knockout animals supports the existence of other endocannabinoid receptors (Begg et al., 2005). AEA is also known to activate the transient receptor potential cation subfamily V member 1 (TRPV1) channel (Di Marzo et al., 2015). Both CB1 and CB2 couple to G protein type $G_{i/o}$ and to a lesser extent G_q. Activation of CB1 can inhibit the opening of voltage-gated calcium channels but increase the opening of inwardly rectifying K^+ channels, typically resulting in a net decrease in neuronal activity and neurotransmitter release.

CB1 is the most abundant GPCR in the human brain (Di Marzo et al., 2015). The neuroanatomical distribution of CB1 has been mapped out in both humans and rodents. The highest densities of CB1 receptors in the human brain are found in the hippocampus, cerebellum, amygdala and basal ganglia, with significantly lower densities throughout the brainstem and thalamus (Fine & Rosenfeld, 2013). A higher density of CB1 receptors is present in the basal ganglia of mice compared to humans (Herkenham et al., 1990), which may provide a basis for the ability of cannabinoids to stimulate locomotor activity to a greater degree in rodents vs. humans. For example, cannabinoids are reported to affect rodent locomotion, though there studies supporting both hypolocomotor and hyperlocomotor responses (Tzschentke, 2007; Zangen, Solinas, Ikemoto, Goldberg, & Wise, 2006). Low densities of receptors in the brainstem, the central regulator of cardiovascular and respiratory functions, might help explain why high doses of natural cannabis do not tend to produce lethality via respiratory depression (Herkenham et al., 1990). Orthologues of the gene for the mammalian CB1 receptor have been found in non-mammalian vertebrates (i.e., zebra finch and newt salamanders) and even some invertebrates (leeches) (Mcpartland, 2004; Mcpartland, Agraval, Gleeson, Heasman, & Glass, 2006). CB2 are expressed primarily in immune cells and are not readily detectable in healthy brains, though these receptors can be expressed in microglia and astrocytes in response to neuroinflammation (Di Marzo et al., 2015).

Cannabinoid signaling is known to modulate common reward pathways by regulating dopamine and opioid systems (Covey, Mateo, Sulzer, Cheer, & Lovinger, 2017; Trezza et al., 2008). While midbrain dopamine neurons do not express CB1 receptors, they synthesize and release 2-AG which acts on CB1 receptors on local presynaptic GABAergic and glutamatergic terminals. The activation of CB1 receptors on GABAergic neurons can increase dopamine levels by eliminating presynaptic inhibition. Human and animal studies have shown that exogenous cannabinoids like THC can have the same effect of increasing DA release, which likely contributes to the experience of euphoria (World Health Organization, 2016). Microdialysis studies have shown that the amount of THC-induced dopamine release in the ventral forebrain (nucleus accumbens (NAc)) is much smaller than the amount induced by cocaine or amphetamines, but the delivery of

THC through inhalation produces the rapid rises in extracellular dopamine (Volkow & Morales, 2015). Endogenous opioid and cannabinoid signaling can also intertwine. For example, CB1 antagonists have been shown to induce classic opiate withdrawal symptoms in morphine-dependent rats, while mu opioid receptor antagonists can disrupt the anxiolytic effects induced by low-dose THC treatments (Berrendero, Mendizabal, Murtra, Kieffer, & Maldonado, 2003; Navarro et al., 1998). Opioid receptors are suspected of forming heterodimers with CB1, based on pharmacological data and a noted reduction in CB signaling in μ-opioid receptor knockout mice (Trezza et al., 2008).

There is evidence implicating the endocannabinoid system in the regulation of normal neural development. In both rodents and humans alike, there are higher densities of functioning CB1 receptors detected in embryonic white matter compared to adults, suggesting a role for CB1 signaling that is specific to early brain development (Trezza et al., 2008). CB receptors are expressed as early as embryonic day 11 in rats and embryonic week 14 in humans (Trezza et al., 2008). The concentrations of endocannabinoid ligands are also higher in embryonic and adolescent humans and rodents compared to the levels found in more mature brains (Trezza et al., 2008). In vitro studies have shown cannabinoids are capable of mediating cell proliferation, migration, axonal pathfinding and synaptogenesis (Trezza et al., 2008). Endocannabinoid signaling is implicated in many functions and behaviors, notably emotion and as discussed below, social behaviors. Dysfunctional endocannabinoid systems have also been implicated in many disorders, including ones with social components like autism (Zamberletti, Gabaglio, & Parolaro, 2017).

3. THE ROLE IN ENDOCANNABINOID SIGNALING IN SOCIAL BEHAVIOR

An ample number of animal models are available to study the endocannabinoid system and its role in modulating natural social behaviors. An understanding of the relationship between endogenous cannabinoid signaling and social behavior may provide insights into how this relationship becomes maladaptive in cannabis abuse. The effects of THC on embryonic and adolescent development might also be better clarified if compared to the natural influence of endocannabinoids on behavior. The core of this literature is based primarily on rodent models, which are innately social species that experience reward from social proximity and are therefore amenable to social testing (Panksepp & Lahvis, 2007). There are many different models in use that engage different aspects of social behaviors.

The three-chamber sociability test, also called the social approach test, examines the willingness of a subject animal to approach a conspecific (Sandi & Haller, 2015). The animals are first habituated to all compartments of the testing arena. On the test day, one side compartment of the testing arena houses a conspecific and the other side compartment either remains empty or houses an inanimate object (Gould et al., 2011, 2012). The conspecific is confined within a mesh barrier, preventing full social interaction. The subject is placed in the center compartment and given free access throughout the entire arena during a 10-min testing interval. A typical rodent will spend significantly more time exploring the compartment containing the conspecific over the other two compartments. Sociability levels are unaltered in CB1 deficient mice, suggesting cannabinoid signaling in not a primary mediator of social approach behavior (Gould et al., 2012). Treatment with a synthetic CB1 agonist also does not enhance sociability in a socially-deficient mouse strain (Gould et al., 2012). Thus, social behaviors measured in this particular paradigm are not particularly sensitive to endocannabinoid system manipulation.

However, treatment with synthetic CB1 agonists impairs social recognition in the social interaction task, which incorporates components of memory and social functioning (Zanettini, 2011). In a typical social recognition task, animals are acquainted with a conspecific during a training phase, and during the testing phase, are presented with the familiar social partner and a novel partner simultaneously. The amount of time the subject spends investigating the new partner is quantified as a measure of sociability (Kaidanovich-beilin, Lipina, Vukobradovic, Roder, & Woodgett, 2011). However, only investigative behaviors are measured (i.e., anogenital exploration, non-anogenital exploration and approach/following), while other social behaviors such as grooming and play behavior are often excluded from analysis (Schneider, Schömig, & Leweke, 2008). Increased time devoted to the exploration of the novel conspecific is interpreted as the subject retaining memory of the partner familiar from training phase, deeming it worthy of less investigation. After chronic, systemic treatment with CB1 agonists, rats behave as though they recognized their familiar social partners, and engaged in fewer social interactions altogether (O'Shea, Mcgregor, & Mallet, 2006; Schneider et al., 2008), suggesting cannabinoid-induced deficits in social interactions. It is important to note that in some reports, the alteration of social behavior was unique to adolescent male rats and could not replicated in the social interactions of adults. Social interactions can also be assessed in a simple, open-field paradigm where two animals are introduced to each other and allowed to move freely. This open meeting allows each animal to initiate prosocial/approach as well as evasive/antisocial behaviors. An alternative method for observing more spontaneous social interactions is the use of a group-housing paradigm. Video recordings can be analyzed for the number and duration of the social encounters.

The open-field setting was also utilized by Trezza and Vanderschuren (2008) to study the role of CB signaling in the play behavior of adolescent rats. Social play behavior was measured as the frequency of pinning and pouncing, with the amount of time engaged in exploratory sniffing recorded for a social non-play control behavior. Interestingly, both the enhancement and attenuation of social play behavior was achieved depending on the conditions under which CB1 receptors were activated. A systemic dose of the direct CB1 agonist, WIN55,212-2, decreased the frequency of play behavior while an indirect agonist, URB597 which inhibits the degradation of AEA via FAAH, increased play behavior. Pretreatment with an opioid receptor antagonist abolished the increase in play behavior induced by URB597, while leaving the decrease induced by WIN55,212-2 unaffected. The same effects occurred when the animals were pretreated with the non-selective dopamine receptor antagonist alpha-flupenthixol. While these drugs significantly impacted pinning and pouncing behavior, no changes were found in other social behaviors such as exploratory sniffing. These data suggest a role for endocannabinoids in prosocial behavior by modulating opioid and dopaminergic activity, as well as deleterious effects of exogenous cannabinoid drugs on sociability.

Using this same play model, additional experiments found increases in the concentration of AEA in the amygdala and NAc of rats immediately following social play behavior (Trezza et al., 2012). These increases were specific to these regions of the reward circuitry, though no increases were observed in the prefrontal cortex. The hippocampus also remained unaffected. Injections of URB597 directly into the NAc and basolateral amygdala (BLA) both independently enhanced social play. However, injections of a CB1 antagonist only reduced systemic URB597-induced play behavior when injected in the BLA (Trezza et al., 2012). Exploratory sniffing behavior remained unaffected by these manipulations. These data support a key role of the BLA in endocannabinoid-mediated social play and add to the mounting evidence supporting CB1 modulation of the BLA-NAc pathway (Ahmad, Sun, Lyons, & Laviolette, 2016).

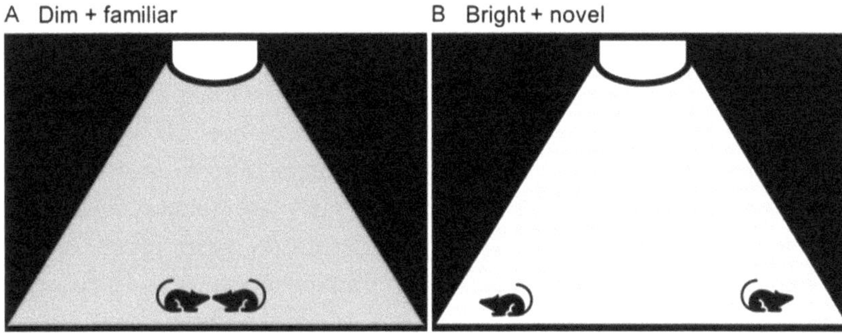

Fig. 3 Schematic representation of the social interaction test #2. (A) Dim lighting and familiarity have anxiolytic properties that promote high levels of social interaction. (B) Bright lights and a novel environment have an anxiogenic effect with limited social interactions.

The endocannabinoid system has also been studied in measures of anxiety (File & Seth, 2003; Genn, Tucci, Marco, Viveros, & File, 2004), using a variation on the social interaction test described above in Fig. 1C. For this version of the social interaction test, rodents are tested in a chamber designed to induce varying levels of anxiety (Fig. 3). Variables such as familiarity of the cage environment and intensity of the lighting can be altered to adjust the level of the provoked anxiety. For example, bright lighting and complete unfamiliarity with the test context are considered the most anxiogenic (Fig. 3A) while dim lighting and familiarity are considered the most anxiolytic (Fig. 3B). Social interactions are more common in the absence of anxiety; hence more interactions are reported under anxiolytic conditions. Animals are tested under a range of conditions (i.e., prior exposure to a stressor, pharmacological or genetic manipulation, etc.) to determine whether a stimulus is anxiolytic or anxiogenic, as measured by the change in social interaction level.

When treated systemically with a potent CB1 agonist, Genn et al. (2004) found that rats significantly decreased the amount of social interactions under anxiolytic conditions in the social interaction test. This result is indicative of CB1 activity having anxiogenic properties, which contradicts other reports utilizing different measures of anxiety. For example, multiple groups have shown CB1 agonists are be anxiolytic in the elevated plus maze (Biscaia et al., 2003). It is plausible that endocannabinoid signaling can have divergent effects at different strengths of antagonism/activation, as well in different environmental contexts that engage different components of anxiety and social interaction.

Many of the commonly used social paradigms in rodent research feature multifaceted tasks to measure complex behaviors involving memory and recognition, investigatory exploration, play, and anxiety-related responses. In general, systemic endocannabinoid agonists have been shown to reduce the number of social encounters and playful interactions that are typical for adolescent rodents. Targeted injections of CB1 antagonists directly into BLA also reduced social play. Endocannabinoid levels are elevated in the amygdala and NAc following social play, implicating the BLA-NAc circuit in CB-mediated social behavior. It is unclear whether any reductions in social behavior are resultant of increases or decreases in the experience of anxiety.

While the culmination of this research helps develop an understanding of the mechanism and social consequences of altering the endogenous cannabinoid system in the brain, further research is needed to examine the therapeutic potential and limitations of cannabinoid drugs. Current studies are examining other therapeutic uses of cannabinoids, such as in treating other drug dependencies. The ability of cannabinoids to act within the same reward circuit as other addictive drugs has been proposed to alleviate cravings, which is one of the main predictors of relapse (Prud'homme, Cata, & Jutras-Aswad, 2015; Socías et al., 2018). Cannabinoid treatments have been shown to reduce the frequency of crack cocaine use in Canada (Socías et al., 2018). Other studies examining the efficacy of cannabis in preventing relapse have demonstrated positive outcomes for cocaine and alcohol addiction, but not heroin addiction (Aharonovich et al., 2005). Expansion on these findings may help facilitate recovery and promote sustained abstinence. Future research is needed that focuses on incorporating these promising results with the dynamic implications of social constructs.

4. SOCIAL INFLUENCES IN HUMAN CANNABIS USE

Social interactions are important in many stages of developing drug dependency. First exposure to substances of abuse typically occurs among peers during adolescence or early adulthood (Neisewander, Peartree, & Pentkowski, 2012). The degree to which an individual finds their first experience using a drug to be positive and rewarding is a good predictor for later dependency; people who reported positive first experiences were more likely to develop habits (Haertzen, Kocher, & Miyasato, 2018). This is true for most abused drugs, including marijuana, alcohol, barbiturates, cocaine, stimulants, hallucinogens, and opiates. The social environment can modulate the reinforcing power of the first drug experience. With nicotine, for example, beginner smokers often credit social motives as their main reason for smoking cigarettes (Baker, Brandon, & Chassin, 2004). Also, having peers who smoke marijuana is strong predictor of initiating cannabis use in adolescents (Bohnert, Bradshaw, & Latkin, 2009). Analysis of self-report questionnaires from over 3000 adolescent students suggests the peer network can account for 65% of variance in marijuana use (Akers, Krohn, Lanza-Kaduce, & Radosevich, 1979).

Social environments can also influence the continuation of drug use (Neisewander et al., 2012). There is consensus in the social network analysis field that a relationship exists between the drug usage of an individual and the substance abuse in the social network to which they belong. Positive reactions from peers may increase rates of marijuana use, even if the marijuana is not intrinsically gratifying to the individual (Kaplan et al., 1986). One epidemiological study regarding marijuana use among 17,592 college students found that students who reported having five or more close student friends on campus were 1.4 times more likely to use marijuana than students who had fewer on campus friends (Bell et al., 1997). It is possible that people prone to substance dependence are more likely to seek friendships with like-minded people, or are more likely to associate with those already dependent on marijuana.

Keyes et al. (2011) studied the relationship between group-level social norms and marijuana use in adolescents. In a 31-year investigation spanning from 1976 to 2007, a total of 986,003 adolescents from grades 8, 10 and 12 were surveyed in hundreds of secondary schools across the United States. The subjects were grouped based on different birth years. Birth cohorts with a higher percentage of students that disapproved of marijuana use overall were less likely to use marijuana, independent of individual feelings toward the drug. This is corroborated by a recent report from the World Health Organization (WHO, 2016), reporting social norms that are tolerant of marijuana use as the main social factor that lead to cannabis consumption in developed countries.

While group norms can reinforce drug intake, a disposition toward social deviance and disregard for societal norms can also be important. Commitment theory suggests that deviant behavior (as marijuana use is conceptualized), is inversely correlated to a person's attachment and adherence to the norms and values of society (Ginsberg & Greenley, 1978; Kaplan et al., 1986). The less an individual care about societal judgment, the more likely he/she is to experiment with marijuana. Bell et al. (1997) demonstrated this theory as well by finding that students who self-reported risky behavior such as smoking cigarettes, having multiple sexual partners, and students who partook in binge drinking were exceedingly more likely to use marijuana than their college counterparts who did not engage in risky (or deviant) behaviors.

Marijuana is colloquially considered to calm people down or "chill them out," and a common expectancy held by marijuana users is that marijuana will alleviate anxiety symptoms (Buckner, Heimberg, & Schmidt, 2011; Gaher & Simons, 2007). In 2010, de Dios et al. investigated marijuana use in a sample of female participants, and a significant correlation between marijuana use and tension reduction expectancies was found. Substance use expectancies are defined as beliefs regarding the anticipated effects from using a substance, and how those anticipated effects facilitate further use (Hayaki et al., 2010). These expectations can relate to how the drug is going to make the person feel, as well as what external consequences they think may occur. It is recognized that substance use expectancies are acquired and maintained through normal learning processes (Shafer & Brown, 1991).

Buckner and Schmidt (2008) looked at marijuana use expectancies and how they related to social anxiety and marijuana use problems. They recruited 337 college participants, the majority of which were Caucasian females, with mean age of 18.8 (SD: 1.17). These investigators hypothesized that social anxiety was related to frequency of marijuana use. They aimed to determine whether particular expectancies were related to an increase in marijuana-related problems, and if individuals that reported higher social anxiety were more likely to predict high-risk marijuana effect expectancies. An example of a high-risk marijuana effect expectancy is using marijuana to cope with social anxiety (Buckner & Schmidt, 2008). They attempted to capture this phenomenon by administering a battery of self-report measures which included: the Marijuana Use Form (MUF), a self-report instrument that assesses marijuana use; Marijuana Problems Scale (MPS), a 19 item measure looking at negative consequences related to marijuana use in the last 90 days; Marijuana Expectancies Questionnaire (MEEQ), a 48-item scale includes six lower-order scales and two higher-order scales measuring positive and negative expectancies; Social Interaction Anxiety Scale (SIAS), a short questionnaire used to measure how social anxiety plays a role in different life situations, and the Beck Depression Inventory (BDI) a 21 item self-report questionnaire that aims to measure the severity of depression. They found a positive correlation between social anxiety and negative expectancies, as well as a negative correlation between social anxiety and tension reduction expectancies. Further, the greater the belief that marijuana would produce cognitive/behavioral impairment, the higher the reported rate of marijuana use (Buckner & Schmidt, 2008).

Social interactions are also relevant to relapse. Interaction with an individual from the drug-taking context is often enough incentive to induce relapse (Neisewander et al., 2012). Individuals have been found to decrease their interactions with drug-using peers after quitting, supporting the theory that social control processes play a role in adult abstinence (Buchanan & Latkin, 2008). A stable social support network is predictive of lower risks for relapse and the ability to maintain abstinence when overcoming drug dependencies (Havassy, Hall, & Wasserman, 1991). Thus, every stage of dependency, from initiation to the termination of drug use, is heavily influenced by social interactions (or lack thereof).

5. ANIMAL MODELS OF CANNABIS ABUSE

Research on the abuse potenital of cannabinoids in animal models has been ongoing since the 1970s and remains a field of much debate (Trezza et al., 2008). In current substance abuse models, THC and other cannabinoid receptor agonists can have both rewarding and aversive effects, depending on the dose and the context (Valjent, Mitchell, Besson, Caboche, & Maldonado, 2002; Wakeford et al., 2017). This is not unlike the human condition, where people often report both positive and negative affective states during marijuana intoxication (Cutler & Mackintosh, 1984). However, THC was initially found to be unrewarding in virtually every animal paradigm tested. For example, in one simple measure of reward, THC produced robust conditioned taste aversion (CTA) (Wakeford & Riley, 2014). In CTA experiments, animals are water deprived and then given free access to an inherently rewarding sweetened solution during conditioning sessions. Immediately afterward, the animals are injected with a drug such as THC so the effects become associated with the consumption of the sugar water. If the drug is considered aversive, then the animals will avoid the sweetened solution during the testing phase, which is precisely what was observed by Wakeford and Riley (2014). The effect is dose-dependent, with higher doses of THC producing stronger aversions. In longitudinal studies with rats, THC exposure in adolescence had no effect on conditioned taste or place preferences in adulthood (Wakeford, Flax, Pomfrey, & Riley, 2016).

CPP is a common paradigm to assess the rewarding properties of a substance. Rodents are classically conditioned to associate neutral environmental stimuli with the effects of a drug, and then are given the choice during a test phase to spend time in the same location or an adjoining neutral chamber. If the animal prefers the drug-paired context, the drug is considered to be rewarding; if the animal avoids the drug-paired context, then the drug is considered aversive. Up until 2000, cannabis primarily produced conditioned place avoidance (CPA). It was unclear whether the cannabis experience was more dysphoric in rodents than it typically is for humans, or whether THC had any rewarding properties in rodents. Research has since proven that, depending on the dosage and the conditioning protocol, cannabis can produce both CPPs and CPAs (Tzschentke, 2007). Priming injections of low-dose THC before conditioning paired with a high number of conditioning sessions for long periods of time are parameters that help facilitate THC CPP (Valjent & Maldonado, 2000). Interestingly, simultaneous administration of cannabidiol can counteract the aversive effects of THC in producing a CPA (Vann et al., 2008).

In experiments targeting specific reward centers in the mesocorticolimbic dopamine system, THC was microinjected into subregions of the ventral tegmental area (VTA) and the NAc. CPP was robust in animals that received this direct treatment in the posterior VTA and the posterior shell of the NAc (Zangen et al., 2006). THC applied to the anterior or dorsal regions of VTA failed to produce CPP, as well as in the core of the NAc (Zangen et al., 2006). Thus, direct stimulation of CB1 receptors in the VTA and NAc is sufficient for experiencing THC-mediated reward. These findings link THC-induced reward to brain regions known to be involved in the reward circuitry for many other drugs of abuse (Zangen et al., 2006).

The CPP paradigm has also been used to examined the effects of prenatal exposure to cannabis on the rewarding properties of other abused drugs (Tzschentke, 2007). Rubio et al. (1998) subjected pregnant rats to daily injections of THC during gestation and lactation periods. When the offspring were later tested, animals that had received exposure to low doses of THC exhibited enhanced CPP responses to morphine. A comparable result was procured by Singh, McGregor, and Mallet (2006), where prenatal exposure to THC produced an enhanced CPP response to heroin in offspring. Interestingly, this enhancing effect does not seem to generalize to THC preexposure in older rodents. Chronic treatment with THC prior to testing, in the form of daily injections for several weeks, had no effect on the CPP response to morphine in adult mice (Jardinaud, Roques, & Noble, 2006; Valverde et al., 2001).

Similar models have been used to study the combined effects of THC and nicotine. This is relevant to human drug intake patterns, as THC and nicotine are often co-used. For example, it is common for people to smoke "spliffs," which are joints comprised of both marijuana and tobacco. However, effects of combined nicotine and THC in rodents have produced inconsistent findings. In adult mice, administering subthreshold doses of nicotine or THC (0.3 mg/kg each) alone failed to produce CPP, but when paired together, robust CPP was observed (Valjent et al., 2002). This suggests an additive effect in the rewarding properties of low doses of nicotine and THC. Conversely, another study found that in adult rats, low doses of THC produced CPPs that were attenuated in rats that received 5 days of nicotine pretreatment (Le Foll, Wiggins, & Goldberg, 2006). THC doses of 1 mg/kg and 2 mg/kg induced conditioned place aversions in the rats pretreated with nicotine (Le Foll et al., 2006). These data suggest nicotine may reduce the rewarding properties and/or accentuate the aversive effects of THC. The parameters of the relationship between nicotine and THC reward need to be further tested.

The traditional gold standard of substance abuse paradigms is a form of operant conditioning called self-administration, where the animal is in control of the amount of drug it receives intravenously or orally through a syringe pump activated by its own lever presses. Cannabis self-administration has been historically difficult for animals to acquire. Much of the pioneer research from the 1970s failed to obtain robust self-administration with THC, meaning the animals did not learn to press a lever in a dose-dependent manner or extinguish lever pressing for THC when it was substituted for saline (Panlilio et al., 2010; van Ree, Slangen, & de Wied, 1978). With perhaps the exception of serotonergic hallucinogens, it is somewhat rare that a rewarding drug for humans is unable to be reliably self-administered in an animal model, and this disconnect was interpreted as marijuana having a low potential for abuse (Panlilio et al., 2010; Tanda, Munzar, & Goldberg, 2000).

In monkeys, THC self-administration was eventually achieved if the animals received prior training with an more readily self-administered drug, like cocaine, and then switched to THC after lever pressing had already been established Even then, less than half of the monkeys would acquire stable THC self-administration (Kaymakçalan, 1981; Tanda et al., 2000). Tanda et al. (2000) noted that the chemical properties of THC made it more difficult to adapt to the self-administration paradigm as compared to other addictive drugs. The lipophilic nature of THC render it difficult to dissolve in an aqueous solution at high doses, which results in animals receiving infusions of a THC suspension. Since particulates in a drug suspension do not stay distributed evenly, the added variability of solution heterogeneity could account for the failure of animals to acquire THC self-administration. It was not until 15 years ago that drug-naïve squirrel monkeys were first shown to self-administer THC, where performance of the animals improved with a fixed-ratio schedule of reinforcement using low but clinically relevant doses of THC fully dissolved in a Tween-80 solution (Justinova, Tanda, Redhi, & Goldberg, 2003).

Successful cannabinoid self-administration has also been shown in rats and mice, though researchers report the same difficulties as in monkeys (Panlilio et al., 2010; Wakeford et al., 2017). The percentage of animals that acquire THC self-administration is low, as is the total amount of THC self-administered compared to other more readily reinforcing drugs such as cocaine. Low doses of THC that are fully dissolved in non-aqueous solvents such as Tween-80 seem to yield the best results. Unlike CPP paradigms, cannabidiol pretreatments were not able to counteract any aversive effects that THC infusions might present (Wakeford et al., 2017). One interesting study utilized intracranial self-administration, where THC infusions were delivered directly to specific brain regions (Zangen et al., 2006). When infused directly into the posterior VTA, animals developed a preference for the active lever after the first training session. Infusions into the NAc shell also produced a similar preference, while rats receiving infusions in the anterior VTA never acquired self-administration. These data support findings from CPP experiments implicating these brain regions in the experience of THC-mediated reward (Zangen et al., 2006).

One study in rodents combined cannabinoid self-administration with a sociability test (Spano, Fadda, Frau, Fattore, & Fratta, 2010). Animals were trained to press a lever to receive intravenous infusions of the synthetic cannabinoid agonist WIN55,212-2. Acquisition and maintenance of self-administration with synthetic cannabinoid agonists that have different pharmacological profiles tends to be more robust than THC (Amchova, Kucerova, Giugliano, Babinska, & Zanda, 2014; Fattore & Cossu, 2001; Fattore et al., 2007; Martellotta, Cossu, Fattore, Gessa, & Fratta, 1998; Wakeford et al., 2017). In the sociability test, animals were scored on time spent engaged in social interactions over a 10-min testing interval. WIN55,212-2 self-administration had no effect on the social interactions of the rats after acute treatment (Spano et al., 2010). Additionally, when phencyclidine (PCP) was administered to the same animals as a psychotomimetic agent, only animals that previously self-administered the cannabinoid agonist showed no change in their social interaction levels, and all cannabinoid-naive animals dramatically decreased their social behaviors. The results of this study suggest that, chronic cannabinoid treatment modulates sensorimotor gating and sociability.

Prior exposure to cannabinoids also enhances the self-administration of other addictive drugs. In one study, it was shown that eight injections of THC throughout adolescence increased heroin self-administration in rats later in adulthood (Ellgren et al., 2007). THC treatment did not predispose the animals to an increased sensitivity for initiating self-administration in the first session, but THC-treated animals showed higher heroin intake during the remaining acquisition and maintenance phases of the experiment. These data were presented as supportive of the gateway drug hypothesis (GDH), a theory that has been associated with marijuana for decades. According to the GDH, marijuana use leads individuals to experiment with "harder" addictive drugs such as cocaine or heroin, and increases the likelihood of developing drug dependency (Ellgren et al., 2007). This phenomenon has been corroborated in human epidemiological studies, as well as a national survey showing that those who started marijuana at age 14 or earlier had a higher percentage of illicit drug dependencies later in life (Agrawal, Neale, Prescott, & Kendler, 2004; Center for Behavioral Health Statistics and Quality, 2015).

6. CONCLUSIONS AND FUTURE DIRECTIONS

Of all the paradigms and experiments aiming to model cannabis abuse in rodents and other non-human species, very few have addressed social components of cannabinoid use. However, as previously discussed, that there are many social components vital to the initiation and maintenance of cannabis use. Social interactions are also known to be very important in animal models of other addictive drugs. Neisewander et al. (2012) reviewed the literature from the alcohol, nicotine, cocaine, amphetamine and opioid addiction fields to develop the model shown in Fig. 4. In this model, every social interaction from familial relationships to connections with peers from birth onward can affect drug use; this results in a massive "sphere" of influential power of social influences. Specific interactions can either enhance or attenuate drug use, depending on whether they occur within or independent of the drug-taking context. Overall, prosocial behavior is protective against drug use outside of the drug-taking environment,

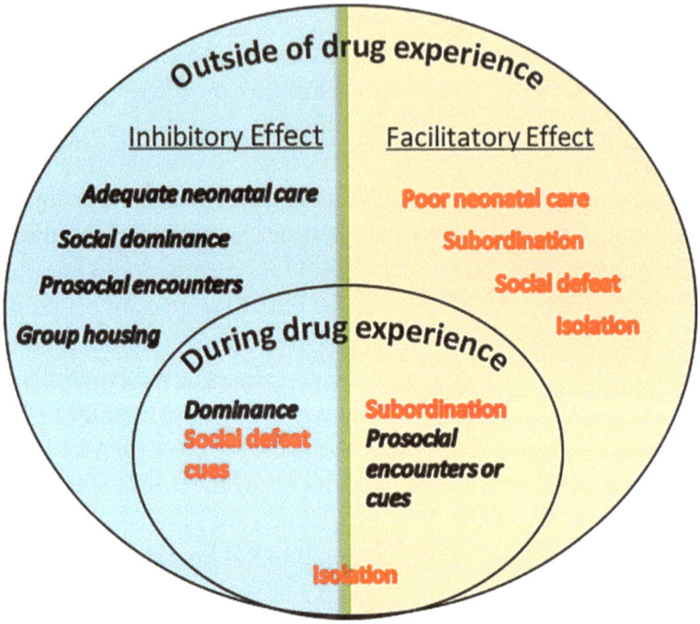

Fig. 4 The effect of social interactions both in and outside the drug context on drug dependency in animal models. The left side (blue) shows interactions that inhibit drug use, while the right side (yellow) show interactions that reinforce drug use. Positive interactions are in black, negative interactions are in red. *Adapted from Neisewander, J. L., Peartree, N. A., & Pentkowski N. S. 2012. Emotional valence and context of social influences on drug abuse-related behavior in animal models of social stress and prosocial interaction,* Psychopharmacology, 224, *33–56 https://doi.org/10.1007/s00213-012-2853-3.*

while a lack of positive interactions or negative interactions can exacerbate use (Neisewander et al., 2012). This is does not appear to be the case with social interactions within the drug-taking environment. The valence of social interactions during the drug experience does not clearly predict the later effects on drug usage; both positive and negative encounters can facilitate use (Neisewander et al., 2012). It should be noted, however, that cannabis-related studies were not considered in the formulation of this model or these conclusions, as literature is on this drug is scarce and requires continued investigation.

Better animal models are needed to reflect these social aspects of the human drug-taking experience. Most animal paradigms have struggled with low motivation for THC, and it would be therefore be of great interest to determine if social interactions facilitate the experience of THC reward and reinforcement. In studies using the CTA paradigm, the presence of a conspecific could be tested as incentive to counteract the aversive effects of the drug and increase sweetened solution consumption. One group utilized social groups in an ethanol CTA study, during which the rats were housed, conditioned and tested in groups of five littermates (Morales, Schatz, Anderson, Spear, & Varlinskaya, 2015). It is important to note that the main goal of this study was to compare different age groups so no isolation control was included, and therefore specific effects of the social context were not examined.

The nicotine field has also explored social interactions in CPP models (Bastle et al., 2016; Thiel, Sanabria, & Neisewander, 2009). Social reward has long been known to produce CPP in rats, as does treatment with a wide range of nicotine doses (Peartree et al., 2012). These researchers found that simultaneous presentation of nicotine and social interactions together during conditioning had greater conditioned rewarding effects, and therefore produced stronger CPP, than the experience of either stimulus alone. A social context has also been incorporated into the nicotine self-administration paradigm (Peartree et al., 2017). In this study, custom operant conditioning chambers were constructed to allow animals to interact with each other through mesh barriers during operant conditioning sessions. Findings of this study showed that with low doses of nicotine, the presence of a conspecific the facilitated the acquisition of nicotine self-administration in male rats. Since various parameters required for THC CPP and self-administration have been established, future studies could be easily adapted to these specialized social context models.

An ideal animal model of addiction would reflect all salient stimuli associated with drug use. For example, important cues involved with marijuana use include smoke or vapor inhalation, which are absent during systemic infusions of a THC solution. Since human marijuana users rarely, if ever, self-administer the drug via the intravenous route, a better model of the human condition of smoking marijuana would be to utilize an aerosol delivery system within a social context. While vapor chambers are often used to administer high doses of ethanol to rodents for modeling alcoholism (Griffin, Lopez, & Becker, 2009; Kufahl, Martin-Fardon, & Weiss, 2011), thus far this has not been performed in a social context Both nicotine and THC have also been passively administered via aerosol to rodents, though without any social component (Laube et al., 2017; Manwell, Ford, Matthews, Heipel, & Mallet, 2014). It would therefore be of great interest to apply this same vapor delivery system to animals to investigate how a social environment affects later self-administration or other measures of THC reward and reinforcement. The potential of the aerosol model to accurately recapitulate social influences and rapid drug delivery simultaneously would be promising for the advancement of research on THC abuse.

REFERENCES

Agrawal, A., Neale, M. C., Prescott, C. A., & Kendler, K. S. (2004). A twin study of early cannabis use and subsequent use and abuse/dependence of other illicit drugs. *Psychological Medicine*, *34*(7), 1227–1237.

Aharonovich, E., Liu, X., Samet, S., Nunes, E., Waxman, R., & Hasin, D. (2005). Post-discharge cannabis use and its relationship to cocaine, alcohol, and heroin use: A prospective study. *The American Journal of Psychiatry*, *162*, 1507–1514.

Ahmad, T., Sun, N., Lyons, D., & Laviolette, S. R. (2016). Bi-directional cannabinoid signalling in the basolateral amygdala controls rewarding and aversive emotional processing via functional regulation of the nucleus accumbens. *Addiction Biology*, *22*, 1218–1231.

Akers, R. L., Krohn, M. D., Lanza-Kaduce, L., & Radosevich, M. (1979). Social learning and deviant behavior: A specific test of a general theory. *American Sociological Review*, *44*(4), 636–655.

Amchova, P., Kucerova, J., Giugliano, V., Babinska, Z., & Zanda, M. T. (2014). Enhanced self-administration of the CB 1 receptor agonist WIN55, 212-2 in olfactory bulbectomized rats: Evaluation of possible serotonergic and dopaminergic underlying mechanisms. *Frontiers in Pharmacology*, *5*, 1–15. https://doi.org/10.3389/fphar.2014.00044.

American Psychiatric Association. (2013). *Diagnostic and statistical manual of mental disorders* (5th ed.). Arlington, VA: American Psychiatric Association.

Anthony, J. C., Warner, L. A., & Kessler, R. C. (1997). Comparative epidemiology of dependence on tobacco, alcohol, controlled substances, and inhalants: Basic findings from the national comorbidity survey. In: G. A. Marlatt & G. R. VandenBos (Eds.), *Addictive Behaviors: Readings on Etiology, Prevention, and Treatment* (pp. 3–39). Washington, DC: American Psychological Association.

Baker, T. B., Brandon, T. H., & Chassin, L. (2004). Motivational influences on cigarette smoking. *Annual Review of Psychology*, *55*(1), 463–491. https://doi.org/10.1146/annurev.psych.55.090902.142054.

Bastle, R. M., Peartree, N. A., Goenaga, J., Hatch, K. N., Henricks, A., Scott, S., et al. (2016). Immediate early gene expression reveals interactions between social and nicotine rewards on brain activity in adolescent male rats. *Behavioural Brain Research*, *313*, 244–254. https://doi.org/10.1016/j.bbr.2016.07.024.

Begg, M., Pacher, P., Bátkai, S., Osei-Hyiaman, D., Offertáler, L., Fong, M. M., et al. (2005). Evidence for novel cannabinoid receptors. *Pharmacology and Therapeutics*, *106*(2), 133–145. https://doi.org/10.1016/j.pharmthera.2004.11.005.

Bell, R., Wechsler, H., & Johnston, L. D. (1997). Correlates of college student marijuana use: Results of a U.S. national survey. *Addiction*, *92*, 571–581.

Berrendero, F., Mendizabal, V., Murtra, P., Kieffer, B. L., & Maldonado, R. (2003). Cannabinoid receptor and WIN 55 212-2-stimulated [^{35}S]-GTPγS binding in the brain of mu-, delta- and kappa-opioid receptor knockout mice. *European Journal of Neuroscience*, *18*, 2197–2202. https://doi.org/10.1046/j.1460-9568.2003.02951.x.

Biscaia, M., Marín, S., Fernµndez, B., Marco, E. M., Rubio, M., Guaza, C., et al. (2003). Chronic treatment with CP55,940 during the peri-adolescent period differentially affects the behavioural responses of male and female rats in adulthood. *Psychopharmacology*, *940*, 301–308. https://doi.org/10.1007/s00213-003-1550-7.

Bohnert, A., Bradshaw, C., & Latkin, C. (2009). A social network perspective on heroin and cocaine use among adults: Evidence of bidirectional influences. *Addiction*, *104*(7), 1210–1218. https://doi.org/10.1111/j.1360-0443.2009.02615.x.A.

Bonnet, U., & Preuss, U. (2017). The cannabis withdrawal syndrome: Current insights. *Substance Abuse and Rehabilitation*, *8*, 9–37. https://doi.org/10.1371/journal.pone.0076635.

Buchanan, A. S., & Latkin, C. A. (2008). Drug use in the social networks of heroin and cocaine users before and after drug cessation. *Drug and Alcohol Dependence*, *96*(3), 286–289. https://doi.org/10.1016/j.drugalcdep.2008.03.008.
Buckner, J. D., Heimberg, R. G., & Schmidt, N. B. (2011). Social anxiety and marijuana-related problems: The role of social avoidance. *Addictive Behaviors*, *36*(1–2), 129–132.
Buckner, J. D., & Schmidt, N. B. (2008). Marijuana effect expectancies: Relations to social anxiety and marijuana use problems. *Addiction Behavior*, *33*, 1477–1483.
Center for Behavioral Health Statistics and Quality. (2015). *Behavioral health trends in the United States: Results from the 2014 National Survey on Drug Use and Health*. (HHS Publication No. SMA 15-4927, NSDUH Series H-50., 64. Retrieved from http://www.samhsa.gov/data/sites/default/files/NSDUH-FRR1-2014/NSDUH-FRR1-2014.pdf%5Cnhttp://www.samhsa.gov/data/.
Covey, D. P., Mateo, Y., Sulzer, D., Cheer, J. F., & Lovinger, D. M. (2017). Neuropharmacology endocannabinoid modulation of dopamine neurotransmission. *Neuropharmacology*, *124*, 52–61. https://doi.org/10.1016/j.neuropharm.2017.04.033.
Crean, R. D., Crane, N. A., & Mason, B. J. (2011). An evidence based review of acute and long-term effects of cannabis use on executive cognitive functions. *Journal of Addiction Medicine*, *5*(1), 1–8. https://doi.org/10.1097/ADM.0b013e31820c23fa.An.
Cutler, M., & Mackintosh, J. H. (1984). Cannabis and delta-9-tetrahydrocannabinol effects on elements of social behavior in mice. *Neuropharmacology*, *23*(9), 1091–1097.
de Dios, M. A., Hagerty, C. E., Herman, D. S., Hayaki, J., Anderson, B. J., Budney, A. J., et al. (2010). General anxiety disorder symptoms, tension reduction, and marijuana use among young adult females. *Journal of Women's Health (2002)*, *19*(9), 1635–1642. https://doi.org/10.1089/jwh.2010.1973.
Di Marzo, V., Stella, N., & Zimmer, A. (2015). Endocannabinoid signalling and the deteriorating brain. *Nature Reviews Neuroscience*, *16*(1), 30–42. https://doi.org/10.1038/nrn3876.
Ellgren, M., Spano, S. M., & Hurd, Y. L. (2007). Adolescent cannabis exposure alters opiate intake and opioid limbic neuronal populations in adult rats. *Neuropharmacology*, *32*, 607–615. https://doi.org/10.1038/sj.npp.1301127.
Fattore, L., & Cossu, G. (2001). Intravenous self-administration of the cannabinoid CB1 receptor agonist WIN55,212-2 in rats. *Psychopharmacology*, *156*, 410–416. https://doi.org/10.1007/s002130100734.
Fattore, L., Spano, M. S., Altea, S., Angius, F., Fadda, P., & Fratta, W. (2007). Cannabinoid self-administration in rats: Sex differences and the influence of ovarian function. *British Journal of Pharmacology*, *152*, 795–804. https://doi.org/10.1038/sj.bjp.0707465.
File, S. E., & Seth, P. (2003). A review of 25 years of the social interaction test. *European Journal of Pharmacology*, *463*, 35–53. https://doi.org/10.1016/S0014-2999(03)01273-1.
Fine, P. G., & Rosenfeld, M. J. (2013). The endocannabinoid system, cannabinoid system, and pain. *Rambam Maimonides Medical Journal*, *4*(4), 1–15. https://doi.org/10.5041/RMMJ.10129.
Gaher, R. M., & Simons, J. S. (2007). Evaluations and expectancies of alcohol and marijuana problems among college students. *Psychology of Addictive Behaviors*, *21*, 545–554.
Genn, R. F., Tucci, S., Marco, E. M., Viveros, M. P., & File, S. E. (2004). Unconditioned and conditioned anxiogenic effects of the cannabinoid receptor agonist CP 55,940 in the social interaction test. *Pharmacology Biochemistry and Behavior*, 77(3), 567–573. https://doi.org/10.1016/j.pbb.2003.12.019.
Ginsberg, I. J., & Greenley, J. R. (1978). Competing theories of marijuana use: A longitudinal study. *Journal of Health and Social Behavior*, *19*, 22–34.
Gostin, L. O., Hodge, J. G., & Wetter, S. A. (2018). Enforcing federal drug laws in states where medical Marijuana is lawful. *JAMA*, *319*(14), 1435–1436. https://doi.org/10.1001/jama.2018.1083.

Gould, G. G., Hensler, J. G., Burke, T. F., Benno, R. H., Onaivi, E. S., & Daws, L. C. (2011). Density and function of central serotonin (5-HT) transporters, 5-HT1A and 5-HT2A receptors, and effects of their targeting on BTBR T + tf/J mouse social behavior. *Journal of Neurochemistry*, *116*(2), 291–303. https://doi.org/10.1111/j.1471-4159.2010.07104.x.

Gould, G. G., Seillier, A., Weiss, G., Giuffrida, A., Burke, T. F., Hensler, J. G., et al. (2012). Acetaminophen differentially enhances social behavior and cortical cannabinoid levels in inbred mice. *Progress in Neuro-psychopharmacology & Biological Psychiatry*, *38*(2), 1–24. https://doi.org/10.1016/j.pnpbp.2012.04.011.Acetaminophen.

Griffin, W. C., Lopez, M. F., & Becker, H. C. (2009). Intensity and duration of chronic ethanol exposure is critical for subsequent escalation of voluntary ethanol drinking in mice. *Alcoholism Clinical and Experimental Research*, *33*(11), 1893–1900. https://doi.org/10.1111/j.1530-0277.2009.01027.x.

Haertzen, C. A., Kocher, T. R., & Miyasato, K. (2018). Reinforcements from the first drug experience can predict later drug habits and/or addiction: Results with coffee, cigarettes, alcohol, barbiturates, minor and major tranquilizers, stimulants, marijuana, hallucinogens, heroin, opiates and cocaine. *Drug and Alcohol Dependence*, *11*(2), 147–165. https://doi.org/10.1016/0376-8716(83)90076-5.

Han, B., Compton, W. M., Blanco, C., & Jones, C. M. (2018). Trends in and correlates of medical marijuana use among adults in the United States. *Drug and Alcohol Dependence*, *186*, 120–129. https://doi.org/10.1016/j.drugalcdep.2018.01.022.

Havassy, B., Hall, S., & Wasserman, D. (1991). Social support and relapse: Commonalities among alcoholics, opiate users, and cigarette smokers. *Addictive Behaviors*, *16*, 235–246.

Hayaki, J., Hagerty, C. E., Herman, D. S., de Dios, M. A., Anderson, B. J., & Stein, M. D. (2010). Expectancies and marijuana use frequency and severity among young females. *Addictive Behaviors*, *35*, 995–1000.

Herkenham, M., Lynn, A. B., Litrle, M. D., Johnsont, M. R., Melvin, L. S., Costa, B. R. D. E., et al. (1990). Cannabinoid receptor localization in brain. *Proceedings of the National Academy of Sciences of the United States of America*, *87*, 1932–1936.

Hill, K., & Palastro, M. (2017). Medical cannabis for the treatment of chronic pain and other disorders: Misconceptions and facts. *Polish Archives of Internal Medicine*, *127*(11), 785–789. http://doi.org/10.20452/pamw.4123.

Jardinaud, F., Roques, B. P., & Noble, F. (2006). Tolerance to the reinforcing effects of morphine in delta9-tetrahydrocannabinol treated mice. *Behavioural Brain Research*, *173*(2), 255–261. https://doi.org/10.1016/j.bbr.2006.06.027.

Justinova, Z., Tanda, G., Redhi, G. H., & Goldberg, S. R. (2003). Self-administration of delta-9-tetrahydrocannabinol (THC) by drug naive squirrel monkeys. *Psychopharmacology*, *169*, 135–140. https://doi.org/10.1007/s00213-003-1484-0.

Kaidanovich-beilin, O., Lipina, T., Vukobradovic, I., Roder, J., & Woodgett, J. R. (2011). Assessment of social interaction behaviors. *Journal of Visualized Experiments*, (48), 1–6. https://doi.org/10.3791/2473.

Kaymakçalan, S. (1981). The addictive potential of cannabis. *Bulletin on Narcotics*, *33*(2), 21–31.

Keyes, K. M., Schulenberg, J. E., O'Malley, P. M., Johnston, L. D., Bachman, J. G., Li, G., et al. (2011). The social norms of birth cohorts and adolescent marijuana use in the United States, 1976–2007. *Addiction*, *106*(10), 1790–1800.

Kufahl, P., Martin-Fardon, R., & Weiss, F. (2011). Enhanced sensitivity to attenuation of conditioned reinstatement by the mGluR2/3 agonist LY379268 and increased functional activity of mGluR2/3 in rats with a history of ethanol dependence. *Neuropsychopharmacology*, *36*, 2762–2773. https://doi.org/10.1038/npp.2011.174.

Laube, B. L., Afshar-mohajer, N., Koehler, K., Chen, G., Collaco, J. M., Mcgrath-morrow, S. A., et al. (2017). Acute and chronic in vivo effects of exposure to nicotine and propylene glycol from an E-cigarette on mucociliary clearance in a murine model. *Inhalation Toxicology*, *29*(5), 197–205. https://doi.org/10.1080/08958378.2017.1336585.

Le Foll, B., Wiggins, M., & Goldberg, S. R. (2006). Nicotine pre-exposure does not potentiate the locomotor or rewarding effects of delta-9-tetrahydrocannabinol in rats. *Behavioral Pharmacology*, *17*, 195–199.

Lichtman, A. H., & Martin, B. R. (2005). Cannabinoid tolerance and dependence. *Handbook of Experimental Pharmacology*, *168*, 691–717.

Lopez-Quintero, C., de lo Cobos, J., Hasin, D. S., Okuda, S. W., Grant, B., & Blanco, C. (2011). Probability and predictors of transition from first use to dependence on nicotine, alcohol, cannabis, and cocaine: Results of the National Epidemiologic Survey on Alcohol and Related Conditions (NESARC). *Drug and Alcohol Dependence*, *115*, 120–130.

Lorenzetti, V., Lubman, D. I., Whittle, S., Solowij, N., & Yücel, M. (2010). Structural MRI findings in long-term cannabis users: What do we know? *Substance Use and Misuse*, *45*(11), 1787–1808. https://doi.org/10.3109/10826084.2010.482443.

Manwell, L. A., Ford, B., Matthews, B. A., Heipel, H., & Mallet, P. E. (2014). A vapourized Δ 9-tetrahydrocannabinol (Δ9-THC) delivery system part II: Comparison of behavioural effects of pulmonary versus parenteral cannabinoid exposure in rodents. *Journal of Pharmacological and Toxicological Methods*, *70*(1), 112–119. https://doi.org/10.1016/j.vascn.2014.06.004.

Martellotta, M. C., Cossu, G., Fattore, L., Gessa, G. L., & Fratta, W. (1998). Self-administration of the cannabinoid receptor agonist WIN 55,212-2 in drug-naive mice. *Neuroscience*, *85*(2), 327–330.

Mcpartland, J. M. (2004). Phylogenomic and chemotaxonomic analysis of the endocannabinoid system. *Brain Research Reviews*, *45*, 18–29. https://doi.org/10.1016/j.brainresrev.2003.11.005.

Mcpartland, J. M., Agraval, J., Gleeson, D., Heasman, K., & Glass, M. (2006). Cannabinoid receptors in invertebrates. *Journal of Evolutionary Biology*, *19*(2001), 366–373. https://doi.org/10.1111/j.1420-9101.2005.01028.x.

Morales, M., Schatz, K., Anderson, R., Spear, L., & Varlinskaya, E. (2015). Conditioned taste aversion to ethanol in a social context: Impact of age and sex. *Behavioural Brain Research*, *261*, 323–327. https://doi.org/10.1016/j.bbr.2013.12.048.Conditioned.

Moreno-Alcázar, A., Gonzalvo, B., Canales-Rodríguez, E. J., Blanco, L., Bachiller, D., Romaguera, A., et al. (2018). Larger gray matter volume in the basal ganglia of heavy cannabis users detected by voxel-based morphometry and subcortical volumetric analysis. *Frontiers in Psychiatry*, *9*(175), 1–9. https://doi.org/10.3389/fpsyt.2018.00175.

Navarro, M., Chowen, C. A. J., Arco, I., Villanúa, M. A., Martin, Y., Roberts, A. J., et al. (1998). CB1 cannabinoid receptor opiate withdrawal in morphine-dependent rats. *Neuropharmacology*, *9*(15), 3397–3402.

Neisewander, J. L., Peartree, N. A., & Pentkowski, N. S. (2012). Emotional valence and context of social influences on drug abuse-related behavior in animal models of social stress and prosocial interaction. *Psychopharmacology*, *224*, 33–56. https://doi.org/10.1007/s00213-012-2853-3.

NIDA. 2014. http://www.drugabuse.gov/publications/drugfacts/marijuana (NIDA site; accessed 4/5/2014).

O'Shea, M., Mcgregor, I. S., & Mallet, P. E. (2006). Repeated cannabinoid exposure during perinatal, adolescent or early adult ages produces similar long-lasting deficits in object recognition in rats. *Journal of Psychopharmacology*, *20*(5), 611–621.

Panksepp, J. B., & Lahvis, G. P. (2007). Social reward among juvenile mice. *Genes, Brain and Behavior*, *6*, 661–671. https://doi.org/10.1111/j.1601-183X.2006.00295.x.

Panlilio, L. V., Justinova, Z., & Goldberg, S. R. (2010). Animal models of cannabinoid reward. *British Journal of Pharmacology, 160*, 499–510. https://doi.org/10.1111/j.1476-5381.2010.00775.x.

Peartree, N. A., Hatch, K. N., Goenaga, J. G., Dado, N. R., Molla, H., Dufwenberg, M. A., et al. (2017). Social context has differential effects on acquisition of nicotine self-administration in male and female rats. *Psychopharmacology, 234*, 1815–1828. https://doi.org/10.1007/s00213-017-4590-0.

Peartree, N. A., Hood, L. E., Thiel, K. J., Sanabria, F., Pentkowski, N. S., Chandler, K. N., et al. (2012). Limited physical contact through a mesh barrier is sufficient for social reward-conditioned place preference in adolescent male rats. *Physiology & Behavior, 105*(3), 749–756. https://doi.org/10.1016/j.physbeh.2011.10.001.

Porath, A. J., & Fried, P. A. (2005). Effects of prenatal cigarette and marijuana exposure on drug use among offspring. *Neurotoxicology and Teratology, 27*, 267–277. https://doi.org/10.1016/j.ntt.2004.12.003.

Prud'homme, M., Cata, R., & Jutras-Aswad, D. (2015). Cannabidiol as an intervention for addictive behaviors: A systematic review of the evidence. *Substance Abuse: Research and Treatment, 9*, 33–38. https://doi.org/10.4137/SART.S25081.

Rubio, P., Rodriguez de Fonseca, F., Martin-Calderon, J. L., Del Arco, I., Bartolome, S., Villanua, M. A., et al. (1998). Maternal exposure to low doses of delta9-tetrahydrocannabinol facilitates morphine-induced place conditioning in adult male offspring. *Pharmacology, Biochemistry and Behavior, 61*(3), 229–238. http://doi.org/S0091305798000999. [pii].

Ryberg, E., Hjorth, S., Hermansson, N., Leonova, J., Elebring, T., Larsson, N., et al. (2007). The orphan receptor GPR55 is a novel cannabinoid receptor. *British Journal of Pharmacology, 152*, 1092–1101. https://doi.org/10.1038/sj.bjp.0707460.

Sandi, C., & Haller, J. (2015). Stress and the social brain: Behavioural effects and neurobiological mechanisms. *Nature Reviews Neuroscience, 16*(5), 290–304. https://doi.org/10.1038/nrn3918.

Schneider, M., Schömig, E., & Leweke, F. M. (2008). Acute and chronic cannabinoid treatment differentially affects recognition memory and social behavior in pubertal and adult rats. *Addiction Biology, 13*(3–4), 345–357. https://doi.org/10.1111/j.1369-1600.2008.00117.x.

Singh, M. E., McGregor, I. S., & Mallet, P. E. (2006). Perinatal exposure to Δ9-tetrahydrocannabinol alters heroin-induced place conditioning and Fos-immunoreactivity. *Neuropsychopharmacology, 31*(1), 58–69. https://doi.org/10.1038/sj.npp.1300770.

Shafer, J., & Brown, S. A. (1991). Marijuana and cocaine effect expectancies and drug use patterns. *Journal of Consulting and Clinical Psychology, 59*, 558–565.

Socías, M. E., Kerr, T., Wood, E., Dong, H., Lake, S., Hayashi, K., et al. (2018). Intentional cannabis use to reduce crack cocaine use in a Canadian setting: A longitudinal analysis. *Addictive Behaviors, 72*, 138–143. https://doi.org/10.1016/j.addbeh.2017.04.006.

Spano, M., Fadda, P., Frau, R., Fattore, L., & Fratta, W. (2010). Cannabinoid self-administration attenuates PCP-induced schizophrenia-like symptoms in adult rats. *European Neuropsychopharmacology, 20*(1), 25–36. https://doi.org/10.1016/j.euroneuro.2009.09.004.

Tanda, G., Munzar, P., & Goldberg, S. R. (2000). Self-administration behavior is maintained by the psychoactive ingredient of marijuana in squirrel monkeys. *Nature Neuroscience, 3*(11), 1073–1074. https://doi.org/10.1038/80577.

Thiel, K. J., Sanabria, F., & Neisewander, J. L. (2009). Synergistic interaction between nicotine and social rewards in adolescent male rats. *Psychopharmacology, 204*, 391–402. https://doi.org/10.1007/s00213-009-1470-2.

Trecki, J., Gerona, R. R., & Schwartz, M. D. (2015). Synthetic cannabinoid–related illnesses and deaths. *New England Journal of Medicine, 373*(2), 103–107. https://doi.org/10.1056/NEJMp1505328.

Trezza, V., Cuomo, V., & Vanderschuren, L. J. M. J. (2008). Cannabis and the developing brain: Insights from behavior. *European Journal of Pharmacology*, *585*, 441–452. https://doi.org/10.1016/j.ejphar.2008.01.058.

Trezza, V., Damsteegt, R., Manduca, A., Petrosino, S., Van Kerkhof, L., Pasteramp, R. J., et al. (2012). Endocannabinoids in amygdala and nucleus accumbens mediate social play reward in adolescent rats. *The Journal of Neuroscience*, *32*(43), 14899–14908. https://doi.org/10.1523/JNEUROSCI.0114-12.2012.Endocannabinoids.

Trezza, V., & Vanderschuren, L. J. M. J. (2008). Bidirectional cannabinoid modulation of social behavior in adolescent rats. *Psychopharmacology*, *197*(2), 217–227.

Tsang, C. C., & Giudice, M. G. (2016). Nabilone for the management of pain. *Pharmacotherapy*, *6*(3), 273–286. https://doi.org/10.1002/phar.1709.

Tzschentke, T. M. (2007). Measuring reward with the conditioned place preference (CPP) paradigm: Update of the last decade. *Addiction Biology*, *12*, 227–462. https://doi.org/10.1111/j.1369-1600.2007.00070.x.

Valjent, E., & Maldonado, R. (2000). A behavioural model to reveal place preference to Δ 9-tetrahydrocannabinol in mice. *Psychopharmacology*, *147*, 436–438.

Valjent, E., Mitchell, J. M., Besson, M., Caboche, J., & Maldonado, R. (2002). Behavioural and biochemical evidence for interactions between D 9-tetrahydrocannabinol and nicotine. *British Journal of Pharmacology*, *135*, 564–578.

Valverde, O., Noble, F., Beslot, F., Daugé, V., Fournié-Zaluski, M. C., & Roques, B. P. (2001). Δ9-Tetrahydrocannabinol releases and facilitates the effects of endogenous enkephalins: Reduction in morphine withdrawal syndrome without change in rewarding effect. *European Journal of Neuroscience*, *13*(9), 1816–1824. https://doi.org/10.1046/j.0953-816X.2001.01558.x.

van Ree, J. M., Slangen, J. L., & de Wied, D. (1978). Intravenous self-administration of drugs in rats. *Journal of Pharmacology and Experimental Therapeutics*, *204*(3), 547–557. Retrieved from http://jpet.aspetjournals.org/content/204/3/547.abstract.

Vann, R. E., Gamage, T. F., Warner, J. A., Marshall, E. M., Nathan, L., Martin, B. R., et al. (2008). Divergent effects of cannabidiol on the discriminative stimulus and place conditioning effects of Δ9-tetrahydrocannabinol. *Drug and Alcohol Dependence*, *94*, 191–198.

Volkow, N. D., & Morales, M. (2015). The brain on drugs: From reward to addiction. *Cell*, *162*(4), 712–725. https://doi.org/10.1016/j.cell.2015.07.046.

Wakeford, A. G. P., Flax, S. M., Pomfrey, R. L., & Riley, A. L. (2016). Adolescent delta-9-tetrahydrocannabinol (THC) exposure fails to affect THC-induced place and taste conditioning in adult male rats. *Pharmacology, Biochemistry and Behavior*, *140*, 75–81. https://doi.org/10.1016/j.pbb.2015.11.004.

Wakeford, A. G. P., & Riley, A. L. (2014). Conditioned taste avoidance induced by Δ9-tetrahydrocannabinol in the Fischer (F344) and Lewis (LEW) rat strains. *Pharmacology, Biochemistry and Behavior*, *116*, 39–44. https://doi.org/10.1016/j.pbb.2013.11.005.

Wakeford, A. G. P., Wetzell, B. B., Pomfrey, R. L., Clasen, M. M., Taylor, W. W., Hempel, B. J., et al. (2017). The effects of cannabidiol (CBD) on Δ9-tetrahydrocannabinol (THC) self-administration in male and female Long-Evans rats. *Experimental and Clinical Psychopharmacology*, *25*(4), 242–248.

World Health Organization. (2016). *The health and social effects of nonmedical cannabis use WHO document production services*. Geneva, Switzerland.

Zamberletti, E., Gabaglio, M., & Parolaro, D. (2017). The endocannabinoid system and autism spectrum disorders: Insights from animal models. *International Journal of Molecular Sciences*, *18*, 1–14. https://doi.org/10.3390/ijms18091916.

Zanettini, C. (2011). Effects of endocannabinoid system modulation on cognitive and emotional behavior. *Frontiers in Behavioral Neuroscience*, *5*, 1–21. https://doi.org/10.3389/fnbeh.2011.00057.

Zangen, A., Solinas, M., Ikemoto, S., Goldberg, S. R., & Wise, R. A. (2006). Two brain sites for cannabinoid reward. *Journal of Neuroscience, 26*(18), 4901–4907. https://doi.org/10.1523/JNEUROSCI.3554-05.2006.

Zawilska, J. B., & Wojcieszak, J. (2014). Spice/K2 drugs-more than innocent substitutes for marijuana. *International Journal of Neuropsychopharmacology*, 17(3), 509–525. https://doi.org/10.1017/S1461145713001247.

Zou, S., & Kumar, U. (2018). Cannabinoid receptors and the endocannabinoid system: Signaling and function in the central nervous system. *International Journal of Molecular Sciences, 19*(3), 1–23. https://doi.org/10.3390/ijms19030833.

CHAPTER EIGHT

Oxytocin and Rodent Models of Addiction

Kah-Chung Leong*, Stewart Cox*, Courtney King*,†, Howard Becker*,†, Carmela M. Reichel*,[1]

*Department of Neurosciences, Medical University of South Carolina, Charleston, SC, United States
†Department of Psychiatry, Medical University of South Carolina, Charleston, SC, United States
[1]Corresponding author: e-mail address: reichel@musc.edu

Contents

Abstract

Interest for the use of oxytocin as a treatment for addiction began over 40 years ago. Better known for its roles in parturition, lactation and pair bonding, oxytocin also has anxiolytic properties, reduces immune and inflammatory responses, and has a role in learning and memory. In this chapter, oxytocin effects on addiction processes are described by highlighting research findings that have used oxytocin within current

International Review of Neurobiology, Volume 140
ISSN 0074-7742
https://doi.org/10.1016/bs.irn.2018.07.007

preclinical animal models of addiction, relapse, or craving. First, we provide a brief background of the endogenous oxytocin system followed by descriptions of the behavioral models used to study addiction, including models of drug taking and seeking. Then we review recent preclinical studies that have used oxytocin as a therapeutic intervention throughout multiple stages of the addiction cycle from a behavioral and neurobiological perspective. These models encompass the entire range of the addiction cycle including acquisition and maintenance of drug taking, withdrawal and craving during periods of drug abstinence, and ultimately relapse. We then posit several theories about how oxytocin interacts with both drug and social reward, as well as presenting a mechanistic account of how specific oxytocin receptor localization may contribute to oxytocin's efficacy as an addiction therapeutic.

1. INTRODUCTION

Over 40 years ago reports emerged that neurohypophyseal peptides interact with memory (De Wied, 1971), alcohol tolerance (Hoffman, Ritzmann, Walter, & Tabakoff, 1978), and opiate addiction (Van Ree & De Wied, 1977) leading to a field of investigation implicating use of the hypothalamic neuropeptide, oxytocin, in opioid, psychostimulant, and alcohol addiction (Kovacs, Sarnyai, & Szabo, 1998; Sarnyai & Kovacs, 2014). Interest in oxytocin and addiction waned for a period while research focused on other features and functions of oxytocin, when it became well known for its role in parturition, lactation and pair bonding (Lee, Macbeth, Pagani, & Young, 2009; Waldherr & Neumann, 2007). The seminal work of Carter, Insel, and Young established that oxytocin takes center stage in areas of social attachment (Carter, Williams, Witt, & Insel, 1992), social memory (Young, Lim, Gingrich, & Insel, 2001), and generating trust in humans (Kosfeld, Heinrichs, Zak, Fischbacher, & Fehr, 2005). In addition to increased trust, oxytocin acts as an anxiolytic, reduces immune and inflammatory responses, and has a role in learning, memory and pain reduction (reviewed in Tops, Koole, H, & Buisman-Pijlman, 2014). Coming full circle, interest in the use of oxytocin as a therapeutic for addiction has risen substantially over the past decade.

In this chapter, we focus on oxytocin's impact on drug taking, seeking, and relapse by highlighting research findings that have used oxytocin within current models of addiction, relapse, or craving. First, we provide a brief background of the endogenous oxytocin system followed by descriptions of the behavioral models used to study addiction, including models of drug taking and seeking. Then we review recent preclinical studies that use oxytocin as a therapeutic intervention throughout multiple stages of the addiction cycle from a behavioral and neurobiological perspective. We then posit several theories about how oxytocin interacts with both drug and social reward, as well as presenting a mechanistic account of how specific oxytocin receptor localization may contribute to oxytocin's efficacy as an addiction therapeutic.

1.1 The Endogenous Central Oxytocin System

The central and peripheral oxytocin systems have been recently reviewed (Bowen & Neumann, 2017; Buisman-Pijlman et al., 2014; Knobloch & Grinevich, 2014), so we will only briefly describe the central system here. The neuropeptide oxytocin, produced by the precursor protein oxytocin/neurophysin I prepropeptide, consists of 9 amino acids. Synthesis within the central nervous system takes place in the paraventricular nucleus (PVN) and the supraoptic nucleus (SON) of the hypothalamus. The majority of oxytocin is transported to the neurohypophysis and thereby released into the bloodstream for peripheral distribution (Leng & Ludwig, 2016; Quirin, Kuhl, & Dusing, 2011). In the periphery, oxytocin release is stimulated by birth, suckling, and sexual stimulation in mammals as well as exposure to physical or psychosocial stress.

Hypothalamic oxytocin projection neurons target the forebrain, limbic and reward systems, including areas of the brain involved in both drug and social reward (Knobloch & Grinevich, 2014). Cells originating in the PVN deliver oxytocin to brain regions such as the amygdala, bed nucleus of the stria terminalis (BNST), lateral septum, hippocampus and nucleus accumbens core (NAcc) (Stoop, 2012). Within these projection areas oxytocin has specific effects on behavior. For example, within the central amygdala oxytocin reduces anxiety and inhibits stress responding (Neumann & Landgraf, 2012; Neumann & Slattery, 2016). Oxytocin within the lateral septum is involved in prosocial behavior including social preference (Lukas, Toth, Veenema, & Neumann, 2013). The impact of central oxytocin on addictive behaviors in rodent models will be discussed in detail in Section 3.

To date, there has been only one oxytocin receptor (OXTR) identified (Busnelli & Chini, 2017). These receptors are abundant throughout the peripheral and central nervous systems and are expressed in a sex dependent and species-specific manner (Dumais & Veenema, 2016; Lee et al., 2009). In the periphery, receptor expression is found in mammary glands, uterus, gastrointestinal tract, heart, and vascular endothelium layer (Gimpl & Fahrenholz, 2001). Centrally, oxytocin receptors are expressed in brain areas that regulate mood, and social and addictive behaviors such as the prefrontal cortex (PFC), ventral tegmental area (VTA), nucleus accumbens (NAc), hippocampus, hypothalamus, ventral pallidum, medial preoptic area, olfactory bulbs, (central) amygdala and brain stem (Gimpl & Fahrenholz, 2001; Sarnyai, 2011). Of these regions, males have increased oxytocin receptor expression mRNA in the NAc, hippocampus, medial preoptic area, and the BNST (an area projecting directly to the amygdala) relative to females (Dumais & Veenema, 2016). However, sex differences are not necessarily consistent across species as female mice show increased oxytocin receptor expression mRNA in the hippocampus and very few sex differences have been observed in other rodent species and humans (reviewed in Dumais & Veenema, 2016).

The OXTR is Gq-coupled and its activation results in changes of activity in phospholipase C and protein kinase C. The oxytocin receptor couples to two different G proteins, Gq/11 at its proximal portion of the C-terminus and to Gi/o depending on brain area and oxytocin concentration (reviewed in Busnelli & Chini, 2017). Activation of these two different signaling pathways may be one explanation for oxytocin's diverse functions (Chini & Fanelli, 2000; Devost, Wrzal, & Zingg, 2008; Gimpl, Reitz, Brauer, & Trossen, 2008; Rimoldi et al., 2003). Supra-physiological levels of oxytocin also bind receptors for vasopressin, a closely related peptide, that also has effects on social mediated behaviors (Bowen & McGregor, 2014; Sala et al., 2011). Lemaire et al. (2002) proposed that the oxytocin and vasopressin V1a receptors are the predominant receptor subtypes in rat brain and spinal cord with specific distribution patterns. However, it is the vasopressin V1b2 receptor that is co-localized with oxytocin receptors in forebrain areas, with oxytocin receptor gene expression being much more robust than that of V1b2 (Vaccari, Lolait, & Ostrowski, 1998).

Characterization of cell-type specific oxytocin receptor expression has been impeded by the lack of specific antibodies recognizing the oxytocin receptor with levels of specificity to rule out vasopressin binding sites. This absence has limited the use of immunohistochemistry to identify the neurochemical and projection phenotype of oxytocin sensitive cells in the brain (see Mouillac, Manning, & Durroux, 2008). Recently, through the use of genetically modified mice, progress has been made to characterize neuronal phenotypes that co-localize with oxytocin receptors in specific brain areas integral to addiction circuitry (Dolen, Darvishzadeh, Huang, & Malenka,

2013; Nakajima, Gorlich, & Heintz, 2014; Peris et al., 2017; Tan et al., 2017). Magnocellular oxytocin neurons in the PVN express high levels of corticotrophin-releasing factor (CRF) type 2 receptors and these CRF receptor positive neurons express oxytocin receptor mRNA as do CRF receptor positive neurons in the bed nucleus of the striatal terminalis (Dabrowska et al., 2011). Oxytocin receptors have been identified on cortical somatostatin-containing interneurons (Li, Nakajima, Ibanez-Tallon, & Heintz, 2016; Nakajima et al., 2014), cortical glutamatergic and GABAergic neurons (Tan et al., 2017), parvalbumin positive interneurons and astrocytes in the NAcc (Dolen et al., 2013), and on glutamate, GABA and dopamine producing neurons in the VTA (Peris et al., 2017). These three brain areas are crucial to the circuitry regulating drug reward and reinforcement as well as relapse to drug use, and oxytocin in each of these areas may uniquely impact drug seeking.

1.2 Oxytocin and the Blood–Brain Barrier

A current research question being investigated is whether oxytocin administration through systemic delivery methods crosses the blood–brain barrier. Recent evidence suggests that it does to a limited extent and mechanisms underlying such penetration into brain are just recently coming into focus (Veening & Olivier, 2013). Central oxytocin synthesized in the SON is secreted into the bloodstream through neurohemal contact in response to reproductive behaviors, social stimuli, and stress (Bowen & Neumann, 2017). However, only 1–2% of oxytocin synthesized and released by peripheral organs (uterine epithelium, ovary, testis, vascular endothelium cells and heart) (Nishimori et al., 2008) and or by the posterior pituitary cross the blood–brain barrier (Ermisch et al., 1985; Landgraf, Ermisch, & Hess, 1979). Veening and Olivier (2013) discuss possible other ways in which oxytocin can pass from peripheral circulation to the brain and back. One mechanism may be through transport via a saturable carrier (McEwen, 2004). Alternatively, oxytocin may be able to cross a damaged BBB more readily. For example, heavy alcohol/drug use may induce leakage of oxytocin from the blood to the cerebrospinal fluid (Kovacs et al., 1998; Nishimori et al., 2008). Stress, hypertension, or disease may also impact blood–brain barrier permeability (Churchland & Winkielman, 2012).

Administration of central oxytocin activates oxytocin-producing cell bodies in the paraventricular nucleus of the hypothalamus (PVN) (Carson, Cornish, Guastella, Hunt, & McGregor, 2010; Carson, Hunt, et al., 2010).

This activation results in local dendritic release of oxytocin and an ensuing positive-feedback effect promoting prolonged activation of the oxytocin system (Ludwig et al., 2002; Rossoni et al., 2008), which is presumed to cause brain wide oxytocin release, including areas associated with addiction. It has been suggested that only a relatively small amount of oxytocin is required to initiate such a feed-forward mechanism to propagate endogenous oxytocin production (Rossoni et al., 2008). Recently, Neumann, Maloumby, Beiderbeck, Lukas, and Landgraf (2013) used microdialysis procedures in rats to demonstrate that systemic (intraperitoneal, ip) administration of oxytocin caused rapid peak levels in brain dialysates 30 min post-injection, providing direct evidence that systemic oxytocin increases central oxytocin levels. Consistent with these results, intravenous oxytocin (5 IU/kg) increased oxytocin levels in cerebrospinal fluid in rhesus macaques, peaking at 15 min post-infusion and gradually returning to baseline by 120 min (Freeman et al., 2016). However, Lee et al. (2018) demonstrated that while oxytocin administered through intranasal and intravenous routes of administration increased oxytocin in cerebrospinal fluid, it did not activate a feed-forward mechanism to elevate endogenous oxytocin. Thus, the mechanism by which exogenous oxytocin delivered in the periphery activates oxytocin signaling in the brain remains unclear. Further studies are needed to address this issue as it is relevant to the potential for oxytocin to serve as a therapeutic for alcohol/drug addiction.

2. PRECLINICAL MODELS OF ADDICTION

The most commonly used behavioral tasks to study alcohol/drug reward and reinforcement include intravenous or oral self-administration involving operant conditioning procedures and place conditioning. Although these two measures have some overlap, the general consensus is that these procedures involve dissociable processes (Bardo & Bevins, 2000). Other models to study alcohol addiction involve free-choice drinking and binge-like drinking procedures. This section will describe effects of oxytocin treatment in studies employing these models as well as procedures commonly used to index social reward.

2.1 Intravenous (IV) Drug Self-administration

Drug self-administration models rely on the basic premises of operant conditioning—a response (e.g., lever press, keypeck, nose poke, etc.) results in an outcome (i.e., a drug infusion typically via an intravenous route of administration). In this model rats or mice are typically trained to press a lever, resulting in an IV infusion of drug into an indwelling jugular catheter. A drug infusion is often paired with the presentation of a stimulus (cue light or a tone), which through repeated pairing can become secondary reinforcers themselves. In addition to an "active" lever that results in drug delivery, a second "inactive" lever is typically available, where responses are recorded but do not result in any consequences. Responses on the "inactive" lever are typically considered a measure of non-differentiated responding and reflective of general (nonspecific) motor activation. Subjects can administer drugs on simple schedules of reinforcement including fixed ratio (FR) or variable ratio (VR) schedules, until stable responding is obtained. Alternatively, more complex schedules may be used, such as a progressive ratio (PR), in which the number of active lever presses required to obtain a reinforcer increases along a logarithm for subsequent reinforcers.

Once stable responding is achieved, animals undergo extinction or abstinence based on specific experimental parameters. The extinction–reinstatement model is considered to have high face and construct validity (Markou, Chiamulera, Geyer, Tricklebank, & Steckler, 2008) as drug-seeking behavior is re-established when the animals are re-exposed to the drug, drug-paired cues, or stress (Bossert, Marchant, Calu, & Shaham, 2013; Feltenstein & See, 2008; Yahyavi-Firouz-Abadi & See, 2009), all of which also induce drug relapse in humans (Childress et al., 1993; Sinha, 2001). Animals undergoing extinction are placed in the drug-paired environment in the absence of drug or drug-paired cues. Extinction results in decreased responding with repeated trials, as reinforcement no longer occurs following a response. Reinstatement occurs when rats resume responding on a previously drug-paired lever, though typically no drug is actually delivered. During drug-primed reinstatement rats typically receive a small dose of the self-administered drug prior to the session and typically results in a robust increase in drug seeking. Cue-induced reinstatement occurs when rats are re-exposed to the previously drug-paired cues. Finally, during stress-induced reinstatement, administration of a physiological stressor or anxiogenic drug, such as footshock or yohimbine, can induce reinstatement.

Adaptations to the above procedures have been incorporated to demonstrate time dependent "incubation" of drug craving without extinction procedures. In this model rats undergo self-administration training as described above. Following the maintenance phase, rats are subject to abstinence for approximately 1–8 weeks. During abstinence rats are placed in their home cages (Grimm, Hope, Wise, & Shaham, 2001), in an alternative environment (Hearing, Schochet, See, & McGinty, 2010), or handled on a daily basis (Moussawi et al., 2011; Reichel, Moussawi, Do, Kalivas, & See, 2011). Rats are never returned to the drug-taking environment during the abstinence period because "relapse" occurs when rats are placed back into the drug-paired context (Reichel & Bevins, 2009). The extent and degree of relapse varies as a function of the time since the last drug experience (Grimm et al., 2001; Tran-Nguyen et al., 1998), and generally increases over time, and thus has been coined "incubation" (Grimm et al., 2001) of drug craving.

2.2 Conditioned Place Preference

In contrast to the behavior-outcome schedules of reinforcement that maintain responding for the drug in self-administration experiments, the place conditioning task relies on the outcome of stimulus–stimulus associations that come to control behavior. In the place conditioning task, one environment or discrete cue (conditioned stimulus, CS) is paired with a biologically and motivationally rewarding stimulus (unconditioned stimulus, US; e.g., cocaine, novelty, water, conspecific), whereas another environment remains unpaired or paired with only a neutral stimulus such as saline administration. On the test day, the animal is given a choice between the two environments. In most place conditioning preparations, exteroceptive cues of the paired environment (CS) become associated with the rewarding aspects of the stimuli of interest (i.e., the US). Accordingly, the paired environment acquires an appetitive value, thus, eliciting approach behaviors (Bardo & Bevins, 2000; Panksepp, Nocjar, Burgdorf, Panksepp, & Huber, 2004; Reichel, Wilkinson, & Bevins, 2010). This approach behavior is expressed as an increase in the amount of time spent in the paired environment on the choice test day.

Compartment preferences motivated by associations with drug reward are also subject to extinction and reinstatement procedures (Mueller & Stewart, 2000). In this case, extinction procedures refer to re-exposure to the drug-associated chamber without the US (i.e., drug effects) (Rescorla, 2004). Accordingly, conditioned responding (i.e., time spent on the cocaine-paired side) decreases on subsequent test days. After conditioned responding is no longer evident, re-exposure US can reinstate conditioned responding (Rescorla, 2004; Shaham, Shalev, Lu, de Wit, & Stewart, 2003). With regards to place conditioning, cocaine-associated cues maintain effectiveness over a considerable length of time (Mueller & Stewart, 2000; Reichel & Bevins, 2008). Notably, this extinguished preference can be reinstated by a priming dose of the drug (Mueller & Stewart, 2000; Reichel & Bevins, 2008).

2.3 Drinking in the Dark

A number of preclinical models have been employed to study the neurobiological consequences of alcohol drinking, however, in many models of voluntary consumption rodents typically consume low levels of alcohol that do not generate blood ethanol concentrations (BECs) considered to be pharmacologically meaningful or the ability to induce dependence. The drinking in the dark (DID) procedure promotes high levels of ethanol drinking in ethanol-preferring strains of mice by taking advantage of time in the animals dark cycle in which activity and nocturnal ingestive behaviors are high (Thiele & Navarro, 2014). A "binge" session, as defined by the National Institutes on Alcohol Abuse and Alcoholism (NIAAA), as a pattern of drinking that produces BECs exceeding 0.08% (80 mg/dL). Using this limited access procedure, mice typically consume enough ethanol to reach BECs >100 mg/dL and resulting in behavioral evidence of intoxication in a short period of time (Sprow & Thiele, 2012).

In the most common variation of the DID procedure, first implemented by Rhodes et al., C57BL/6J mice are presented with a single bottle containing 20% (v/v) ethanol in place of the home cage water bottle for 2 h, 3 h after the beginning of the dark cycle, for 3 consecutive days. On the 4th day, the drinking session is extended to 4 h during which mice exhibit binge-like behavior, consuming significant and physiologically relevant amounts of ethanol (Rhodes, Best, Belknap, Finn, & Crabbe, 2005; Thiele, Crabbe, & Boehm, 2014). Several studies have characterized DID and have shown that BECs of <100 mg/dL produce significant motor incoordination (Moore et al., 2007; Rhodes et al., 2007), tolerance to this effect after repeated binge-like episodes (Linsenbardt, Moore, Griffin, Gigante, & Boehm, 2011) and increases in subsequent voluntary ethanol drinking (Cox, Olney, et al., 2013; Cox, Young, See, & Reichel, 2013). The DID model does not require training prior exposure to ethanol or inclusion of sweet compounds to motivate high levels of ethanol intake. Further, in contrast to other models of binge-like drinking, DID does not require ethanol to be administered by the experimenter as by gavage, injection or ethanol vapor exposure (Becker & Lopez, 2004), or by fluid deprivation or inclusion of ethanol into an animals source of nutrients. Such procedures may cause stress to the animal or other confounding variables, calling into the question the motivation to drink ethanol (Thiele & Navarro, 2014). Thus, DID provides a simple and effective high throughput tool for characterization of potential pharmacotherapies and may provide a useful model for studying binge patterns of consumption and the transition to ethanol dependence.

2.4 Social Interaction and Conditioned Social Reward

Environments that have been associated with social interaction can also engender place preferences. In general, conditioned social reward is evaluated by placing the animal into one compartment of the place conditioning apparatus with their social partner (US), and during the other half of the sessions the animals are placed alone in the alternate compartment (Douglas, Varlinskaya, & Spear, 2004; Thiel, Okun, & Neisewander, 2008; Trezza, Damsteegt, & Vanderschuren, 2009). On test, animals are given a choice to spend time between the two compartments. More time in the side that was associated with social interaction is interpreted as conditioned social reward. Interestingly, conditioned social reward can be extinguished and reinstated by a single re-exposure to the social partner in the social-paired compartment (Trezza et al., 2009).

In mice, cocaine (15 mg/kg) and social interaction (15 min) engender similar preferences for the paired environments (Kummer et al., 2014; Panksepp & Lahvis, 2007) suggesting similar conditioned rewarding value of the US. In a variation of a place conditioning procedure developed by Reichel and Bevins (2008, 2010) competition between conditioned rewards can define the relative competitive value of each reward. Conditioned social reward can directly compete with cocaine conditioned reward when one side was paired with social interaction and the other with cocaine (Fritz et al., 2011). Furthermore, after establishing a place preference to cocaine, extinction of the cocaine CS with social reward conditioning in the alternate side inhibits cocaine-primed reinstatement of the cocaine CR (Fritz et al., 2011).

3. OXYTOCIN EFFECTS ON PRECLINICAL MODELS OF ADDICTION

The previous section described some of the most common behavioral assays employed to study alcohol and drug addiction. Within each of these models, oxytocin decreases the tendency to take drugs and reduces the propensity relapse. This next section will detail the oxytocin effects on methamphetamine, cocaine, heroin and alcohol intake and seeking within the models previously described.

3.1 Methamphetamine

Methamphetamine (*N*-methyl-*O*-phenylisopropylamine; meth) is a highly addictive illicit stimulant. According to the United Nations Office on Drugs and Crime (2017) meth and other amphetamine-like stimulants are second only to opioids in the burden of disease risk. And, within the United States, there is a rising trend of meth-related visits at publicly funded treatment facilities (Gonzales, Mooney, & Rawson, 2010) and emergency departments (Dobkin & Nicosia, 2009). The physiological and cognitive consequences of meth use are detailed in Barr et al. (2006), in which the authors highlight the high abuse potential of meth and its associated impairments in working memory, attention, and executive function. Deficits in long-term memory have also been reported following more chronic meth use (reviewed in Bernheim, See, & Reichel, 2016).

Physiologically, the short-term effects of meth use include enhanced energy, alertness, elation, and elevated heart rate/blood pressure (Rawson, Gonzales, & Brethen, 2002). Meth facilitates rapid release of norepinephrine, serotonin, and dopamine at synapses (Rothman et al., 2001). Meth reverses mechanisms that regulate homeostatic levels of these monoamines leading to a sustained increase in synaptic levels through its action as an indirect agonist (Elkashef et al., 2008). Chronic meth users often have various cardiovascular impairments such as arrhythmias, hypertension, cardiomyopathy, and acute myocardial infarction (Kaye, McKetin, Duflou, & Darke, 2007; Turnipseed, Richards, Kirk, Diercks, & Amsterdam, 2003; Westover, Nakonezny, & Haley, 2008). It is not uncommon for users to suffer from accelerated coronary artery disease and cardiac hypertrophy. Meth users also present with severe mental health problems, skin infections, and dental pathology (Hendrickson, Cloutier, & McConnell, 2008). The abuse potential of meth, taken together with the high abuse rate and subsequent physical and mental health consequences highlights the need for effective pharmacotherapies. Preclinical data suggest that oxytocin is a viable therapeutic target for the treatment of meth addiction at various stages of the addiction cycle.

3.1.1 Oxytocin Effects on Acquisition and Maintenance of Meth Addiction

Significant progress has been made examining the therapeutic effect of oxytocin on the acquisition of meth-seeking behavior. For example, Qi et al. (2009) demonstrated that intracerebroventricular (icv) infusions of oxytocin (2.5 μg) given 30 min before meth impaired acquisition of a place preference in mice. Here the authors infused oxytocin during training in a conditioned place preference (CPP) task and found that oxytocin-injected animals showed less preference for the meth-associated chamber relative to controls. However, when oxytocin was infused after training, prior to testing, there was no effect on meth place preference. Combined this data shows that oxytocin impacts the induction but not the expression of meth-induced CPP (Qi et al., 2009). Consistently, systemic oxytocin administration (0.3 and 1 mg/kg; ip) attenuated active lever presses and meth infusions in self-administering male rats (Carson, Cornish, et al., 2010). However, oxytocin did not block acquisition or maintenance of self-administration behaviors, as all rats self-administered meth and responded on PR schedule of reinforcement. Plasma oxytocin concentrations were also increased 10 days following the last administration of oxytocin, providing evidence that systemic oxytocin administration produces an up-regulation of the endogenous oxytocin system (Carson, Guastella, Taylor, & McGregor, 2013).

3.1.2 Oxytocin Effects on Motivation to Seek Meth

The effect of oxytocin on motivation to seek meth is dependent on experimental procedures, but overall studies indicate that oxytocin decreases motivation to take meth in male and female rats. Our initial report found that peripheral oxytocin administration (1 mg/kg; ip) decreased motivation to seek meth only in females as indicated by reduction of active lever presses, break point (i.e., number of lever presses before reward is obtained), and infusions during a PR test (Cox, Young, et al., 2013). Baseline levels of motivation for meth differed significantly between males and females, with female rats showing greater levels of active lever presses and breakpoints. This is consistent with previous studies showing that females typically display greater motivated meth taking relative to males (Roth & Carroll, 2004). Although oxytocin did not impact male responding on the PR tests, a similar study with males found that oxytocin (1 mg/kg; ip) decreased motivated responding for meth as indicated by lower break points (Carson, Cornish, et al., 2010; Carson, Hunt, et al., 2010). These contrasting results can be attributed to methodological differences. In Cox, Young, et al. (2013), rats were maintained on an FR 5 schedule of reinforcement prior to PR testing. In Carson, Cornish, et al. (2010) and Carson, Hunt, et al. (2010), rats were maintained along a PR schedule of reinforcement and received escalating doses of oxytocin before testing over 5 consecutive days. Consecutive administration of oxytocin may have resulted in increased sensitivity to oxytocin in male rats leading to a greater effect on motivated meth seeking.

Perhaps shedding some light on oxytocin's impact on motivated meth taking comes from a follow up study in our laboratory using a newly developed behavioral economics (BE) paradigm (Cox et al., 2017). Using this task, comprised of a within-session BE procedure (Bentzley, Jhou, & Aston-Jones, 2014), we determined that greater motivation to seek meth was predictive of an increased likelihood to reinstate meth-seeking behavior in both males and females. Furthermore, oxytocin (1 mg/kg; ip) decreased motivation for meth in both male and female rats as indicated by an increase in alpha score (the task-specific measure of motivation). It is important to note that although oxytocin was effectively reduced motivation to seek meth, drug consumption at low effort (consumption when cost to obtain reward requires little to no effort, i.e., "hedonic set point") was not affected. The effects were centrally-driven as icv and intra-NAc infusions of an oxytocin antagonist blocked the attenuating effect of peripheral oxytocin. In general, studies assessing oxytocin's effects on motivation to take meth show decreases in responding for both males and females, which may subsequently affect the likelihood for relapse.

3.1.3 Oxytocin Effects on Extinction of Meth Seeking

Few studies have looked at the direct effect of oxytocin on the extinction of meth-seeking behaviors. In a CPP paradigm, multiple doses of central oxytocin infusions (0.1, 0.5, and 2.5 μg; icv) enhanced the extinction of meth-induced conditioned place preference (Qi et al., 2009). Mice went through CPP and were subsequently extinguished after 10 days. During extinction, mice were centrally infused with oxytocin before chamber placement in the absence of meth. Extinction criteria was set as a lack of significant difference between exploration in either the previously meth-paired chamber and control chamber. Oxytocin facilitated extinction indexed by less time spent in the meth associated side by the 7th day of extinction training. While no studies to our knowledge have investigated the effect of oxytocin on extinction following meth self-administration, Carson, Cornish, et al. (2010) and Carson, Hunt, et al. (2010) report that multiple injections of oxytocin during the self-administration period do not impact lever responding during extinction.

3.1.4 Oxytocin Effects on Reinstated Meth-Seeking Behavior

A number of preclinical studies have shown that oxytocin is a viable therapeutic target for the treatment of reinstated meth seeing in response to conditioned cues, a drug or a stress prime. Meth-primed reinstatement is reduced by systemic oxytocin (1 mg/kg; ip) in both male and female rats (Carson, Cornish, et al., 2010; Carson, Hunt, et al., 2010; Cox, Olney, et al., 2013; Cox, Young, et al., 2013). When directly compared, females in general respond to a greater extent to a priming injection of meth, but oxytocin is effective in both sexes (Cox, Young, et al., 2013). Repeated treatment in adolescent females also has an inhibitory effect on meth-primed reinstatement during adulthood (Hicks, Cornish, Baracz, Suraev, & McGregor, 2016). Oxytocin directly applied to the NAcc blocked meth-induced CPP (Baracz et al., 2012) and attenuated meth-primed reinstatement (Baracz, Everett, & Cornish, 2015; Baracz, Everett, McGregor, & Cornish, 2016). Co-administration of an oxytocin receptor antagonist into the NAcc only partially blocked the attenuating effect of intra-NAc oxytocin (Baracz et al., 2016; Everett, McGregor, Baracz, & Cornish, 2018) in contrast to a V1a antagonist, which sufficiently blocked the attenuating effect of systemic and intra-NAc oxytocin on meth-primed reinstatement (Everett et al., 2018). In regard to meth-primed reinstatement, the ability of oxytocin to reduce responding may be through vasopressin receptors driving the therapeutic effect. In contrast, NAcc oxytocin receptor antagonism blocked the attenuating effect of systemic oxytocin on motivation to seek meth and in response to meth associated cues (Cox et al., 2017). Taken together, it appears that oxytocin receptors in the NAcc may have a different roll depending on the phase of addiction pathology or the events that precipitate relapse.

In regard to cued reinstatement of meth seeking, oxytocin decreases responding in males and females (Bernheim, Leong, Berini, & Reichel, 2017; Cox et al., 2017; Cox, Young, et al., 2013). This decrease may involve additional behavioral parameters because in one study (Cox, Young, et al., 2013), rats were reinstated on an FR5 schedule of reinstatement and responding was greater female rats. When rats were reinstated on an FR1, both sexes robustly reinstated on the active lever and oxytocin decreased this reinstatement (Bernheim et al., 2017; Cox et al., 2017). Recently, we have proposed that the ability of oxytocin to reduce cued reinstatement is through interactions with the glutamatergic system in the NAcc (discussed in Section 4.1).

Consistent with other reinstatement modalities, oxytocin is also equally effective in treating stress-induced meth-reinstatement. A pharmacological stressor yohimbine, which acts an alpha-2-adrenoceptor antagonist, facilitated meth seeking after extinction to a greater degree in females relative to males (Cox, Young, et al., 2013). Systemic oxytocin administered before yohimbine prevented reinstatement (Cox, Young, et al., 2013). Oxytocin also decreased reinstatement in response to an ethologically-relevant predator odor (trimethylthiazoline; TMT) (Ferland, Reichel, & McGinty, 2016). Male rats underwent exposure to TMT (or a neutral odor) in locomotor chambers for 5 days before self-administration training. During testing rats were re-exposed to TMT in the testing chamber, and rats reinstated meth seeking. An acute injection of oxytocin given before reinstatement testing was sufficient to attenuate meth seeking in both conditions. As well, in a separate experiment, repeated oxytocin doses following TMT pre-exposure but prior to self-administration training blocked TMT-induced reinstatement, suggesting a protective effect of oxytocin administration on future meth-seeking behavior (Ferland et al., 2016; Hicks et al., 2016). Centrally administered oxytocin (0.1, 0.5, and 2.5 μg; icv) reduced the effect of restraint stress on reinstatement of meth CPP but did not block meth-primed CPP (Qi et al., 2009).

3.2 Cocaine

The number of cocaine users in the United States rose exponentially in the mid-1990s, where there was an 82% increase within a 4 year period. As of 2012, there were 4.7 million individuals above the age of 12 reported current or past use of cocaine within the United States (Palamar, Davies, Ompad, Cleland, & Weitzman, 2015). Approximately 5–10% of all emergency room visits in the country can be attributed to cocaine usage, resulting in an estimated cost of $83 million dollars a year for treatment of cocaine users (Maraj, Figueredo, & Lynn Morris, 2010). Some of the acute toxic effects of cocaine use include cardiovascular complications, anxiety, panic, and paranoia. Chronic use of cocaine can result in fatigue, blood borne disease, stroke, seizures, and sudden death. Despite the prevalence of cocaine as a prominent drug of abuse, and its associated negative health outcomes, there are no well-established pharmacological treatment options for cocaine addiction. Preclinical studies have found that oxytocin might be a viable therapeutic target (Sarnyai & Kovacs, 2014) through blockade of cocaine-associated behaviors such as hypervocalization (Kovacs, Sarnyai, Barbarczi, Szabo, & Telegdy, 1990) and cocaine-induced stereotypy (Sarnyai & Kovacs, 1994). Here we will cover the effect of oxytocin administration on various stages of the cocaine addiction cycle and its potential mechanisms.

3.2.1 Oxytocin Effects on Cocaine Acquisition and Maintenance

One of the earliest studies investigating the effect of oxytocin administration on cocaine self-administration determined that a systemic injection of oxytocin reduced the amount of lever presses during the maintenance phase of cocaine self-administration (Sarnyai & Kovacs, 1994). Other preclinical studies have found that oxytocin dose dependently (0.3, 1, and 3 mg/kg; ip) reduced active lever pressing and cocaine intake on both an FR1 and FR5 schedule of reinforcement in male rats (Zhou et al., 2014). In females, similar doses of systemic oxytocin (0.3, 1, and 3 mg/kg; ip) attenuated lever pressing and cocaine intake on an FR1 schedule of reinforcement (Leong, Zhou, Ghee, See, & Reichel, 2016). Interestingly, despite receiving multiple oxytocin injections over the course of self-administration, female rats displayed no sustained reduction in overall cocaine intake between self-administration test days (Leong et al., 2016).

3.2.2 Oxytocin Effects on Motivation of Cocaine Seeking

Consistent with meth, oxytocin decreases motivation to take cocaine in male and female rats. Zhou et al. (2014) discovered that peripheral administration of oxytocin (1 mg/kg) was effective in reducing active lever presses and break points in a PR test of cocaine seeking in male rats. This suggests that, in males, oxytocin is effective in suppressing the motivational limit of cocaine. These results were further supported in a recent study that employed a within-session model of economic demand of cocaine (Bentzley et al., 2014). Here, oxytocin reduced demand and motivation for cocaine in male rats with a history of cocaine self-administration. Through this novel BE paradigm, the effectiveness of oxytocin in suppressing demand was proportional to their initial baseline demand, which provides translational value for the use of oxytocin as a pharmacological treatment option of cocaine use. In females, oxytocin proved to be effective in reducing breakpoint in a PR test of cocaine seeking in female rats (Fig. 1). However, in contrast to Zhou et al. (2014) which found that oxytocin attenuates motivation in male rats, oxytocin did not impact break point values in male rats in our unpublished study (Fig. 1). One explanation for this sex difference is that

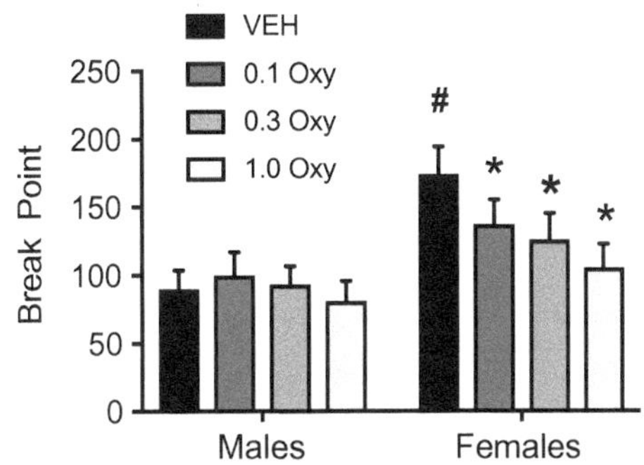

Fig. 1 Oxytocin decreased the breakpoint for cocaine self-administration in female rats. # Females significantly greater than males. *Significantly reduced relative to vehicle-treated animals of the same sex vehicle.

oxytocin has sex-specific effects on motivation to take cocaine. However, the difference could also be due to innate differences in motivational properties between males and females assessed with a PR procedure. Females displayed significantly greater motivation to take cocaine leading to a floor effect in male rats that might render oxytocin ineffective relative to females. Indeed, oxytocin did reduce motivation for cocaine in the behavior economics study that accounts for both the demand and motivation for drug (Bentzley et al., 2014).

3.2.3 Oxytocin Effects on Cocaine Extinction

To our knowledge, few studies have investigated the effect of oxytocin on cocaine extinction behavior. A recent study revealed that oxytocin reduces lever responding on the first day of extinction (Ext Day 1) in male rats (Bentzley et al., 2014). However, it is uncertain whether this single dose of oxytocin on Day 1 of extinction is sufficient to produce long-lasting changes on subsequent extinction sessions or on measures of relapse. We gave oxytocin to male rats 30 min before daily extinction sessions (Oxy Pre) or immediately after the session (Oxy Post). A separate group of rats received vehicle (Fig. 2). There were no differences in lever responding over 8 days, all rats then underwent cue-induced reinstatement tests with oxytocin or vehicle injected 30 min before chamber placement. Oxytocin attenuated lever responses in Veh and Oxy Pre-rats relative to vehicle injected rats in the same group. Interestingly, Oxy Post rats given vehicle had reduced reinstatement responding consistent with rats that received oxytocin on test. Despite not having an effect of extinction responding, subsequent reinstatement was blunted when oxytocin was given after the

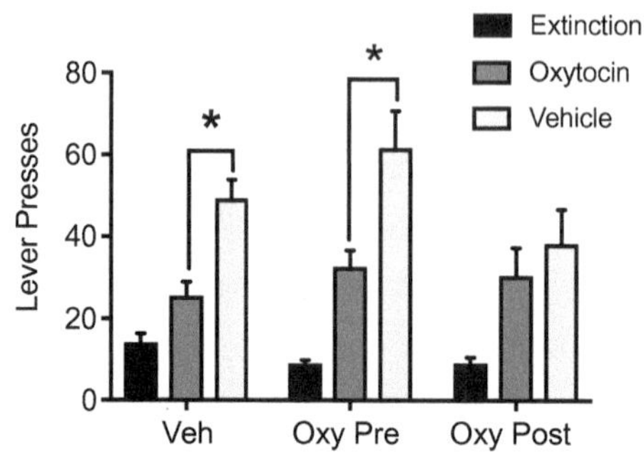

Fig. 2 Daily oxytocin administered following extinction sessions deceased cue-induced reinstatement of cocaine seeking. Vehicle-treated animals received daily saline injections, Oxy Pre animals received oxytocin (1 mg/kg) before the extinction session and Oxy Post animals had oxytocin after the session. On test days rats were injected with vehicle or oxytocin. *Significant difference between oxytocin and vehicle treatments.

daily session, future work will need to determine mechanisms by which oxytocin enhanced learning about the extinction session. Early studies have found that oxytocin infused icv or directly into the hippocampus facilitated extinction of an active avoidance task (Bohus, Urban, van Wimersma Greidanus, & de Wied, 1978; Ibragimov, 1990). However, future work will need to determine mechanisms by which oxytocin enhanced learning about the extinction session within a self-administration paradigm.

3.2.4 Oxytocin Effects on Cocaine Relapse/Reinstatement

Overwhelming preclinical evidence suggests that oxytocin may have therapeutic potential for the treatment of cocaine addiction through preclinical assessment of reinstatement of cocaine seeking. In general, peripheral oxytocin administration attenuates lever pressing during a cocaine-primed reinstatement test and cue-induced reinstatement tests in male (Bentzley et al., 2014; Zhou et al., 2014) and female rats and this reduction is not dependent on cycle (Leong et al., 2016; Leong et al., 2017; Weber et al., 2018). Central infusions of oxytocin (icv) also attenuate cue-induced cocaine-seeking behavior in both sexes (Leong et al., 2017; Morales-Rivera et al., 2014). Recent studies have begun to highlight specific neural regions that may be mediating oxytocin's central effect on cocaine-seeking behavior.

Few studies have directly compared oxytocin's effect on neuronal function following cocaine self-administration. Recently, we demonstrated that cue-induced *fos* activation in various brain regions was normalized (returned to extinction values) by a systemic injection of oxytocin (Leong et al., 2017). Specifically, in the medial prefrontal cortex (mPFC), NAc, and subthalamic nucleus (STN) *fos* levels were elevated in response to a cocaine-associated cue and oxytocin brought this expression back down to control levels. In contrast, oxytocin enhanced central amygdala (CeA) *fos* expression. The CeA is a region critical in mediating emotional responses and the mPFC, NAc, and STN are critical components of the reward circuit. Direct oxytocin infusions into the aforementioned brain areas result in regionally specific changes in cued reinstatement. Intra-accumbal oxytocin reduced whereas intra-PFC oxytocin increased reinstatement of cocaine seeking (Weber et al., 2018). The STN, a downstream projection of the NAc, is also an important region mediating the effect of oxytocin on drug seeking (Baracz & Cornish, 2016). Fig. 3 shows that intra-STN infusions of oxytocin reduced cued reinstatement of cocaine seeking, but not sucrose seeking. The involvement of the STN in mediating oxytocin's effect on

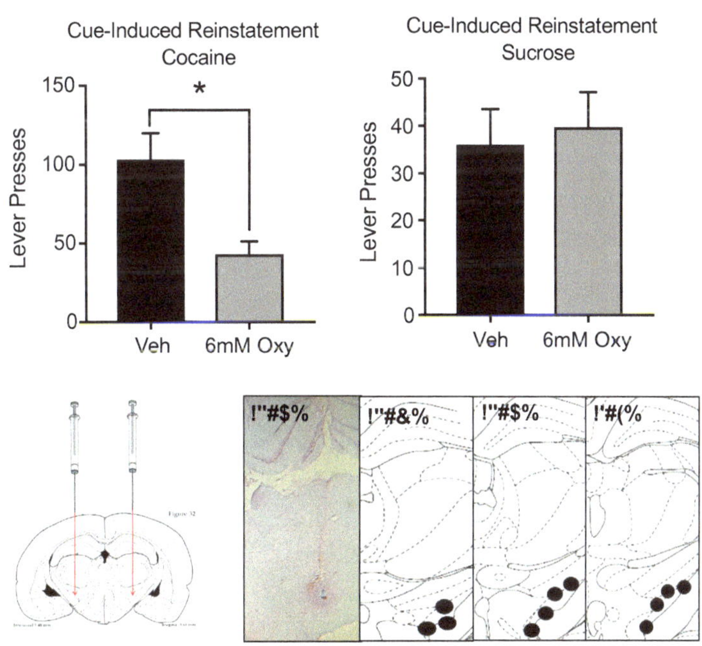

Fig. 3 Oxytocin infused into the STN deceased cued-induced reinstatement of cocaine but not sucrose seeking. Vehicle or oxytocin was infused into the STN before reinstatement testing. Images depict the terminal point of the injectors used to infuse compounds, with numbers next to brain sections detailing the distance in mm to Bregma. *Significantly decreased compared to vehicle.

cocaine seeking is similar to its involvement in mediating oxytocin's effect on meth-seeking behavior (Baracz et al., 2012; Baracz et al., 2015).

Another similarity with meth, a concurrent intra-NAcc infusion of an mGluR2/3 antagonist blocked the oxytocin reduction of cued reinstatement suggesting presynaptic mechanisms in the NAcc (Bernheim et al., 2017; Weber et al., 2018). Similar findings were uncovered with peripheral administration of both compounds. This suggests that the mechanisms underlying oxytocin's therapeutic effect on relapse behavior are pharmacologically and neuroanatomically similar in both cocaine and meth addiction. Overall, there is strong evidence that oxytocin is a viable pharmacological therapeutic target for the treatment of cocaine addiction throughout various stages of the addiction cycle. Recent studies have provided insight into the mechanisms underlying oxytocin's impact on cocaine-seeking behavior. Importantly, few sex differences have been documented thus far, at the behavioral and neuronal level suggesting that it might be equally as effective in both males and females.

3.3 Heroin

There has been a substantial increase in prevalence of heroin and prescription opioid use in the United States over the last few years, with users ranging from affluent suburban backgrounds to low-income minority populations (Cicero, Ellis, Surratt, & Kurtz, 2014). Nationwide, there has been a stark increase in documented overdose-related hospitalizations within the past 10 years (Unick, Rosenblum, Mars, & Ciccarone, 2013). The rise in heroin users is attributed to the increase in prescription opioid abuse in the last 20 years (Unick et al., 2013), accelerated by the frequently prescribed oxycodone in the 1990s (Siegal, Carlson, Kenne, & Swora, 2003). In 2014 alone, opioid overdose accounted for 28,647 deaths in the United States (Rudd, Rudd, Seth, David, & Scholl, 2016). Various medical complications are associated with heroin use including cardiac infections, kidney and liver disease, autoimmune disorders, and depression. Furthermore, the threat of heroin overdose is immensely high as large doses of heroin can quickly cause respiratory arrest.

Despite the availability of drugs to counteract the immediate effects of heroin overdose, such as the opioid receptor antagonist naloxone and buprenorphine (Sung & Conry, 2006), few other pharmacological options exist to combat heroin's significant addictive properties. Preliminary findings suggest that oxytocin might act as a potential therapeutic target or adjunct treatment for heroin addiction at various stages of the addiction cycle. Chronic opioid administration reduced oxytocin immunoreactivity in the hippocampus and decreased oxytocin mRNA levels in the hypothalamus and SON, the primary area of oxytocinergic innervation within the brain (for review see Zanos et al., 2017). It is possible that endogenous administration of oxytocin can provide therapeutic value through restoration of oxytocin homeostasis.

3.3.1 Oxytocin Effects on Heroin Self-Administration and Maintenance

Most early studies investigating the effect of oxytocin on heroin addiction have focused on its effect during self-administration. In one study, daily peripheral oxytocin administration diminished acquisition of heroin self-administration in heroin-tolerant male rats (Kovacs, Borthaiser, & Telegdy, 1985). Oxytocin had no effect on heroin-naïve animals during self-administration. Similarly, oxytocin reduced heroin intake in heroin-tolerant rats but not in naïve rats. Site specific application intra-hippocampal and intra-accumbal oxytocin impaired acquisition of lever pressing responses (Ibragimov, Kovacs, Szabo, & Telegdy, 1987). These findings suggest that oxytocin attenuated the development of heroin tolerance. Support for this idea comes from a follow up study showing that oxytocin pretreatment diminished heroin tolerance in mice as measured by its antinociceptive effects (Kovacs, Faludi, & Telegdy, 1985).

3.3.2 Oxytocin Effects on Heroin Reinstatement

Oxytocin's effect on reinstatement of heroin seeking has not been as extensively studied relative to psychostimulants. To our knowledge, no studies have investigated the impact of oxytocin during the extinction phase of heroin addiction. Furthermore, no studies have been published showing that oxytocin impacts heroin reinstatement. Related, carbetocin, an oxytocin analogue, reduced stress-induced (Zanos et al., 2014) and drug-primed (Georgiou et al., 2015) reinstatement of morphine-seeking. Preliminary data from our laboratory (Fig. 4) shows that oxytocin also blocks cued reinstatement of heroin seeking. Male and female rats self-administered heroin for 14 days, followed by extinction, and subsequent cued reinstatement testing. So, consistent with cocaine and methamphetamine, oxytocin and/or its analogs decrease stress, drug, and cued reinstatement of opioid seeking.

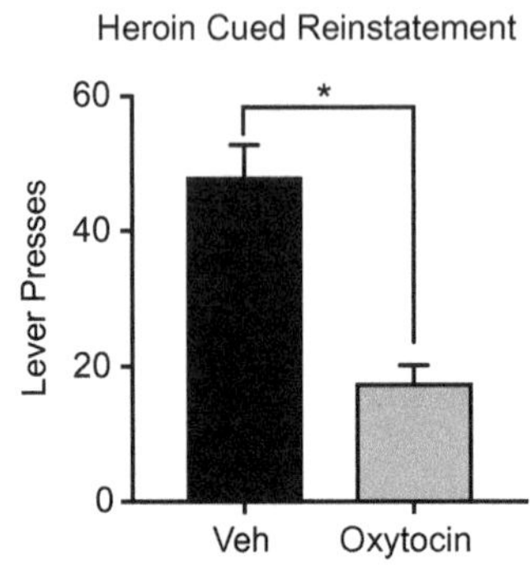

Fig. 4 Oxytocin decreased cued reinstatement of heroin seeking. *Significant difference between vehicle and oxytocin treatments.

3.4 Alcohol

Alcohol use disorder is a substantial and disabling public health concern, with lifetime prevalence rates of 29.1% reported in the United States (Grant et al., 2015). Excessive alcohol consumption is linked to a number of diseases, including cancers, cardiovascular disease, liver cirrhosis, neuropsychiatric disorders and fetal alcohol syndrome (Rehm et al., 2009). The Center for Disease Control places excessive alcohol use/abuse as the third leading cause of preventable deaths in the United States (Mokdad, Marks, Stroup, & Gerberding, 2004). Economically, alcohol misuse is estimated to cost the U.S. roughly $223.5 billion per year. Despite the significant personal, social and economic burden alcohol use disorders present, there are few effective pharmacotherapies available to help curb alcohol addiction. Preclinical evidence indicates that oxytocin influences a number of behavioral and physiological effects of alcohol (Lee & Weerts, 2016). Here we discuss the findings that support oxytocin as a promising novel therapeutic target in the treatment of alcohol use disorders.

3.4.1 Oxytocin Effects on Alcohol Self-Administration Behavior

The early seminal body of research in this area focused on the effect of oxytocin on tolerance to the physiological effects of alcohol. Repeated systemic administration (ip) of oxytocin prevented the development of tolerance to the hypothermic, hypnotic, and ataxic effects of alcohol (Jodogne, Tirelli, Klingbiel, & Legros, 1991; Kovacs et al., 1998; Szabo, Kovacs, Szekeli, & Telegdy, 1985; Szabo, Kovacs, & Telegdy, 1989). Oxytocin treatment was also found to modulate the severity of alcohol withdrawal symptoms (Szabo, Kovacs, Szekeli, Balaspiri, & Telegdy, 1987). Additionally, central administration of oxytocin (icv) was shown to block rapid tolerance to the effects of alcohol (Szabo et al., 1989). Central administration was found to be more efficacious than peripheral administration, pointing to a centrally mediated mechanism. In this regard, interactions with dopamine and serotonin neurotransmission were postulated to mediate this effect (Szabo et al., 1989).

In more recent reports, systemic administration of oxytocin has been shown to reduce alcohol preference and intake in a variety of voluntary drinking models in rats (Bowen, Carson, Spiro, Arnold, & McGregor, 2011; MacFadyen et al., 2016; McGregor & Bowen, 2012) and mice (King et al., 2017; Peters, Slattery, Flor, Neumann, & Reber, 2013). Specifically, McGregor and Bowen (2012) found that a single dose of oxytocin (1 mg/kg) produced a long-lasting reduction in preference for an alcohol-containing solution compared with a nonalcoholic sweet solution. Additionally, treatment with oxytocin for 2 weeks prior to the induction of a two-bottle free-choice paradigm resulted in lower alcohol preference in oxytocin-treated rats compared to controls. MacFadyen et al. (2016) reported that systemic administration of a lower dose range of oxytocin (0.1–0.5 mg/kg; ip) reduced operant alcohol self-administration in rats. Further, in alcohol-preferring C57BL/6J mice, systemic administration of oxytocin decreased alcohol self-administration and binge-like alcohol consumption in a dose-related manner (0.3, 1, 3 or 10 mg/kg; ip), as well as reduced motivation for alcohol reward as measured by responding under a PR schedule (King et al., 2017). Peripheral oxytocin treatment in a similar dose range (1–10 mg/kg) reduced alcohol consumption in male and female prairie voles that had two-bottle choice free-access to 15% alcohol (Stevenson, Wenner, et al., 2017). Additionally, in a continuous access paradigm, systemic oxytocin administration decreased alcohol intake in the first hour after treatment but had no significant effect on consumption over a 24-h period (Stevenson, Wenner, et al., 2017).

While there is strong preclinical evidence to suggest a role for oxytocin in alcohol use disorders, the mechanisms by which oxytocin reduces alcohol consumption is not fully understood. In a recent study, direct (icv) administration of oxytocin reduced alcohol consumption and alcohol-induced dopamine efflux in the NAc (Peters, Bowen, Bohrer, McGregor, & Neumann, 2017). Additionally, oxytocin receptors have been identified on neurons in the ventral tegmental area projecting to the NAc (Peris et al., 2017). Thus, it is possible that oxytocin reduces alcohol self-administration by altering mesolimbic dopamine activity that ordinarily signals alcohol-related reward. However, few studies have examined the role of oxytocin receptors in mediating the neuropeptide's effect on the motivational actions of alcohol. Recent reports showed that viral-mediated overexpression of oxytocin receptors in NAc reduced alcohol-induced conditioned place preference and alcohol consumption in mice (Bahi, 2015; Bahi, Al Mansouri, & Al Maamari, 2016). Additionally, the ability of oxytocin to reduce binge-like alcohol drinking was blocked by pre-treatment with an oxytocin receptor antagonist, suggesting that oxytocin reduced alcohol consumption via interaction with its own receptor (King et al., 2017). While the oxytocin receptor appears to be involved in the rewarding effects of alcohol, it is important to note that Bowen and colleagues showed that attenuation of sedative and ataxic effects by central oxytocin administration was mediated by $GABA_A$ receptors (Bowen et al., 2015). Although additional studies are necessary to further investigate the conditions and mechanisms by which oxytocin reduces alcohol consumption, the growing body of literature investigating the effect of oxytocin on the neurobehavioral response to alcohol points to the oxytocin system as a promising target for therapeutic intervention of alcohol use disorders.

3.4.2 Oxytocin Effects on Alcohol Relapse/Reinstatement Behaviors

To our knowledge, only one published study has investigated the effects of oxytocin administration on relapse-like behavior in animal models. Following repeated cycles of chronic intermittent alcohol vapor exposure to induce dependence in rats, Hansson and colleagues injected (icv) oxytocin in alcohol-dependent and non-dependent rats trained to orally self-administer alcohol using operant conditioning procedures. Central administration of oxytocin reduced cue-induced alcohol relapse-like behavior in alcohol-dependent rats, but not in non-dependent rats (Hansson et al., 2017). Additionally, the authors demonstrated that oxytocin mRNA was reduced in the NAc by acute alcohol and during early withdrawal, but increased after prolonged (3 weeks) abstinence, suggesting dynamic dependence-related neuroadaptations of the oxytocin system. After 3 weeks of abstinence, dependent rats showed increases in oxytocin receptor mRNA and protein levels in PFC, striatal, amygdala and hippocampal regions, and a reduction in oxytocin mRNA and peptide expression in hypothalamic nuclei. In congruence with this, a loss of hypothalamic oxytocin neurons after prolonged alcohol intake has been reported in other studies (Stevenson, Young, et al., 2017).

4. OXYTOCIN AS A TREATMENT FOR ADDICTION

The mechanisms by which oxytocin provides therapeutic value for the treatment of addiction are just now beginning to be identified. This section will present a mechanistic account that illustrates oxytocin's interactions with other neurotransmitter systems within the addiction circuitry. Following, we present prominent theories hypothesizing that oxytocin attenuates drug-seeking behavior through its anxiolytic and prosocial properties.

4.1 Interactions Within Addiction Circuitry

As mentioned in a previous section, oxytocin receptors are ubiquitous in the brain and in particular throughout the addiction circuitry. Within this circuitry, oxytocin directly interacts with dopamine, glutamate and GABA neurotransmission. Oxytocin projection neurons within the PVN synapse on dopaminergic cells within the NAc (Knobloch & Grinevich, 2014) and oxytocin receptors are also present on dopamine neurons projecting from the VTA to the NAc and mPFC (Knobloch & Grinevich, 2014; Peris et al., 2017). Intra-NAc oxytocin infusions block cocaine-induced increases in dopamine utilization (Kovacs et al., 1998) and ICV infusion of oxytocin inhibited meth-induced dopamine turnover (Qi et al., 2008). Similarly, ICV oxytocin blocks dopamine release induced by ethanol administration within the NAc (Peters et al., 2017), which corresponds with reduced ethanol seeking behavior. Recent studies have characterized the presence of oxytocin/dopamine receptor complexes (OXTR-D2R) within the NAc in which oxytocin acts as an allosteric agonist to increase D2R affinity (de la Mora et al., 2016; Fuxe et al., 2012; Romero-Fernandez, Borroto-Escuela, Agnati, & Fuxe, 2013). Interestingly, activation of D2R reduces drug-seeking behavior and this receptor subtype is down-regulated following chronic drug exposure (for review see Koob & Mason, 2016; Volkow & Morales, 2015). Therefore, oxytocin may reduce drug-seeking behavior via interactions with D2Rs in the NAc.

Alternatively, oxytocin may impact drug seeking and dopaminergic signaling through GABAergic interneurons. A number of studies have found that excitatory Gq-coupled oxytocin receptors are localized on GABAergic interneurons in the NAc (Dolen et al., 2013), hippocampus (Zaninetti & Raggenbass, 2000), and PFC (Li et al., 2016; Nakajima et al., 2014). Oxytocin signaling in these neurons would increase inhibitory tone within these regions. Along these lines, an oxytocin receptor agonist suppressed hippocampal pyramidal cell firing via GABAergic interneurons within the CA1 layer of the hippocampus (Owen et al., 2013; Zaninetti & Raggenbass, 2000). Given that oxytocin receptors are also present on parvalbumin-containing GABA interneurons in the NAc (Dolen et al., 2013) and that these interneurons regulate the expression of psychostimulant-induced behavioral adaptations (Wang et al., 2018), it is possible that oxytocin impacts drug-seeking behavior directly through interactions with its own receptor on GABAergic interneurons in areas that are critical to dopamine signaling and addiction processes.

A recent model from our laboratory postulates that oxytocin receptors located on astrocytes (Dolen et al., 2013) within the NAc modulate drug-seeking behavior (Bernheim et al., 2017; Weber et al., 2018). More specifically, oxytocin activation of Gq-coupled oxytocin receptors on astrocytes results in astrocytic glutamate release. Extra-synaptic glutamate can thereby activate type 2 and 3 metabotropic (mGlu2/3) glutamate receptors located presynaptically on cortical glutamatergic axons, inhibiting synaptic release of glutamate thus, restoring glutamatergic tone (Moussawi & Kalivas, 2010). Scofield et al. (2015) demonstrated that chemogenetic activation of astrocytes in NAc core decreased cue reinstatement of cocaine (Scofield et al., 2015) seeking through a mGlu2/3 receptor dependent mechanism. Support for this notion, comes from work showing concurrent mGlu2/3 receptor antagonism reversed the attenuating effect of oxytocin on cocaine and meth-seeking behavior (Bernheim et al., 2017; Weber et al., 2018).

4.2 Relief of Stress

Due to the strong link between stress, drug use and relapse, the ability of oxytocin to modulate stress processes has generated growing interest in its potential as a treatment for substance use disorders. Stress is recognized as a significant risk factor that promotes the development of addictive behavior, and in particular, triggering relapse (Sinha, 2001, 2008, 2012), and may be a motivator to use drugs (Koob, 2014). However, drugs themselves can also serve as stressors by activating the hypothalamic–pituitary–adrenocortical (HPA) axis, the primary mediator of neuroendocrine response to stress. This includes stimulating the release of CRF from the hypothalamus (Sinha, 2001; Smith & Vale, 2006), norepinephrine from the locus coeruleus (Koob, 2008) as well as other stress-related neuropeptides in extrahypothalamic stress circuits in the brain (Koob, 2009; Koob & Mason, 2016). Repeated drug exposure followed by withdrawal can lead to profound dysregulation of the HPA axis as well as increased CRF activity within brain stress-reward pathways, which contribute to negative affective states related to withdrawal (e.g., anxiety, dysphoria, irritability) along with increased drive to use or drink (Koob & Volkow, 2010).

Oxytocin is known to exert potent anti-stress and anxiolytic effects (Jurek et al., 2015; Peters, Slattery, Uschold-Schmidt, Reber, & Neumann, 2014; Slattery & Neumann, 2010). Oxytocin administration has been shown to dampen increases in corticosterone (CORT) levels and stress-induced HPA-axis activation, as well as reduce behavioral responses in animal models of anxiety and depression (Neumann, Kromer, Toschi, & Ebner, 2000; Windle et al., 2004; Windle, Shanks, Lightman, & Ingram, 1997). Conversely, mice genetically engineered to lack the gene for oxytocin (OTKO) display more anxiety-related behavior and greater CORT response following exposure to stress (Amico, Mantella, Vollmer, & Li, 2004; Mantella, Vollmer, & Amico, 2005). Further, high levels of oxytocin and oxytocin receptor mRNA expression are localized to forebrain regions, such as the extended amygdala, that are critically involved in anxiety and stress-responsiveness (Dabrowska et al., 2011; Gimpl & Fahrenholz, 2001; Veinante & Freund-Mercier, 1997), thereby enabling oxytocin to contribute to the regulation of anxiety, stress and reward-related behaviors.

While endogenous oxytocin appears to be intricately involved in the homeostatic regulation of stress responses (Lee & Weerts, 2016), oxytocin signaling in brain has been shown to be altered by both acute and chronic exposure to alcohol and/or drugs of abuse. Thus, exogenous oxytocin administration may help to attenuate the effects of stress associated with heavy drinking and drug use. In support of this hypothesis, Peters and colleagues demonstrated that oxytocin administration attenuated stress-induced increase in alcohol consumption in rats (Peters et al., 2013). Additionally, systemic administration of the oxytocin analogue carbetocin reversed anxiety- and depressive-like behaviors in morphine-withdrawn mice to levels similar to control groups (Zanos et al., 2014). Clinically, a single dose of intranasal oxytocin was shown to reduce stress-induced craving and anxiety in cannabis-dependent individuals (McRae-Clark, Baker, Maria, & Brady, 2013). Further, intranasal oxytocin treatment decreased alcohol withdrawal symptoms in treatment-seeking patients compared to placebo and significantly reduced anxiety in dependent subjects following cessation of drinking (Pedersen, 2017; Pedersen et al., 2013). Taken together, oxytocin treatment may be a useful therapeutic to reduce stress-related physiological responses (e.g., anxiety, craving) relevant to drug and alcohol use and relapse.

4.3 Social Affiliation

Oxytocin may amplify the reinforcing or rewarding properties of social engagement. Often people begin using drugs or alcohol in social contexts and the reinforcing value of some drugs are enhanced when used in the presence of others (McGregor & Bowen, 2012). Social factors can also have a profound impact on maintaining abstinence from drug use for some people. It is suggested that parallel brain circuits mediate social vs. drug reward-related behaviors and that oxytocin may act to redirect behaviors between these reward circuits (McGregor & Bowen, 2012), shift the preference for novel situations and experiences to one for familiarity (Tops et al., 2014), and rebalances the addicted brain (Bowen & Neumann, 2017).

McGregor and Bowen (2012) posit that oxytocin administration can break the behavioral "loops" that maintain the cyclical patterns of drug use–abstinence–relapse. Specifically, they hypothesize that excessive dopaminergic signaling within basal ganglia circuitry resulting from high dose stimulant use lead to a bias in behaviors directed toward object-oriented rewards (food, drugs, alcohol) and the execution of these behavioral loops to seek the object (McGregor & Bowen, 2012). Oxytocin may interfere with this biased behavior by shifting a focus from object-oriented behavioral seeking to one of preference for social interaction. As a treatment for addiction, oxytocin may increase the proclivity towards social rewards thereby strengthening social bonds that may be essential for maintaining abstinence.

Along these same lines Tops et al. (2014) postulate that oxytocin involves overlapping mechanisms between attachment formation and stress by shifting preferences from novel experiences and reward to one of familiarity by facilitating a ventral-to-dorsal shift in activation of cortical striatal loops. This hypothesis is in line with a ventral-to-dorsal shift in striatal control of behavior in addiction. That is, early in drug use, behavior is driven by dopamine release in the ventral striatum mediated by the acute rewarding effects of the drug. With repeated use habitual responding for drugs comes under control of the dorsal striatum (Belin & Everitt, 2008; Gabriele, Pacchioni, & See, 2012). Perhaps oxytocin given as an adjunct for addiction increases the preference for familiarity thereby enhancing social bonds that may help maintain abstinence.

5. OXYTOCIN, SOCIAL REWARD, AND ADDICTION

Prosocial interactions are necessary for the promotion and enhancement of cooperation between group members in many diverse animal species. Although these behaviors serve a critical function, the underlying neural circuitry is understudied. Mounting evidence suggests prosocial behaviors are reinforcing (Song, Borland, Larkin, O'Malley, & Albers, 2016) and the reinforcing properties of prosocial behavior are mediated by the activation of the dopaminergic mesocorticolimbic (MCL) circuit (Insel, 2003). Over the past few decades, oxytocin has attracted interest for its role in prosocial behaviors (Dolen et al., 2013; Hung et al., 2017). As discussed previously in this chapter, both the MCL and oxytocin systems play crucial roles in encoding the rewarding and reinforcing properties of drugs of abuse. In the following sections, the current understanding of the neural circuitry of social rewards will be discussed, focusing in particular on oxytocin's regulation of these behaviors. In addition, the literature describing the bidirectional modulation of social reward and drugs of abuse will be described with a special interest in the role oxytocin may play in these competing reward signals.

5.1 The Neural Circuitry of Social Reward and Oxytocin

Prosocial behaviors, such as pair bonding and social interaction, have rewarding properties that are encoded by the MCL pathway (Becker, Rudick, & Jenkins, 2001; Insel, 2003), similar to drugs of abuse. For example, dopamine is released in the NAc following pup exposure in maternal female rats (Hansen, Bergvall, & Nyiredi, 1993), and lesions in either the VTA or the NAc will disrupt maternal approach and interaction with pups (Lee, Li, Watchus, & Fleming, 1999; Numan & Smith, 1984). The dopaminergic reinforcement of social reward alone is an incomplete picture of the complexity of these behaviors. Oxytocin has gained growing interest for its role in social reward since Pedersen and Prange (1979) demonstrated that oxytocin induced maternal behavior in virgin female rats. Since this seminal study, it has been established that oxytocin modulates social interaction (Young & Wang, 2004) and pair bonding (Carter et al., 1992), as well as enhanced social memory (Ferguson et al., 2000). The interplay between the MCL and oxytocin systems that was first observed in studies of monogamous prairie voles, concluded oxytocin receptors in the NAc are required for pair bonding formation (Young & Wang, 2004). Similar to the rat model, dopamine is released in the NAc during social rewards in prairie voles. More specifically, D2-specific activity induces mating-induced partner preference in female prairie voles (Young & Wang, 2004). Moreover, Young et al. (2001) demonstrated an oxytocin receptor antagonist directly infused into the NAc of these voles reduced D2-mediated partner preference.

While high densities of oxytocin receptor have been observed in the NAc, they have also been detected in other areas related to reward signaling. For example, oxytocin receptors have been discovered in high density in the PFC of female prairie voles, and direct infusion of an oxytocin receptor antagonist into this brain region blocked partner preference (Young et al., 2001). The role of oxytocin in social reward in non-monogamous species is also becoming more apparent. Oxytocin projections to the PFC have been identified in rats (reviewed in Buisman-Pijlman et al., 2014) but there is no study to date that has examined these projections in relation to social reward. Further, dopamine administration induces central release of oxytocin, while oxytocin administration potentiates central dopamine in the rat (Melis et al., 2007). Rodents that do not demonstrate monogamous mating behavior seem to have similar levels of circulating oxytocin as that of prairie voles, but oxytocin receptor densities vary across species (Buisman-Pijlman et al., 2014), which may contribute to the differences in their mating preferences (Young & Wang, 2004). Regardless of social and mating structure, oxytocin modulates reward signals within the NAc as direct administration of an oxytocin receptor antagonist eliminated social reward preference in a social CPP test in mice without altering cocaine CPP (Dolen et al., 2013).

Dolen et al. (2013) additionally discovered oxytocin within the NAc induced long-term depression (LTD) on presynaptic excitatory synapses of medium spiny neurons (MSNs) in a mouse model. The oxytocin-induced LTD was significantly attenuated if the mice had been previously socially conditioned. The prosocial behavior did not preferentially act on oxytocin-induced LTD within the D1 or D2 MSN pathways, suggesting another possible mechanistic difference between monogamous and non-monogamous rodents. Finally, social reward and oxytocin-induced LTD requires inputs into the NAc from the dorsal raphe nucleus (DRN). The pertinent afferents from the DRN were shown to be serotonergic (5-HT), acting on 5-HT-1b receptors on glutamate neurons. These DRN afferents have oxytocin receptors that mediate 5-HT release within the NAc. During social reward, oxytocin is released into the NAc via PVN neurons, which binds oxytocin receptor on serotonergic inputs from the

DRN, causing a release of 5-HT. Subsequently, it binds to 5-HT-1b receptors on presynaptic excitatory synapses onto MSNs, inducing LTD. Taken together, these experiments eloquently demonstrate the mechanistic impact of oxytocin within NAc during prosocial behaviors.

The VTA is another region that has long been characterized within the MCL pathway as a critical brain region for reward signals related to drugs of abuse. Recently, studies have implicated the VTA in social reward cues (Song et al., 2016) via increases in dopaminergic NAc-projecting VTA neuron activity following social interactions in mice (Gunaydin et al., 2014). Exciting research from Hung et al. (2017) explored the role oxytocin plays in this change in dopamine concentrations following social reward. They provide evidence that there are oxytocin projections from the PVN to the VTA in mice. These oxytocin projections within the VTA were often co-localized to oxytocin receptor-containing dopamine neurons. Mice lacking oxytocin receptor specifically on dopaminergic neurons within the VTA did not express social CPP behavior compared to controls. These mice did, however, express CPP in response to cocaine. They also determined *c-fos* levels in VTA-projecting PVN oxytocin neurons were significantly potentiated early on during a bout of social interaction, but activation of these oxytocin neurons alone was not rewarding. Instead, temporally-dependent oxytocin release into the VTA during social reward seems to directly enhance the excitability of NAc-projecting VTA neurons, potentiating dopamine levels within the NAc, thereby reinforcing the prosocial behavior (Hung et al., 2017). In summary, there is clear evidence from these rodent models for the concomitant role of the dopaminergic MCL and oxytocin systems in the control and regulation of social reward. In addition, a growing body of research indicates oxytocin gates dopamine's reinforcing effects of social reward via indirect modulation of other neurotransmitters including, but not limited to, 5-HT.

5.2 The Overlap Between Oxytocin Effects on Social Reward and Drug Reward

Numerous studies described previously in this chapter have worked to explain how oxytocin is able to affect addictive processes associated with drugs of abuse. The overlap between oxytocin's role in the shared neural circuitry of addiction and social reward has been postulated (Insel, 2003), and there is evidence to suggest oxytocin may facilitate the rewarding effects of prosocial behavior at the expense of drug-related rewards (Sarnyai & Kovacs, 2014; Thompson, Callaghan, Hunt, Cornish, & McGregor, 2007).

The method by which oxytocin does this, however, is still under investigation. In this section, the current hypotheses of the role of oxytocin in mediating the differing reward signals will be discussed.

It has long been hypothesized that drug-mediated and social reward signals compete within the reward circuitry of the brain (Buisman-Pijlman et al., 2014; Insel, 2003). Several rodent studies have shown that social rewards, such as conspecific interaction, have the ability to completely reverse cocaine CPP (Fritz et al., 2011; Kummer et al., 2014). In a comparative CPP paradigm, mice that were given a choice between cocaine and social interaction with a conspecific did not develop a significant preference for either compared to mice that underwent cocaine or social interaction CPP alone (Liu, Wang, Zhan, & Cheng, 2016). Interestingly, when the experimenters evaluated oxytocin expression in the PVN, CPP of cocaine alone, social interaction alone, and social interaction and cocaine combined all had elevated levels compared to controls. However, while social interaction alone had the largest increase and cocaine had the lowest, the animals forced to choose between the two rewards had oxytocin expression in the PVN that fell in between the levels quantified in the two groups that only experienced a single reward (Liu et al., 2016). It is evident the brain's oxytocin system is extremely plastic, especially in response to these differing rewards. Drugs of abuse may cause lasting downregulation in oxytocin expression or function, which explains social behavior deficits observed in animals repeatedly exposed to them (McGregor & Bowen, 2012). Conversely, during prosocial behavior, oxytocin may promote reward value associated with prosocial behavior to compete with that of the drugs (Campbell, 2008).

Addiction has long been understood as a neuroadaptive process of "pathological learning" to drugs of abuse (Koob & Volkow, 2016). For example, they can alter or induce long-term potentiation (LTP), a cellular prerequisite for learning and memory, by altering the strength of functional neuronal connections (Berretta, Nisticò, Bernardi, & Mercuri, 2008; Sarnyai & Kovacs, 2014). Additionally, drugs cause dramatic changes in synaptic density within the MCL regions of the brain, which play a critical role in motivation and learning (Koehl & Abrous, 2011). Social reward may negate or reverse these changes through an oxytocin-mediated neuroplasticity in some of the same brain regions. The LTD induced in the NAc by oxytocin during social reward (Dolen et al., 2013) may help reverse drug-induced LTP and shift the addictive bias from habitual behavior loops associated with drug towards social stimuli (McGregor & Bowen, 2012).

Further, Dölen and colleague's study established a critical interaction between oxytocin and serotonergic transmission in the NAc. Serotonin release in the NAc has been implicated in the regulation of psychostimulant drug-related memories (Cunningham & Anastasio, 2014) and the reinforcing effects of methamphetamine (Chiu & Schenk, 2012; Sarnyai & Kovacs, 2014). It may be possible that oxytocinergic modulation of this serotonergic pathway can reduce these pathological changes in addiction. Further research is needed to examine whether endogenous changes to oxytocin via prosocial activity is sufficient to significantly alter these drug-related changes in learning and memory.

The hippocampus is a critical brain structure in addiction, providing contextual learning components to drugs of abuse (Robbins & Everitt, 2002) and oxytocin was shown to attenuate drug-mediated learning by specifically targeting fast-spiking inhibitory interneurons within the hippocampus to sculpt its inhibitory signals (Owen et al., 2013). It is hypothesized this alteration in hippocampal excitability may, in turn, influence regions in the MCL circuit, such as the NAc and PFC, thereby mediating drug-induced neuroplasticity (Sarnyai & Kovacs, 2014). A recent study has also established that oxytocin receptors within the hippocampus (notably the anterior dentate gyrus, and anterior CA2/CA3) of mice are necessary for social stimuli discrimination and social recognition (Raam, McAvoy, Besnard, Veenema, & Sahay, 2017). It is possible, therefore, that prosocial memories, mediated by oxytocin in the hippocampus, inhibit the drug-related neuroplasticity, or alter the engrams associated with drug cues and reward. Despite these advancements, further evidence evaluating the competing influence social and drug reward on neuroplasticity are still needed.

6. CONCLUSIONS

Very little research has focused on the effect of prosocial behavior on neural changes associated with drugs of abuse. From these few studies, it is clear that a potentiation in oxytocin activity during social reward attenuates or reverses the reinforcing effects of drugs of abuse (Fritz et al., 2011; Kummer et al., 2014; Liu et al., 2016). Yet, important questions still remain. At present, there is no definitive understanding of the causality of these changes in neuroplasticity. In our previous sections we have presented strong preclinical evidence from our laboratories and others to suggest that exogenous administration of oxytocin reduces drug seeking and self-administration, tolerance, and reinstatement. Importantly, these effects occur across multiple drug classes and impact multiple neurotransmitter systems (dopamine, GABA, serotonin) indicating that oxytocin may be rebalancing an addicted brain through multiple dependent and interdependent mechanisms (Bowen & Neumann, 2017). Additionally, burgeoning research has exemplified the impact of oxytocin on social reward alone (Dolen et al., 2013; Hung et al., 2017). Future studies must work to elucidate the direct interaction of these two reward systems in hopes of understanding how prosocial behaviors aid in the recovery of those with substance abuse disorders. Adaptations to current methodologies to provide new information on the neurobiology or function of oxytocin signaling should include several different approaches. First, continued development of oxytocin receptor specific pharmaceuticals to determine specificity of the neuropeptide will aid understanding the neurobiology and treatment approaches. Second, adaptation of intranasal oxytocin as route of administration will add consistency with addiction studies in humans, which will directly increase the translational relevance of the basic science models (Ditzen et al., 2009; Guastella et al., 2010; Pedersen et al., 2013). And, third, recent technological advances in viral mediated gene transfer approaches will increase understanding brain region specific roles (through receptor over expression or knockdown) and functional circuitry (through optogenetics and/or chemogenetics) of oxytocin from a basic science level.

REFERENCES

Amico, J. A., Mantella, R. C., Vollmer, R. R., & Li, X. (2004). Anxiety and stress responses in female oxytocin deficient mice. *Journal of Neuroendocrinology*, *16*(4), 319–324. https://doi.org/10.1111/j.0953-8194.2004.01161.x.

Bahi, A. (2015). The oxytocin receptor impairs ethanol reward in mice. *Physiology & Behavior*, *139*, 321–327. https://doi.org/10.1016/j.physbeh.2014.11.046.

Bahi, A., Al Mansouri, S., & Al Maamari, E. (2016). Nucleus accumbens lentiviral-mediated gain of function of the oxytocin receptor regulates anxiety- and ethanol-related behaviors in adult mice. *Physiology & Behavior*, *164*(Pt. A), 249–258. https://doi.org/10.1016/j.physbeh.2016.06.009.

Baracz, S. J., & Cornish, J. L. (2016). The neurocircuitry involved in oxytocin modulation of methamphetamine addiction. *Frontiers in Neuroendocrinology*, *43*, 1–18. https://doi.org/10.1016/j.yfrne.2016.08.001.

Baracz, S. J., Everett, N. A., & Cornish, J. L. (2015). The involvement of oxytocin in the subthalamic nucleus on relapse to methamphetamine-seeking behaviour. *PLoS One*, *10*(8), e0136132. https://doi.org/10.1371/journal.pone.0136132.

Baracz, S. J., Everett, N. A., McGregor, I. S., & Cornish, J. L. (2016). Oxytocin in the nucleus accumbens core reduces reinstatement of methamphetamine-seeking behaviour in rats. *Addiction Biology*, *21*(2), 316–325. https://doi.org/10.1111/adb.12198.

Baracz, S. J., Rourke, P. I., Pardey, M. C., Hunt, G. E., McGregor, I. S., & Cornish, J. L. (2012). Oxytocin directly administered into the nucleus accumbens core or subthalamic

nucleus attenuates methamphetamine-induced conditioned place preference. *Behavioural Brain Research*, *228*(1), 185–193. https://doi.org/10.1016/j.bbr.2011.11.038.
Bardo, M. T., & Bevins, R. A. (2000). Conditioned place preference: What does it add to our preclinical understanding of drug reward? *Psychopharmacology*, *153*(1), 31–43.
Barr, A. M., Panenka, W. J., MacEwan, G. W., Thornton, A. E., Lang, D. J., Honer, W. G., et al. (2006). The need for speed: An update on methamphetamine addiction. *Journal of Psychiatry & Neuroscience*, *31*(5), 301–313.
Becker, H. C., & Lopez, M. F. (2004). Increased ethanol drinking after repeated chronic ethanol exposure and withdrawal experience in C57BL/6 mice. *Alcoholism, Clinical and Experimental Research*, *28*(12), 1829–1838.
Becker, J. B., Rudick, C. N., & Jenkins, W. J. (2001). The role of dopamine in the nucleus accumbens and striatum during sexual behavior in the female rat. *The Journal of Neuroscience*, *21*(9), 3236–3241.
Belin, D., & Everitt, B. J. (2008). Cocaine seeking habits depend upon dopamine-dependent serial connectivity linking the ventral with the dorsal striatum. *Neuron*, *57*(3), 432–441. https://doi.org/10.1016/j.neuron.2007.12.019.
Bentzley, B. S., Jhou, T. C., & Aston-Jones, G. (2014). Economic demand predicts addiction-like behavior and therapeutic efficacy of oxytocin in the rat. *Proceedings of the National Academy of Sciences of the United States of America*, *111*(32), 11822–11827. https://doi.org/10.1073/pnas.1406324111.
Bernheim, A., Leong, K. C., Berini, C., & Reichel, C. M. (2017). Antagonism of mGlu2/3 receptors in the nucleus accumbens prevents oxytocin from reducing cued methamphetamine seeking in male and female rats. *Pharmacology, Biochemistry, and Behavior*, *161*, 13–21. https://doi.org/10.1016/j.pbb.2017.08.012.
Bernheim, A., See, R. E., & Reichel, C. M. (2016). Chronic methamphetamine self-administration disrupts cortical control of cognition. *Neuroscience and Biobehavioral Reviews*, *69*, 36–48. https://doi.org/10.1016/j.neubiorev.
Berretta, N., Nisticò, R., Bernardi, G., & Mercuri, N. B. (2008). Synaptic plasticity in the basal ganglia: A similar code for physiological and pathological conditions. *Progress in Neurobiology*, *84*(4), 343–362. https://doi.org/10.1016/j.pneurobio.2007.12.004.
Bohus, B., Urban, I., van Wimersma Greidanus, T. B., & de Wied, D. (1978). Opposite effects of oxytocin and vasopressin on avoidance behaviour and hippocampal theta rhythm in the rat. *Neuropharmacology*, *17*(4–5), 239–247.
Bossert, J. M., Marchant, N. J., Calu, D. J., & Shaham, Y. (2013). The reinstatement model of drug relapse: Recent neurobiological findings, emerging research topics, and translational research. *Psychopharmacology*, *229*(3), 453–476. https://doi.org/10.1007/s00213-013-3120-y.
Bowen, M. T., Carson, D. S., Spiro, A., Arnold, J. C., & McGregor, I. S. (2011). Adolescent oxytocin exposure causes persistent reductions in anxiety and alcohol consumption and enhances sociability in rats. *PLoS One*, *6*(11), e27237. https://doi.org/10.1371/journal.pone.0027237.
Bowen, M. T., & McGregor, I. S. (2014). Oxytocin and vasopressin modulate the social response to threat: A preclinical study. *The International Journal of Neuropsychopharmacology*, *17*(10), 1621–1633. https://doi.org/10.1017/s1461145714000388.
Bowen, M. T., & Neumann, I. D. (2017). Rebalancing the addicted brain: Oxytocin interference with the neural substrates of addiction. *Trends in Neurosciences*, *40*(12), 691–708. https://doi.org/10.1016/j.tins.2017.10.003.
Bowen, M. T., Peters, S. T., Absalom, N., Chebib, M., Neumann, I. D., & McGregor, I. S. (2015). Oxytocin prevents ethanol actions at delta subunit-containing GABAA receptors and attenuates ethanol-induced motor impairment in rats. *Proceedings of the National Academy of Sciences of the United States of America*, *112*(10), 3104–3109. https://doi.org/10.1073/pnas.1416900112.

Buisman-Pijlman, F. T., Sumracki, N. M., Gordon, J. J., Hull, P. R., Carter, C. S., & Tops, M. (2014). Individual differences underlying susceptibility to addiction: Role for the endogenous oxytocin system. *Pharmacology, Biochemistry, and Behavior*, *119*, 22–38. https://doi.org/10.1016/j.pbb.2013.09.005.

Busnelli, M., & Chini, B. (2017). Molecular basis of oxytocin receptor signalling in the brain: What we know and what we need to know. *Current Topics in Behavioral Neurosciences*, *35*, 3–29. https://doi.org/10.1007/7854_2017_6.

Campbell, A. (2008). Attachment, aggression and affiliation: The role of oxytocin in female social behavior. *Biological Psychology*, 77(1), 1–10. https://doi.org/10.1016/j.biopsycho.2007.09.001.

Carson, D., Cornish, J., Guastella, A., Hunt, G., & McGregor, I. (2010). Oxytocin decreases methamphetamine self-administration, methamphetamine hyperactivity, and relapse to methamphetamine-seeking behaviour in rats. *Neuropharmacology*, *58*(1), 38–43. https://doi.org/10.1016/j.neuropharm.2009.06.018.

Carson, D. S., Guastella, A. J., Taylor, E. R., & McGregor, I. S. (2013). A brief history of oxytocin and its role in modulating psychostimulant effects. *Journal of Psychopharmacology*, *27*(3), 231–247. https://doi.org/10.1177/0269881112473788.

Carson, D. S., Hunt, G. E., Guastella, A. J., Barber, L., Cornish, J. L., Arnold, J. C., et al. (2010). Systemically administered oxytocin decreases methamphetamine activation of the subthalamic nucleus and accumbens core and stimulates oxytocinergic neurons in the hypothalamus. *Addiction Biology*, *15*(4), 448–463. https://doi.org/10.1111/j.1369-1600.2010.00247.x.

Carter, C. S., Williams, J. R., Witt, D. M., & Insel, T. R. (1992). Oxytocin and social bonding. *Annals of the New York Academy of Sciences*, *652*, 204–211.

Childress, A. R., Hole, A. V., Ehrman, R. N., Robbins, S. J., McLellan, A. T., & O'Brien, C. P. (1993). Cue reactivity and cue reactivity interventions in drug dependence. *NIDA Research Monograph*, *137*, 73–95.

Chini, B., & Fanelli, F. (2000). Molecular basis of ligand binding and receptor activation in the oxytocin and vasopressin receptor family. *Experimental Physiology*, *85*, 59s–66s. Spec No.

Chiu, V. M., & Schenk, J. O. (2012). Mechanism of action of methamphetamine within the catecholamine and serotonin areas of the central nervous system. *Current Drug Abuse Reviews*, *5*(3), 227–242.

Churchland, P. S., & Winkielman, P. (2012). Modulating social behavior with oxytocin: How does it work? What does it mean? *Hormones and Behavior*, *61*(3), 392–399. https://doi.org/10.1016/j.yhbeh.2011.12.003.

Cicero, T. J., Ellis, M. S., Surratt, H. L., & Kurtz, S. P. (2014). The changing face of heroin use in the United States: A retrospective analysis of the past 50 years. *JAMA Psychiatry*, *71*(7), 821–826. https://doi.org/10.1001/jamapsychiatry.2014.366.

Cox, B., Bentzley, B., Regen-Tuero, H., See, R., Reichel, C., & Aston-Jones, G. (2017). Oxytocin acts in nucleus accumbens to attenuate methamphetamine seeking and demand. *Biological Psychiatry*, *81*(11), 949–958. https://doi.org/10.1016/j.biopsych.2016.11.011.

Cox, B. R., Olney, J. J., Lowery-Gionta, E. G., Sprow, G. M., Rinker, J. A., Navarro, M., et al. (2013). Repeated cycles of binge-like ethanol (EtOH)-drinking in male C57BL/6J mice augments subsequent voluntary EtOH intake but not other dependence-like phenotypes. *Alcoholism, Clinical and Experimental Research*, *37*(10), 1688–1695. https://doi.org/10.1111/acer.12145.

Cox, B., Young, A., See, R., & Reichel, C. (2013). Sex differences in methamphetamine seeking in rats: Impact of oxytocin. *Psychoneuroendocrinology*, *38*(10), 2343–2353. https://doi.org/10.1016/j.psyneuen.2013.05.005.

Cunningham, K. A., & Anastasio, N. C. (2014). Serotonin at the nexus of impulsivity and cue reactivity in cocaine addiction. *Neuropharmacology*, *76*, 460–478. https://doi.org/10.1016/j.neuropharm.2013.06.030.

Dabrowska, J., Hazra, R., Ahern, T. H., Guo, J. D., McDonald, A. J., Mascagni, F., et al. (2011). Neuroanatomical evidence for reciprocal regulation of the corticotrophin-releasing factor and oxytocin systems in the hypothalamus and the bed nucleus of the stria terminalis of the rat: Implications for balancing stress and affect. *Psychoneuroendocrinology*, *36*(9), 1312–1326. https://doi.org/10.1016/j.psyneuen.2011.03.003.

de la Mora, M. P., Perez-Carrera, D., Crespo-Ramirez, M., Tarakanov, A., Fuxe, K., & Borroto-Escuela, D. O. (2016). Signaling in dopamine D2 receptor-oxytocin receptor heterocomplexes and its relevance for the anxiolytic effects of dopamine and oxytocin interactions in the amygdala of the rat. *Biochimica et Biophysica Acta*, *1862*(11), 2075–2085. https://doi.org/10.1016/j.bbadis.2016.07.004.

De Wied, D. (1971). Long term effect of vasopressin on the maintenance of a conditioned avoidance response in rats. *Nature*, *232*(5305), 58–60.

Devost, D., Wrzal, P., & Zingg, H. H. (2008). Oxytocin receptor signalling. *Progress in Brain Research*, *170*, 167–176. https://doi.org/10.1016/s0079-6123(08)00415-9.

Ditzen, B., Schaer, M., Gabriel, B., Bodenmann, G., Ehlert, U., & Heinrichs, M. (2009). Intranasal oxytocin increases positive communication and reduces cortisol levels during couple conflict. *Biological Psychiatry*, *65*(9), 728–731.

Dobkin, C., & Nicosia, N. (2009). The war on drugs: Methamphetamine, public health, and crime. *The American Economic Review*, *99*(1), 324–349. https://doi.org/10.1257/aer.99.1.324.

Dolen, G., Darvishzadeh, A., Huang, K. W., & Malenka, R. C. (2013). Social reward requires coordinated activity of nucleus accumbens oxytocin and serotonin. *Nature*, *501*(7466), 179–184. https://doi.org/10.1038/nature12518.

Douglas, L. A., Varlinskaya, E. I., & Spear, L. P. (2004). Rewarding properties of social interactions in adolescent and adult male and female rats: Impact of social versus isolate housing of subjects and partners. *Developmental Psychobiology*, *45*(3), 153–162. https://doi.org/10.1002/dev.20025.

Dumais, K. M., & Veenema, A. H. (2016). Vasopressin and oxytocin receptor systems in the brain: Sex differences and sex-specific regulation of social behavior. *Frontiers in Neuroendocrinology*, *40*, 1–23. https://doi.org/10.1016/j.yfrne.2015.04.003.

Elkashef, A., Vocci, F., Hanson, G., White, J., Wickes, W., & Tiihonen, J. (2008). Pharmacotherapy of methamphetamine addiction: An update. *Substance Abuse*, *29*(3), 31–49. https://doi.org/10.1080/08897070802218554.

Ermisch, A., Barth, T., Ruhle, H. J., Skopkova, J., Hrbas, P., & Landgraf, R. (1985). On the blood-brain barrier to peptides: Accumulation of labelled vasopressin, DesGlyNH2 vasopressin and oxytocin by brain regions. *Endocrinologia Experimentalis*, *19*(1), 29–37.

Everett, N. A., McGregor, I. S., Baracz, S. J., & Cornish, J. L. (2018). The role of the vasopressin V1A receptor in oxytocin modulation of methamphetamine primed reinstatement. *Neuropharmacology*, *133*, 1–11. https://doi.org/10.1016/j.neuropharm.2017.12.036.

Feltenstein, M., & See, R. (2008). The neurocircuitry of addiction: An overview. *British Journal of Pharmacology*, *154*(2), 261–274. https://doi.org/10.1038/bjp.2008.51.

Ferguson, J. N., Young, L. J., Hearn, E. F., Matzuk, M. M., Insel, T. R., & Winslow, J. T. (2000). Social amnesia in mice lacking the oxytocin gene. *Nature Genetics*, *25*, 284. https://doi.org/10.1038/77040.

Ferland, C. L., Reichel, C. M., & McGinty, J. F. (2016). Effects of oxytocin on methamphetamine-seeking exacerbated by predator odor pre-exposure in rats. *Psychopharmacology*, *233*(6), 1015–1024. https://doi.org/10.1007/s00213-015-4184-7.

Freeman, S. M., Samineni, S., Allen, P. C., Stockinger, D., Bales, K. L., Hwa, G. G., et al. (2016). Plasma and CSF oxytocin levels after intranasal and intravenous oxytocin in awake macaques. *Psychoneuroendocrinology*, *66*, 185–194. https://doi.org/10.1016/j.psyneuen.2016.01.014.

Fritz, M., El Rawas, R., Salti, A., Klement, S., Bardo, M. T., Kemmler, G., et al. (2011). Reversal of cocaine-conditioned place preference and mesocorticolimbic Zif268 expression by social interaction in rats. *Addiction Biology, 16*(2), 273–284. https://doi.org/10.1111/j.1369-1600.2010.00285.x.

Fuxe, K., Borroto-Escuela, D. O., Romero-Fernandez, W., Ciruela, F., Manger, P., Leo, G., et al. (2012). On the role of volume transmission and receptor-receptor interactions in social behaviour: Focus on central catecholamine and oxytocin neurons. *Brain Research, 1476*, 119–131. https://doi.org/10.1016/j.brainres.2012.01.062.

Gabriele, A., Pacchioni, A. M., & See, R. E. (2012). Dopamine and glutamate release in the dorsolateral caudate putamen following withdrawal from cocaine self-administration in rats. *Pharmacology, Biochemistry, and Behavior, 103*(2), 373–379. https://doi.org/10.1016/j.pbb.2012.09.015.

Georgiou, P., Zanos, P., Garcia-Carmona, J. A., Hourani, S., Kitchen, I., Kieffer, B. L., et al. (2015). The oxytocin analogue carbetocin prevents priming-induced reinstatement of morphine-seeking: Involvement of dopaminergic, noradrenergic and MOPr systems. *European Neuropsychopharmacology, 25*(12), 2459–2464. https://doi.org/10.1016/j.euroneuro.2015.09.015.

Gimpl, G., & Fahrenholz, F. (2001). The oxytocin receptor system: Structure, function, and regulation. *Physiological Reviews, 81*(2), 629–683.

Gimpl, G., Reitz, J., Brauer, S., & Trossen, C. (2008). Oxytocin receptors: Ligand binding, signalling and cholesterol dependence. *Progress in Brain Research, 170*, 193–204. https://doi.org/10.1016/s0079-6123(08)00417-2.

Gonzales, R., Mooney, L., & Rawson, R. A. (2010). The methamphetamine problem in the United States. *Annual Review of Public Health, 31*, 385–398. https://doi.org/10.1146/annurev.publhealth.012809.103600.

Grant, B. F., Goldstein, R. B., Smith, S. M., Jung, J., Zhang, H., Chou, S. P., et al. (2015). The Alcohol Use Disorder and Associated Disabilities Interview Schedule-5 (AUDADIS-5): Reliability of substance use and psychiatric disorder modules in a general population sample. *Drug and Alcohol Dependence, 148*, 27–33. https://doi.org/10.1016/j.drugalcdep.2014.11.026.

Grimm, J. W., Hope, B. T., Wise, R. A., & Shaham, Y. (2001). Neuroadaptation. Incubation of cocaine craving after withdrawal. *Nature, 412*(6843), 141–142. https://doi.org/10.1038/35084134.

Guastella, A. J., Einfeld, S. L., Gray, K. M., Rinehart, N. J., Tonge, B. J., Lambert, T. J., et al. (2010). Intranasal oxytocin improves emotion recognition for youth with autism spectrum disorders. *Biological Psychiatry, 67*(7), 692–694.

Gunaydin, L. A., Grosenick, L., Finkelstein, J. C., Kauvar, I. V., Fenno, L. E., Adhikari, A., et al. (2014). Natural neural projection dynamics underlying social behavior. *Cell, 157*(7), 1535–1551. https://doi.org/10.1016/j.cell.2014.05.017.

Hansen, S., Bergvall, A. H., & Nyiredi, S. (1993). Interaction with pups enhances dopamine release in the ventral striatum of maternal rats: A microdialysis study. *Pharmacology, Biochemistry, and Behavior, 45*(3), 673–676.

Hansson, A. C., Koopmann, A., Uhrig, S., Buhler, S., Domi, E., Kiessling, E., et al. (2017). Oxytocin reduces alcohol cue-reactivity in alcohol-dependent rats and humans. *Neuropsychopharmacology, 43*(6), 1235–1246. https://doi.org/10.1038/npp.2017.257.

Hearing, M. C., Schochet, T. L., See, R. E., & McGinty, J. F. (2010). Context-driven cocaine-seeking in abstinent rats increases activity-regulated gene expression in the basolateral amygdala and dorsal hippocampus differentially following short and long periods of abstinence. *Neuroscience, 170*(2), 570–579. https://doi.org/10.1016/j.neuroscience.2010.07.027.

Hendrickson, R. G., Cloutier, R., & McConnell, K. J. (2008). Methamphetamine-related emergency department utilization and cost. *Academic Emergency Medicine, 15*(1), 23–31. https://doi.org/10.1111/j.1553-2712.2007.00006.x.

Hicks, C., Cornish, J. L., Baracz, S. J., Suraev, A., & McGregor, I. S. (2016). Adolescent pre-treatment with oxytocin protects against adult methamphetamine-seeking behavior in female rats. *Addiction Biology*, *21*(2), 304–315. https://doi.org/10.1111/adb.12197.

Hoffman, P. L., Ritzmann, R. F., Walter, R., & Tabakoff, B. (1978). Arginine vasopressin maintains ethanol tolerance. *Nature*, *276*(5688), 614–616.

Hung, L. W., Neuner, S., Polepalli, J. S., Beier, K. T., Wright, M., Walsh, J. J., et al. (2017). Gating of social reward by oxytocin in the ventral tegmental area. *Science*, *357*(6358), 1406–1411. https://doi.org/10.1126/science.aan4994.

Ibragimov, R. (1990). Influence of neurohypophyseal peptides on the formation of active avoidance conditioned reflex behavior. *Neuroscience and Behavioral Physiology*, *20*(3), 189–193.

Ibragimov, R., Kovacs, G. L., Szabo, G., & Telegdy, G. (1987). Microinjection of oxytocin into limbic-mesolimbic brain structures disrupts heroin self-administration behavior: A receptor-mediated event? *Life Sciences*, *41*(10), 1265–1271.

Insel, T. R. (2003). Is social attachment an addictive disorder? *Physiology & Behavior*, *79*(3), 351–357.

Jodogne, C., Tirelli, E., Klingbiel, P., & Legros, J. J. (1991). Oxytocin attenuates tolerance not only to the hypothermic but also to the myorelaxant and akinesic effects of ethanol in mice. *Pharmacology, Biochemistry, and Behavior*, *40*(2), 261–265.

Jurek, B., Slattery, D. A., Hiraoka, Y., Liu, Y., Nishimori, K., Aguilera, G., et al. (2015). Oxytocin regulates stress-induced Crf gene transcription through CREB-regulated transcription coactivator 3. *The Journal of Neuroscience*, *35*(35), 12248–12260. https://doi.org/10.1523/JNEUROSCI.1345-14.2015.

Kaye, S., McKetin, R., Duflou, J., & Darke, S. (2007). Methamphetamine and cardiovascular pathology: A review of the evidence. *Addiction*, *102*(8), 1204–1211. https://doi.org/10.1111/j.1360-0443.2007.01874.x.

King, C. E., Griffin, W. C., Luderman, L. N., Kates, M. M., McGinty, J. F., & Becker, H. C. (2017). Oxytocin reduces ethanol self-administration in mice. *Alcoholism, Clinical and Experimental Research*, *41*(5), 955–964. https://doi.org/10.1111/acer.13359.

Knobloch, H. S., & Grinevich, V. (2014). Evolution of oxytocin pathways in the brain of vertebrates. *Frontiers in Behavioral Neuroscience*, *8*, 31. https://doi.org/10.3389/fnbeh.2014.00031.

Koehl, M., & Abrous, D. N. (2011). A new chapter in the field of memory: Adult hippocampal neurogenesis. *The European Journal of Neuroscience*, *33*(6), 1101–1114. https://doi.org/10.1111/j.1460-9568.2011.07609.x.

Koob, G. F. (2008). A role for brain stress systems in addiction. *Neuron*, *59*(1), 11–34. https://doi.org/10.1016/j.neuron.2008.06.012.

Koob, G. F. (2009). Neurobiological substrates for the dark side of compulsivity in addiction. *Neuropharmacology*, *56*(Suppl. 1), 18–31. https://doi.org/10.1016/j.neuropharm.2008.07.043.

Koob, G. F. (2014). Neurocircuitry of alcohol addiction: Synthesis from animal models. *Handbook of Clinical Neurology*, *125*, 33–54. https://doi.org/10.1016/B978-0-444-62619-6.00003-3.

Koob, G. F., & Mason, B. J. (2016). Existing and future drugs for the treatment of the dark side of addiction. *Annual Review of Pharmacology and Toxicology*, *56*, 299–322. https://doi.org/10.1146/annurev-pharmtox-010715-103143.

Koob, G. F., & Volkow, N. D. (2010). Neurocircuitry of addiction. *Neuropsychopharmacology*, *35*(1), 217–238. https://doi.org/10.1038/npp.2009.110.

Koob, G. F., & Volkow, N. D. (2016). Neurobiology of addiction: A neurocircuitry analysis. *The Lancet. Psychiatry*, *3*(8), 760–773. https://doi.org/10.1016/s2215-0366(16)00104-8.

Kosfeld, M., Heinrichs, M., Zak, P. J., Fischbacher, U., & Fehr, E. (2005). Oxytocin increases trust in humans. *Nature*, *435*(7042), 673–676. https://doi.org/10.1038/nature03701.

Kovacs, G. L., Borthaiser, Z., & Telegdy, G. (1985). Oxytocin reduces intravenous heroin self-administration in heroin-tolerant rats. *Life Sciences*, *37*(1), 17–26.

Kovacs, G. L., Faludi, M., & Telegdy, G. (1985). Oxytocin diminishes heroin tolerance in mice. *Psychopharmacology*, *86*(3), 377–379.

Kovacs, G. L., Sarnyai, Z., Barbarczi, E., Szabo, G., & Telegdy, G. (1990). The role of oxytocin-dopamine interactions in cocaine-induced locomotor hyperactivity. *Neuropharmacology*, *29*(4), 365–368.

Kovacs, G. L., Sarnyai, Z., & Szabo, G. (1998). Oxytocin and addiction: A review. *Psychoneuroendocrinology*, *23*(8), 945–962.

Kummer, K. K., Hofhansel, L., Barwitz, C. M., Schardl, A., Prast, J. M., Salti, A., et al. (2014). Differences in social interaction- vs. cocaine reward in mouse vs. rat. *Frontiers in Behavioral Neuroscience*, *8*, 363. https://doi.org/10.3389/fnbeh.2014.00363.

Landgraf, R., Ermisch, A., & Hess, J. (1979). Indications for a brain uptake of labelled vasopressin and ocytocin and the problem of the blood-brain barrier. *Endokrinologie*, *73*(1), 77–81.

Lee, A., Li, M., Watchus, J., & Fleming, A. S. (1999). Neuroanatomical basis of maternal memory in postpartum rats: Selective role for the nucleus accumbens. *Behavioral Neuroscience*, *113*(3), 523–538.

Lee, H. J., Macbeth, A. H., Pagani, J. H., & Young, W. S., 3rd (2009). Oxytocin: The great facilitator of life. *Progress in Neurobiology*, *88*(2), 127–151. https://doi.org/10.1016/j.pneurobio.2009.04.001.

Lee, M. R., Scheidweiler, K. B., Diao, X. X., Akhlaghi, F., Cummins, A., Huestis, M. A., et al. (2018). Oxytocin by intranasal and intravenous routes reaches the cerebrospinal fluid in rhesus macaques: Determination using a novel oxytocin assay. *Molecular Psychiatry*, *23*(1), 115–122. https://doi.org/10.1038/mp.2017.27.

Lee, M. R., & Weerts, E. M. (2016). Oxytocin for the treatment of drug and alcohol use disorders. *Behavioural Pharmacology*, *27*(8), 640–648. https://doi.org/10.1097/FBP.0000000000000258.

Lemaire, W., O'Brien, J. A., Burno, M., Chaudhary, A. G., Dean, D. C., Williams, P. D., et al. (2002). A nonpeptide oxytocin receptor antagonist radioligand highly selective for human receptors. *European Journal of Pharmacology*, *450*(1), 19–28.

Leng, G., & Ludwig, M. (2016). Intranasal oxytocin: Myths and delusions. *Biological Psychiatry*, *79*(3), 243–250. https://doi.org/10.1016/j.biopsych.2015.05.003.

Leong, K., Freeman, L., Berini, C., Ghee, S., See, R., & Reichel, C. (2017). Oxytocin reduces cocaine cued Fos activation in a regionally specific manner. *The International Journal of Neuropsychopharmacology*, *20*(10), 844–854. https://doi.org/10.1093/ijnp/pyx058.

Leong, K., Zhou, L., Ghee, S., See, R., & Reichel, C. (2016). Oxytocin decreases cocaine taking, cocaine seeking, and locomotor activity in female rats. *Experimental and Clinical Psychopharmacology*, *24*(1), 55–64. https://doi.org/10.1037/pha0000058.

Li, K., Nakajima, M., Ibanez-Tallon, I., & Heintz, N. (2016). A cortical circuit for sexually dimorphic oxytocin-dependent anxiety behaviors. *Cell*, *167*(1). 60–72.e11 https://doi.org/10.1016/j.cell.2016.08.067.

Linsenbardt, D. N., Moore, E. M., Griffin, K. D., Gigante, E. D., & Boehm, S. L., 2nd (2011). Tolerance to ethanol's ataxic effects and alterations in ethanol-induced locomotion following repeated binge-like ethanol intake using the DID model. *Alcoholism, Clinical and Experimental Research*, *35*(7), 1246–1255. https://doi.org/10.1111/j.1530-0277.2011.01459.x.

Liu, C., Wang, J., Zhan, B., & Cheng, G. (2016). Neuronal activity and the expression of hypothalamic oxytocin and vasopressin in social versus cocaine conditioning. *Behavioural Brain Research*, *310*, 84–92. https://doi.org/10.1016/j.bbr.2016.05.010.

Ludwig, M., Sabatier, N., Bull, P. M., Landgraf, R., Dayanithi, G., & Leng, G. (2002). Intracellular calcium stores regulate activity-dependent neuropeptide release from dendrites. *Nature, 418*(6893), 85–89. https://doi.org/10.1038/nature00822.

Lukas, M., Toth, I., Veenema, A. H., & Neumann, I. D. (2013). Oxytocin mediates rodent social memory within the lateral septum and the medial amygdala depending on the relevance of the social stimulus: Male juvenile versus female adult conspecifics. *Psychoneuroendocrinology, 38*(6), 916–926. https://doi.org/10.1016/j.psyneuen.2012.09.018.

MacFadyen, K., Loveless, R., DeLucca, B., Wardley, K., Deogan, S., Thomas, C., et al. (2016). Peripheral oxytocin administration reduces ethanol consumption in rats. *Pharmacology, Biochemistry, and Behavior, 140*, 27–32. https://doi.org/10.1016/j.pbb.2015.10.014.

Mantella, R. C., Vollmer, R. R., & Amico, J. A. (2005). Corticosterone release is heightened in food or water deprived oxytocin deficient male mice. *Brain Research, 1058*(1–2), 56–61. https://doi.org/10.1016/j.brainres.2005.07.062.

Maraj, S., Figueredo, V. M., & Lynn Morris, D. (2010). Cocaine and the heart. *Clinical Cardiology, 33*(5), 264–269. https://doi.org/10.1002/clc.20746.

Markou, A., Chiamulera, C., Geyer, M. A., Tricklebank, M., & Steckler, T. (2008). Removing obstacles in neuroscience drug discovery: The future path for animal models. *Neuropsychopharmacology, 34*(1), 74–89.

McEwen, B. B. (2004). Closing remarks: Review and commentary on selected aspects of the roles of vasopressin and oxytocin in memory processing. *Advances in Pharmacology, 50*(593–654), 655–708. https://doi.org/10.1016/s1054-3589(04)50015-7.

McGregor, I. S., & Bowen, M. T. (2012). Breaking the loop: Oxytocin as a potential treatment for drug addiction. *Hormones and Behavior, 61*(3), 331–339. https://doi.org/10.1016/j.yhbeh.2011.12.001.

McRae-Clark, A. L., Baker, N. L., Maria, M. M., & Brady, K. T. (2013). Effect of oxytocin on craving and stress response in marijuana-dependent individuals: A pilot study. *Psychopharmacology, 228*(4), 623–631. https://doi.org/10.1007/s00213-013-3062-4.

Melis, M. R., Melis, T., Cocco, C., Succu, S., Sanna, F., Pillolla, G., et al. (2007). Oxytocin injected into the ventral tegmental area induces penile erection and increases extracellular dopamine in the nucleus accumbens and paraventricular nucleus of the hypothalamus of male rats. *The European Journal of Neuroscience, 26*(4), 1026–1035. https://doi.org/10.1111/j.1460-9568.2007.05721.x.

Mokdad, A. H., Marks, J. S., Stroup, D. F., & Gerberding, J. L. (2004). Actual causes of death in the United States, 2000. *JAMA, 291*(10), 1238–1245. https://doi.org/10.1001/jama.291.10.1238.

Moore, E. M., Serio, K. M., Goldfarb, K. J., Stepanovska, S., Linsenbardt, D. N., & Boehm, S. L., 2nd (2007). GABAergic modulation of binge-like ethanol intake in C57BL/6J mice. *Pharmacology, Biochemistry, and Behavior, 88*(1), 105–113. https://doi.org/10.1016/j.pbb.2007.07.011.

Morales-Rivera, A., Hernandez-Burgos, M. M., Martinez-Rivera, A., Perez-Colon, J., Rivera, R., Montalvo, J., et al. (2014). Anxiolytic effects of oxytocin in cue-induced cocaine seeking behavior in rats. *Psychopharmacology, 231*, 4145–4155. https://doi.org/10.1007/s00213-014-3553-y.

Mouillac, B., Manning, M., & Durroux, T. (2008). Fluorescent agonists and antagonists for vasopressin/oxytocin G protein-coupled receptors: Usefulness in ligand screening assays and receptor studies. *Mini Reviews in Medicinal Chemistry, 8*(10), 996–1005.

Moussawi, K., & Kalivas, P. W. (2010). Group II metabotropic glutamate receptors (mGlu2/3) in drug addiction. *European Journal of Pharmacology, 639*(1–3), 115–122. https://doi.org/10.1016/j.ejphar.2010.01.030.

Moussawi, K., Zhou, W., Shen, H., Reichel, C., See, R., Carr, D., et al. (2011). Reversing cocaine-induced synaptic potentiation provides enduring protection from relapse. *Proceedings of the National Academy of Sciences of the United States of America, 108*(1), 385–390. https://doi.org/10.1073/pnas.1011265108.

Mueller, D., & Stewart, J. (2000). Cocaine-induced conditioned place preference: Reinstatement by priming injections of cocaine after extinction. *Behavioural Brain Research, 115*(1), 39–47.

Nakajima, M., Gorlich, A., & Heintz, N. (2014). Oxytocin modulates female sociosexual behavior through a specific class of prefrontal cortical interneurons. *Cell, 159*(2), 295–305. https://doi.org/10.1016/j.cell.2014.09.020.

Neumann, I. D., Kromer, S. A., Toschi, N., & Ebner, K. (2000). Brain oxytocin inhibits the (re)activity of the hypothalamo-pituitary-adrenal axis in male rats: Involvement of hypothalamic and limbic brain regions. *Regulatory Peptides, 96*(1–2), 31–38.

Neumann, I. D., & Landgraf, R. (2012). Balance of brain oxytocin and vasopressin: Implications for anxiety, depression, and social behaviors. *Trends in Neurosciences, 35*(11), 649–659. https://doi.org/10.1016/j.tins.2012.08.004.

Neumann, I. D., Maloumby, R., Beiderbeck, D. I., Lukas, M., & Landgraf, R. (2013). Increased brain and plasma oxytocin after nasal and peripheral administration in rats and mice. *Psychoneuroendocrinology, 38*(10), 1985–1993. https://doi.org/10.1016/j.psyneuen.2013.03.003.

Neumann, I. D., & Slattery, D. A. (2016). Oxytocin in general anxiety and social fear: A translational approach. *Biological Psychiatry, 79*(3), 213–221. https://doi.org/10.1016/j.biopsych.2015.06.004.

Nishimori, K., Takayanagi, Y., Yoshida, M., Kasahara, Y., Young, L. J., & Kawamata, M. (2008). New aspects of oxytocin receptor function revealed by knockout mice: Sociosexual behaviour and control of energy balance. *Progress in Brain Research, 170*, 79–90. https://doi.org/10.1016/s0079-6123(08)00408-1.

Numan, M., & Smith, H. G. (1984). Maternal behavior in rats: Evidence for the involvement of preoptic projections to the ventral tegmental area. *Behavioral Neuroscience, 98*(4), 712–727.

Owen, S. F., Tuncdemir, S. N., Bader, P. L., Tirko, N. N., Fishell, G., & Tsien, R. W. (2013). Oxytocin enhances hippocampal spike transmission by modulating fast-spiking interneurons. *Nature, 500*(7463), 458–462. https://doi.org/10.1038/nature12330.

Palamar, J. J., Davies, S., Ompad, D. C., Cleland, C. M., & Weitzman, M. (2015). Powder cocaine and crack use in the United States: An examination of risk for arrest and socioeconomic disparities in use. *Drug and Alcohol Dependence, 149*, 108–116. https://doi.org/10.1016/j.drugalcdep.2015.01.029.

Panksepp, J. B., & Lahvis, G. P. (2007). Social reward among juvenile mice. *Genes, Brain, and Behavior, 6*(7), 661–671. https://doi.org/10.1111/j.1601-183X.2006.00295.x.

Panksepp, J., Nocjar, C., Burgdorf, J., Panksepp, J. B., & Huber, R. (2004). The role of emotional systems in addiction: A neuroethological perspective. *Nebraska Symposium on Motivation, 50*, 85–126.

Pedersen, C. A. (2017). Oxytocin, tolerance, and the dark side of addiction. *International Review of Neurobiology, 136*, 239–274. https://doi.org/10.1016/bs.irn.2017.08.003.

Pedersen, C. A., & Prange, A. J., Jr. (1979). Induction of maternal behavior in virgin rats after intracerebroventricular administration of oxytocin. *Proceedings of the National Academy of Sciences of the United States of America, 76*(12), 6661–6665.

Pedersen, C. A., Smedley, K. L., Leserman, J., Jarskog, L. F., Rau, S. W., Kampov-Polevoi, A., et al. (2013). Intranasal oxytocin blocks alcohol withdrawal in human subjects. *Alcoholism, Clinical and Experimental Research, 37*(3), 484–489. https://doi.org/10.1111/j.1530-0277.2012.01958.x.

Peris, J., MacFadyen, K., Smith, J. A., de Kloet, A. D., Wang, L., & Krause, E. G. (2017). Oxytocin receptors are expressed on dopamine and glutamate neurons in the mouse ventral tegmental area that project to nucleus accumbens and other mesolimbic targets. *The Journal of Comparative Neurology*, *525*(5), 1094–1108. https://doi.org/10.1002/cne.24116.

Peters, S. T., Bowen, M. T., Bohrer, K., McGregor, I. S., & Neumann, I. D. (2017). Oxytocin inhibits ethanol consumption and ethanol-induced dopamine release in the nucleus accumbens. *Addiction Biology*, *22*(3), 702–711. https://doi.org/10.1111/adb.12362.

Peters, S., Slattery, D. A., Flor, P. J., Neumann, I. D., & Reber, S. O. (2013). Differential effects of baclofen and oxytocin on the increased ethanol consumption following chronic psychosocial stress in mice. *Addiction Biology*, *18*(1), 66–77. https://doi.org/10.1111/adb.12001.

Peters, S., Slattery, D. A., Uschold-Schmidt, N., Reber, S. O., & Neumann, I. D. (2014). Dose-dependent effects of chronic central infusion of oxytocin on anxiety, oxytocin receptor binding and stress-related parameters in mice. *Psychoneuroendocrinology*, *42*, 225–236. https://doi.org/10.1016/j.psyneuen.2014.01.021.

Qi, J., Yang, J. Y., Song, M., Li, Y., Wang, F., & Wu, C. F. (2008). Inhibition by oxytocin of methamphetamine-induced hyperactivity related to dopamine turnover in the mesolimbic region in mice. *Naunyn-Schmiedeberg's Archives of Pharmacology*, *376*(6), 441–448. https://doi.org/10.1007/s00210-007-0245-8.

Qi, J., Yang, J. Y., Wang, F., Zhao, Y. N., Song, M., & Wu, C. F. (2009). Effects of oxytocin on methamphetamine-induced conditioned place preference and the possible role of glutamatergic neurotransmission in the medial prefrontal cortex of mice in reinstatement. *Neuropharmacology*, *56*(5), 856–865. https://doi.org/10.1016/j.neuropharm.2009.01.010.

Quirin, M., Kuhl, J., & Dusing, R. (2011). Oxytocin buffers cortisol responses to stress in individuals with impaired emotion regulation abilities. *Psychoneuroendocrinology*, *36*(6), 898–904. https://doi.org/10.1016/j.psyneuen.2010.12.005.

Raam, T., McAvoy, K. M., Besnard, A., Veenema, A. H., & Sahay, A. (2017). Hippocampal oxytocin receptors are necessary for discrimination of social stimuli. *Nature Communications*, *8*(1), 2001. https://doi.org/10.1038/s41467-017-02173-0.

Rawson, R. A., Gonzales, R., & Brethen, P. (2002). Treatment of methamphetamine use disorders: An update. *Journal of Substance Abuse Treatment*, *23*(2), 145–150.

Rehm, J., Mathers, C., Popova, S., Thavorncharoensap, M., Teerawattananon, Y., & Patra, J. (2009). Global burden of disease and injury and economic cost attributable to alcohol use and alcohol-use disorders. *The Lancet*, *373*(9682), 2223–2233. https://doi.org/10.1016/S0140-6736(09)60746-7.

Reichel, C. M., & Bevins, R. A. (2008). Competition between the conditioned rewarding effects of cocaine and novelty. *Behavioral Neuroscience*, *122*(1), 140–150. https://doi.org/10.1037/0735-7044.122.1.140.

Reichel, C. M., & Bevins, R. A. (2009). Forced abstinence model of relapse to study pharmacological treatments of substance use disorder. *Current Drug Abuse Reviews*, *2*(2), 184–194.

Reichel, C. M., & Bevins, R. A. (2010). Competition between novelty and cocaine conditioned reward is sensitive to drug dose and retention interval. *Behavioral Neuroscience*, *124*(1), 141–151.

Reichel, C. M., Moussawi, K., Do, P. H., Kalivas, P. W., & See, R. E. (2011). Chronic N-acetylcysteine during abstinence or extinction following cocaine self-administration produces enduring reductions in drug-seeking. *The Journal of Pharmacology and Experimental Therapeutics*, *337*(2), 487–493. https://doi.org/10.1124/jpet.111.179317.

Reichel, C. M., Wilkinson, J. L., & Bevins, R. A. (2010). Reference place conditioning procedure with cocaine: Increased sensitivity for measuring associatively motivated choice behavior in rats. *Behavioural Pharmacology*, *21*(4), 323–331. https://doi.org/10.1097/FBP.0b013e32833b110b.

Rescorla, R. A. (2004). Spontaneous recovery. *Learning & Memory*, *11*(5), 501–509. https://doi.org/10.1101/lm.77504.

Rhodes, J. S., Best, K., Belknap, J. K., Finn, D. A., & Crabbe, J. C. (2005). Evaluation of a simple model of ethanol drinking to intoxication in C57BL/6J mice. *Physiology & Behavior*, *84*(1), 53–63. https://doi.org/10.1016/j.physbeh.2004.10.007.

Rhodes, J. S., Ford, M. M., Yu, C. H., Brown, L. L., Finn, D. A., Garland, T., Jr., et al. (2007). Mouse inbred strain differences in ethanol drinking to intoxication. *Genes, Brain, and Behavior*, *6*(1), 1–18. https://doi.org/10.1111/j.1601-183X.2006.00210.x.

Rimoldi, V., Reversi, A., Taverna, E., Rosa, P., Francolini, M., Cassoni, P., et al. (2003). Oxytocin receptor elicits different EGFR/MAPK activation patterns depending on its localization in caveolin-1 enriched domains. *Oncogene*, *22*(38), 6054–6060. https://doi.org/10.1038/sj.onc.1206612.

Robbins, T. W., & Everitt, B. J. (2002). Limbic-striatal memory systems and drug addiction. *Neurobiology of Learning and Memory*, *78*(3), 625–636.

Romero-Fernandez, W., Borroto-Escuela, D. O., Agnati, L. F., & Fuxe, K. (2013). Evidence for the existence of dopamine D2-oxytocin receptor heteromers in the ventral and dorsal striatum with facilitatory receptor-receptor interactions. *Molecular Psychiatry*, *18*(8), 849–850. https://doi.org/10.1038/mp.2012.103.

Rossoni, E., Feng, J., Tirozzi, B., Brown, D., Leng, G., & Moos, F. (2008). Emergent synchronous bursting of oxytocin neuronal network. *PLoS Computational Biology*, *4*(7), e1000123. https://doi.org/10.1371/journal.pcbi.1000123.

Roth, M. E., & Carroll, M. E. (2004). Sex differences in the acquisition of IV methamphetamine self-administration and subsequent maintenance under a progressive ratio schedule in rats. *Psychopharmacology*, *172*(4), 443–449. https://doi.org/10.1007/s00213-003-1670-0.

Rothman, R. B., Baumann, M. H., Dersch, C. M., Romero, D. V., Rice, K. C., Carroll, F. I., et al. (2001). Amphetamine-type central nervous system stimulants release norepinephrine more potently than they release dopamine and serotonin. *Synapse*, *39*(1), 32–41. https://doi.org/10.1002/1098-2396(20010101)39:1<32::Aid-syn5>3.0.Co;2-3.

Rudd, R. A., Rudd, R. A., Seth, P., David, F., & Scholl, L. (2016). Increases in drug and opioid-involved overdose deaths—United States, 2010–2015. *MMWR. Morbidity and Mortality Weekly Report*, *65*, 1445–1452.

Sala, M., Braida, D., Lentini, D., Busnelli, M., Bulgheroni, E., Capurro, V., et al. (2011). Pharmacologic rescue of impaired cognitive flexibility, social deficits, increased aggression, and seizure susceptibility in oxytocin receptor null mice: A neurobehavioral model of autism. *Biological Psychiatry*, *69*(9), 875–882. https://doi.org/10.1016/j.biopsych.2010.12.022.

Sarnyai, Z. (2011). Oxytocin as a potential mediator and modulator of drug addiction. *Addiction Biology*, *16*(2), 199–201. https://doi.org/10.1111/j.1369-1600.2011.00332.x.

Sarnyai, Z., & Kovacs, G. L. (1994). Role of oxytocin in the neuroadaptation to drugs of abuse. *Psychoneuroendocrinology*, *19*(1), 85–117.

Sarnyai, Z., & Kovacs, G. L. (2014). Oxytocin in learning and addiction: From early discoveries to the present. *Pharmacology, Biochemistry, and Behavior*, *119*, 3–9. https://doi.org/10.1016/j.pbb.2013.11.019.

Scofield, M. D., Boger, H. A., Smith, R. J., Li, H., Haydon, P. G., & Kalivas, P. W. (2015). Gq-DREADD selectively initiates glial glutamate release and inhibits cue-induced cocaine seeking. *Biological Psychiatry*, *78*(7), 441–451. https://doi.org/10.1016/j.biopsych.2015.02.016.

Shaham, Y., Shalev, U., Lu, L., de Wit, H., & Stewart, J. (2003). The reinstatement model of drug relapse: History, methodology and major findings. *Psychopharmacology, 168*(1–2), 3–20. https://doi.org/10.1007/s00213-002-1224-x.

Siegal, H. A., Carlson, R. G., Kenne, D. R., & Swora, M. G. (2003). Probable relationship between opioid abuse and heroin use. *American Family Physician, 67*(5), 942. 945.

Sinha, R. (2001). How does stress increase risk of drug abuse and relapse? *Psychopharmacology, 158*(4), 343–359. https://doi.org/10.1007/s002130100917.

Sinha, R. (2008). Chronic stress, drug use, and vulnerability to addiction. *Annals of the New York Academy of Sciences, 1141*, 105–130. https://doi.org/10.1196/annals.1441.030.

Sinha, R. (2012). How does stress lead to risk of alcohol relapse? *Alcohol Research: Current Reviews, 34*(4), 432–440.

Slattery, D. A., & Neumann, I. D. (2010). Chronic icv oxytocin attenuates the pathological high anxiety state of selectively bred Wistar rats. *Neuropharmacology, 58*(1), 56–61. https://doi.org/10.1016/j.neuropharm.2009.06.038.

Smith, S. M., & Vale, W. W. (2006). The role of the hypothalamic-pituitary-adrenal axis in neuroendocrine responses to stress. *Dialogues in Clinical Neuroscience, 8*(4), 383–395.

Song, Z., Borland, J. M., Larkin, T. E., O'Malley, M., & Albers, H. E. (2016). Activation of oxytocin receptors, but not arginine-vasopressin V1a receptors, in the ventral tegmental area of male Syrian hamsters is essential for the reward-like properties of social interactions. *Psychoneuroendocrinology, 74*, 164–172. https://doi.org/10.1016/j.psyneuen.2016.09.001.

Sprow, G. M., & Thiele, T. E. (2012). The neurobiology of binge-like ethanol drinking: Evidence from rodent models. *Physiology & Behavior, 106*(3), 325–331. https://doi.org/10.1016/j.physbeh.2011.12.026.

Stevenson, J. R., Wenner, S. M., Freestone, D. M., Romaine, C. C., Parian, M. C., Christian, S. M., et al. (2017). Oxytocin reduces alcohol consumption in prairie voles. *Physiology & Behavior, 179*, 411–421. https://doi.org/10.1016/j.physbeh.2017.07.021.

Stevenson, J. R., Young, K. A., Bohidar, A. E., Francomacaro, L. M., Fasold, T. R., Buirkle, J. M., et al. (2017). Alcohol consumption decreases oxytocin neurons in the anterior paraventricular nucleus of the hypothalamus in prairie voles. *Alcoholism, Clinical and Experimental Research, 41*(8), 1444–1451. https://doi.org/10.1111/acer.13430.

Stoop, R. (2012). Neuromodulation by oxytocin and vasopressin. *Neuron, 76*(1), 142–159. https://doi.org/10.1016/j.neuron.2012.09.025.

Sung, S., & Conry, J. M. (2006). Role of buprenorphine in the management of heroin addiction. *The Annals of Pharmacotherapy, 40*(3), 501–505. https://doi.org/10.1345/aph.1G276.

Szabo, G., Kovacs, G. L., Szekeli, S., Balaspiri, L., & Telegdy, G. (1987). C-terminal fragments of oxytocin (prolyl-leucyl-glycinamide and Z-prolyl-D-leucine) attenuate the development of tolerance to ethanol. *Acta Physiologica Hungarica, 69*(1), 115–122.

Szabo, G., Kovacs, G. L., Szekeli, S., & Telegdy, G. (1985). The effects of neurohypophyseal hormones on tolerance to the hypothermic effect of ethanol. *Alcohol, 2*(4), 567–574.

Szabo, G., Kovacs, G. L., & Telegdy, G. (1989). Intraventricular administration of neurohypophyseal hormones interferes with the development of tolerance to ethanol. *Acta Physiologica Hungarica, 73*(1), 97–103.

Tan, Y., Singhal, S., Hiller, H., Harden, S., Nguyen, D., Colon-Perez, C., et al. (2017). *Optogenetic excitation of neurons in the prefrontal cortex that express oxytocin receptors eliminates preference for social novelty in male mice.* Society for Neurosciece Abstract. 2017.

Thiel, K. J., Okun, A. C., & Neisewander, J. L. (2008). Social reward-conditioned place preference: A model revealing an interaction between cocaine and social context rewards in rats. *Drug and Alcohol Dependence, 96*(3), 202–212. https://doi.org/10.1016/j.drugalcdep.2008.02.013.

Thiele, T. E., Crabbe, J. C., & Boehm, S. L., 2nd. (2014). "Drinking in the dark" (DID): A simple mouse model of binge-like alcohol intake. *Current Protocols in Neuroscience, 68.* 9.49.1–9.49-12 https://doi.org/10.1002/0471142301.ns0949s68.

Thiele, T. E., & Navarro, M. (2014). "Drinking in the dark" (DID) procedures: A model of binge-like ethanol drinking in non-dependent mice. *Alcohol*, *48*(3), 235–241. https://doi.org/10.1016/j.alcohol.2013.08.005.

Thompson, M. R., Callaghan, P. D., Hunt, G. E., Cornish, J. L., & McGregor, I. S. (2007). A role for oxytocin and 5-HT1A receptors in the prosocial effects of 3,4 methylenedioxymethamphetamine ("ecstasy"). *Neuroscience*, *146*(2), 509–514. https://doi.org/10.1016/j.neuroscience.2007.02.032.

Tops, M., Koole, S. L., H, I. J., & Buisman-Pijlman, F. T. (2014). Why social attachment and oxytocin protect against addiction and stress: Insights from the dynamics between ventral and dorsal corticostriatal systems. *Pharmacology, Biochemistry, and Behavior*, *119*, 39–48. https://doi.org/10.1016/j.pbb.2013.07.015.

Tran-Nguyen, L. T., Fuchs, R. A., Coffey, G. P., Baker, D. A., O'Dell, L. E., & Neisewander, J. L. (1998). Time-dependent changes in cocaine-seeking behavior and extracellular dopamine levels in the amygdala during cocaine withdrawal. *Neuropsychopharmacology*, *19*(1), 48–59.

Trezza, V., Damsteegt, R., & Vanderschuren, L. J. (2009). Conditioned place preference induced by social play behavior: Parametrics, extinction, reinstatement and disruption by methylphenidate. *European Neuropsychopharmacology*, *19*(9), 659–669. https://doi.org/10.1016/j.euroneuro.2009.03.006.

Turnipseed, S. D., Richards, J. R., Kirk, J. D., Diercks, D. B., & Amsterdam, E. A. (2003). Frequency of acute coronary syndrome in patients presenting to the emergency department with chest pain after methamphetamine use. *The Journal of Emergency Medicine*, *24*(4), 369–373.

Unick, G. J., Rosenblum, D., Mars, S., & Ciccarone, D. (2013). Intertwined epidemics: National demographic trends in hospitalizations for heroin- and opioid-related overdoses, 1993-2009. *PLoS One*, *8*(2), e54496. https://doi.org/10.1371/journal.pone.0054496.

United Nations Office on Drugs and Crime. 2017. World Drug Report, June 16. Vienna.

Vaccari, C., Lolait, S. J., & Ostrowski, N. L. (1998). Comparative distribution of vasopressin V1b and oxytocin receptor messenger ribonucleic acids in brain. *Endocrinology*, *139*(12), 5015–5033. https://doi.org/10.1210/endo.139.12.6382.

Van Ree, J. M., & De Wied, D. (1977). Modulation of heroin self-administration by neurohypophyseal principles. *European Journal of Pharmacology*, *43*(2), 199–202.

Veening, J. G., & Olivier, B. (2013). Intranasal administration of oxytocin: Behavioral and clinical effects, a review. *Neuroscience and Biobehavioral Reviews*, *37*(8), 1445–1465. https://doi.org/10.1016/j.neubiorev.2013.04.012.

Veinante, P., & Freund-Mercier, M. J. (1997). Distribution of oxytocin- and vasopressin-binding sites in the rat extended amygdala: A histoautoradiographic study. *The Journal of Comparative Neurology*, *383*(3), 305–325.

Volkow, N. D., & Morales, M. (2015). The brain on drugs: From reward to addiction. *Cell*, *162*(4), 712–725. https://doi.org/10.1016/j.cell.2015.07.046.

Waldherr, M., & Neumann, I. D. (2007). Centrally released oxytocin mediates mating-induced anxiolysis in male rats. *Proceedings of the National Academy of Sciences of the United States of America*, *104*(42), 16681–16684. https://doi.org/10.1073/pnas.0705860104.

Wang, X., Gallegos, D. A., Pogorelov, V. M., O'Hare, J. K., Calakos, N., Wetsel, W. C., et al. (2018). Parvalbumin interneurons of the mouse nucleus accumbens are required for amphetamine-induced locomotor sensitization and conditioned place preference. *Neuropsychopharmacology*, *43*(5), 953–963. https://doi.org/10.1038/npp.2017.178.

Weber, R. A., Logan, C. N., Leong, K. C., Peris, J., Knackstedt, L., & Reichel, C. M. (2018). Regionally specific effects of oxytocin on reinstatement of cocaine seeking in male and female rats. *The International Journal of Neuropsychopharmacology*, *21*(7), 677–686. https://doi.org/10.1093/ijnp/pyy025.

Westover, A. N., Nakonezny, P. A., & Haley, R. W. (2008). Acute myocardial infarction in young adults who abuse amphetamines. *Drug and Alcohol Dependence*, *96*(1–2), 49–56. https://doi.org/10.1016/j.drugalcdep.2008.01.027.

Windle, R. J., Kershaw, Y. M., Shanks, N., Wood, S. A., Lightman, S. L., & Ingram, C. D. (2004). Oxytocin attenuates stress-induced c-fos mRNA expression in specific forebrain regions associated with modulation of hypothalamo-pituitary-adrenal activity. *The Journal of Neuroscience*, *24*(12), 2974–2982. https://doi.org/10.1523/JNEUROSCI.3432-03.2004.

Windle, R. J., Shanks, N., Lightman, S. L., & Ingram, C. D. (1997). Central oxytocin administration reduces stress-induced corticosterone release and anxiety behavior in rats. *Endocrinology*, *138*(7), 2829–2834. https://doi.org/10.1210/endo.138.7.5255.

Yahyavi-Firouz-Abadi, N., & See, R. E. (2009). Anti-relapse medications: Preclinical models for drug addiction treatment. *Pharmacology & Therapeutics*, *124*(2), 235–247. https://doi.org/10.1016/j.pharmthera.2009.06.014.

Young, L. J., Lim, M. M., Gingrich, B., & Insel, T. R. (2001). Cellular mechanisms of social attachment. *Hormones and Behavior*, *40*(2), 133–138. https://doi.org/10.1006/hbeh.2001.1691.

Young, L. J., & Wang, Z. (2004). The neurobiology of pair bonding. *Nature Neuroscience*, 7, 1048. https://doi.org/10.1038/nn1327.

Zaninetti, M., & Raggenbass, M. (2000). Oxytocin receptor agonists enhance inhibitory synaptic transmission in the rat hippocampus by activating interneurons in stratum pyramidale. *The European Journal of Neuroscience*, *12*(11), 3975–3984.

Zanos, P., Georgiou, P., Weber, C., Robinson, F., Kouimtsidis, C., Niforooshan, R., et al. (2017). Oxytocin and opioid addiction revisited: Old drug, new applications. *British Journal of Pharmacology*, *175*(14), 2809–2824. https://doi.org/10.1111/bph.13757.

Zanos, P., Georgiou, P., Wright, S. R., Hourani, S. M., Kitchen, I., Winsky-Sommerer, R., et al. (2014). The oxytocin analogue carbetocin prevents emotional impairment and stress-induced reinstatement of opioid-seeking in morphine-abstinent mice. *Neuropsychopharmacology*, *39*(4), 855–865. https://doi.org/10.1038/npp.2013.285.

Zhou, L., Sun, W. L., Young, A. B., Lee, K., McGinty, J. F., & See, R. E. (2014). Oxytocin reduces cocaine seeking and reverses chronic cocaine-induced changes in glutamate receptor function. *The International Journal of Neuropsychopharmacology*, *18*(1), 677–686. https://doi.org/10.1093/ijnp/pyu009.

CHAPTER NINE

Effects of Sexual Experience on Psychostimulant- and Opiate-Induced Behavior and Neural Plasticity in the Mesocorticolimbic Pathway

Lauren N. Beloate*, Lique M. Coolen*,†,1
*Department of Neurobiology and Anatomical Sciences, University of Mississippi Medical Center, Jackson, MS, United States
†Department of Physiology and Biophysics, University of Mississippi Medical Center, Jackson, MS, United States
[1]Corresponding author: e-mail address: lcoolen@umc.edu

Contents

International Review of Neurobiology, Volume 140
ISSN 0074-7742
https://doi.org/10.1016/bs.irn.2018.07.008

Abstract

Many factors, including social elements, influence drug addiction in humans and can be modeled in laboratory rodents. In general, the presence of social reward is protective against drug abuse and the absence or removal of social reward in both humans and rodents increases vulnerability to drug addiction. The current review chapter is focused on studies from our lab that have examined the effects of sociosexual behavior in male rats on drug-induced behaviors, including changes in both psychostimulant and opiate behavior. Furthermore, we review the underlying neural mechanisms by which these effects occur. Together, these results may help elucidate the neural mechanisms underlying the interaction between social and drug rewards and the mechanisms by which a loss of social rewards increase the vulnerability to drug addiction development.

1. INTRODUCTION

While many factors influence the vulnerability, development, and maintenance of human drug use and addiction, social elements, including environment and parental, peer, and romantic relationships, play a particularly important role (Andrews, Tildesley, Hops, & Li, 2002; Bahr, Hoffmann, & Yang, 2005; Beckmeyer, 2015; Zeinali, Sharifi, Enayati, Asgari, & Pasha, 2011). These social elements are studied in laboratory rodents by manipulating social variables, such as the utilization of social isolation (group versus individual housing), social and environmental enrichment (versus standard or poor enrichment), pair bonding and sexual behavior (reviewed in Beloate & Coolen, 2017). In rodent models of drug use and addiction, the presence of and experience with social behaviors are typically protective against drug reward and "seeking" (Adams, Klug, Quast, & Stairs, 2013; Aragona, Detwiler, & Wang, 2007; Beloate & Coolen, 2017; Gipson, Beckmann, El-Maraghi, Marusich, & Bardo, 2011; Solinas, Chauvet, Thiriet, Rawas, & Jaber, 2008; Thiel, Sanabria, Pentkowski, & Neisewander, 2009), while their absence or removal increases drug-seeking behavior and enhances vulnerability to drug addiction (Nader et al., 2014; Pitchers, Balfour, et al., 2010; Pitchers et al., 2013).

This review will focus on studies that have examined the effects of sociosexual behavior on drug-induced behaviors, including changes in both psychostimulant and opiate behavior. In these studies, sexual behavior of male rats was selected as the social behavior to examine effects of social experience, for several reasons: 1. it is rewarding and reinforcing; 2. measures of sexual motivation, performance and reward are well validated; 3. the behavior occurs during a relatively short time; 4. it does not require changes in housing conditions; and 5. the neurobiology of sexual performance is well studied. Importantly, sexual experience in male rats leads to cross-sensitization of locomotor response to D-amphetamine (Amph) and enhanced Amph conditioned place preference (CPP) or cross-sensitization of drug reward (Pitchers, Balfour, et al., 2010; Pitchers et al., 2013). The latter effect is dependent on a period of sexual abstinence following sexual experience, or loss of sexual reward (Pitchers et al., 2013). Additionally, sexually-experienced male rats display a cross-tolerance to morphine CPP, although it remains unknown if a period of abstinence is required (Pitchers et al., 2014).

These studies on the effects of sociosexual behavior on drug behavior also identified the underlying neural mechanisms by which these effects occurred. The interaction between sexual and drug reward is mediated, at least in part, by changes in the mesocorticolimbic pathway. Sexual behavior and the removal of sexual behavior leads to neuroplasticity that is similar to what is seen following experience with drugs of abuse, such as long-lasting increases in deltaFosB expression in the nucleus accumbens (NAc), medial prefrontal cortex (mPFC), basolateral amygdala (BLA) and ventral tegmental area (VTA) in both male and female rodents (Meisel & Mullins, 2006; Pitchers, Frohmader, et al., 2010). Sexual experience in male rodents also leads to transient increases in dendritic arborization and number of spines in NAc (Pitchers, Frohmader, et al., 2010; Pitchers et al., 2013), decreases in α-amino-3-hydroxy-5-methyl-4-isoxazolepropionic acid/*N*-methyl-D-aspartate (AMPA/NMDA) ratios in the NAc (Pitchers et al., 2012), changes in NMDA (GluN) and AMPA (GluA) receptors in NAc (Pitchers et al., 2012), and a reduction in soma size of VTA dopamine neurons (Beloate, Omrani, Adan, Webb, & Coolen, 2016; Pitchers et al., 2014). Furthermore, both VTA dopamine activity and NAc dopamine and glutamate receptor activation during sexual experience regulates sex-induced changes in drug-related behaviors and neural alterations (Beloate, Omrani, et al., 2016; Beloate, Weems, Casey, Webb, & Coolen, 2016; Pitchers et al., 2013).

Below, in further detail, laboratory measures of rodent sex- and drug-related behaviors and the influence of sexual experience on both psychostimulant and opiate behavior are described. Furthermore, the influence of sexual behavior-induced neuroplasticity on rodent drug behavior and the functional role of such neural changes will be discussed.

2. MEASURES OF MALE RAT SEXUAL BEHAVIOR

Sexual behavior is a social behavior for which parameters of appetitive and consummatory elements are well-defined and neural circuitry has been studied extensively (Beloate & Coolen, 2017; Heijkoop, Huijgens, & Snoeren, 2017). Furthermore, it is robustly rewarding and reinforcing (Beloate & Coolen, 2017; Frohmader, Pitchers, Balfour, & Coolen, 2010; Hernandez-Gonzalez, Prieto-Beracoechea, Arteaga-Silva, & Guevara, 2007; Pitchers, Frohmader, et al., 2010; Tenk, Wilson, Zhang, Pitchers, & Coolen, 2009). For the current chapter, we focus on specific neural mechanisms that mediate the interaction between sex and drug reward. For detailed discussions of the neural circuitry underlying sexual performance behavior in male and female rodents, see Veening and Coolen (2014) and Veening, Coolen, and Gerrits (2014).

2.1 Sexual Motivation and Initiation of Consummatory Behaviors

The appetitive phase of sexual behavior consists of approach and investigation of the female. The initiation of the consummatory phase of sexual behavior in rodents is typically defined by the display of the first mount or intromission. While mounts and intromissions are sometimes difficult to distinguish for novice experimenters, in rats an intromission can be discerned by a noticeable hip thrust and vaginal insertion by the male rat, followed by grooming of the penis. Upon display of a series of mounts and intromissions, males display an ejaculation. In male rats, this is accompanied by a stereotypic posture, characterized by a stillness of behavior while reared on the hind legs with front legs lifted upward. Furthermore, the male will emit ultrasonic vocalizations and will not engage with the female for approximately 5 min, referred to as the post-ejaculatory interval (McIntosh, Barfield, & Thomas, 1984). Following the post-ejaculatory interval, the male will again initiate mounts and intromissions and complete a copulatory series until sexual satiation. Typically, male rats are able to display five to eight ejaculations before reaching satiety but will not engage in sexual behavior for the following multiple days (Phillips-Farfan & Fernandez-Guasti, 2009).

2.2 Reinforcement: Sexual Experience and Long-Term Facilitation

With the acquisition of sexual experience, the behaviors in both the appetitive and consummatory phases change, which has been described in detail for male rats (for a review, see Veening & Coolen, 2014). As experience is gained, male rats spend less time on the investigation of the female prior to initiation of consummatory behaviors. The latency to initiation of consummatory behavior, or latency to first mount or intromission, thereby decreases. Males will typically reduce the numbers of mounts prior to ejaculation and thus increase copulation efficiency, and latencies to ejaculation are reduced (Beloate, Omrani, et al., 2016; Pitchers et al., 2014, 2012). This effect of sexual experience on decreased latency to mount, intromission and ejaculation, i.e., facilitation of sexual behavior, is evident even after only one mating experience, but typically requires several sessions to reach reliable detection of statistically significant differences (Beloate, Omrani, et al., 2016; Beloate, Weems, et al., 2016; Pitchers et al., 2014, 2012, 2013; Pitchers, Di Sebastiano, & Coolen, 2016; Pitchers, Frohmader, et al., 2010). Finally, experience-induced facilitation of sexual behavior is expressed in a long-term manner, as it is maintained for at least a month after the last mating session (Pitchers et al., 2014, 2012).

2.3 Reinforcement: Non-Sexual Tasks

Access to a sexually-receptive female can facilitate completion in non-sexual tasks as well. For instance, it enhances learning in paradigms such as mazes (reviewed in Frohmader, Pitchers, et al., 2010). Specifically, sexually-experienced male rats execute a T maze task faster when the goal was to reach a sexually-receptive female (Hernandez-Gonzalez et al., 2007). Furthermore, sexually-experienced male rats will learn to jump over heightened partitions and even dangerous obstacles to reach a sexually-receptive female (Bai, Belin, Zheng, Liu, & Zhang, 2017).

2.4 Reward: Conditioned Place Preference

Sexual behavior is rewarding to male rats, as is typically demonstrated using a CPP paradigm (Tenk et al., 2009). In this task, display of sexual activity is paired with one of two chambers in the CPP apparatus (sex-paired chamber) that is distinguishable from the unpaired chamber by different tactile, visual and/or olfactory contextual cues. The unpaired chamber is typically paired with a control manipulation such as handling. The change in time spent in the sex-paired chamber is determined during a posttest and used as an indicator of sexual reward-seeking behavior. Notably in sex CPP, sexual activity can be experienced inside the paired chamber, as is typically done for food- or drug-related CPP tasks (Tzschentke, 2007). However, CPP for sexual behavior is also formed when males are allowed to display sexual behavior in a separate cage outside the CPP apparatus but are transferred to the paired chamber shortly afterward. Hence, sex-induced CPP is not dependent on display of the sexual activity inside the paired chamber. Moreover, a single conditioning session is sufficient for sex-induced CPP to be detected (Pitchers et al., 2014; Tenk et al., 2009). Finally, CPP for sexual behavior is influenced by sexual experience. Specifically, both sexually-experienced and naïve males will develop a robust CPP for ejaculation. However, only sexually-naïve, but not experienced males, also develop a CPP for sexual activity that only includes mounts or intromissions, but not ejaculation (Tenk et al., 2009).

3. MEASURES OF RODENT DRUG-RELATED BEHAVIOR

3.1 Locomotor Sensitization

A widely-used approach to measure behavior related to drugs of abuse is by examining the effects of non-contingent, or experimenter-administered, drugs of abuse on locomotor behavior, often using automated activity tracking apparatuses. While these types of experiments do not directly model human drug use, they allow for experimenter control, are relatively easy to use, and can be used to determine effects of a single or repeated administration of drugs. Repeated exposure to psychostimulants, such as cocaine and Amph, leads to a sensitized locomotor response, evident as an increase in locomotion to a lower dose of the psychostimulant compared to drug-naïve animals (Robinson & Becker, 1986). Note that locomotor cross-sensitization can also occur across drug classes (Cunningham, Finn, & Kelley, 1997).

3.2 Conditioned Place Preference

As mentioned above, CPP is a procedure that measures reward, defined as a conditioned preference for an environment that was paired with a stimulus, such as drugs of abuse and social rewards (Napier, Herrold, & de Wit, 2013; Tenk et al., 2009). An animal is said to have formed a CPP for a drug if they spend more time in the drug-paired environment post-conditioning compared to pre-conditioning. This assay is particularly useful because it requires little drug exposure and animals are tested in a drug-free state during a post-test, which models human addicts' response to drug-associated cue relapse (Napier et al., 2013). In fact, CPP is also used in human studies (Napier et al., 2013). For extensive discussion on various design variations and considerations and elements of learning behaviors underlying the assay, see Tzschentke (2007) and Huston, Silva, Topic, and Muller (2013).

3.3 Self-Administration

Although the studies discussed in the present chapter focus on locomotor sensitization and CPP, we will also briefly discuss the self-administration paradigm, which is considered to be the gold standard for contingent, or subject-administered, rodent drug model. In this model, animals are trained to perform an action (lever press, nose poke, etc.) in order to receive a reward (i.e., drug) that is typically paired with auditory, visual and/or olfactory cues. Drug cessation can be modeled in the self-administration paradigm by withdrawal/abstinence (i.e., animal remaining in home cage for a period of time) or extinction (i.e., animal is placed in chamber for a period of time but receives no drug) and can be followed by a reinstatement session, during which pressing on the drug-paired lever will be reinstated by stimuli, such as stress, drug-associated cues or drug itself. Like CPP, there are many variations to self-administration, depending on the purpose of the study (Bossert, Marchant, Calu, & Shaham, 2013; O'Brien, Childress, McLellan, & Ehrman, 1992; Sinha, Shaham, & Heilig, 2011). It has been argued that self-administration better reflects the human aspects of addiction because of the contingent drug intake, and both the self-administration and CPP paradigms can be used to study different stages of the drug addiction process (i.e., initial use, withdrawal/abstinence and relapse) (Bossert et al., 2013; Napier et al., 2013).

4. INFLUENCE OF SEXUAL EXPERIENCE ON DRUG-RELATED BEHAVIOR

4.1 Psychostimulants

4.1.1 Locomotor Sensitization

In drug addiction studies, it is common to utilize consecutive daily sessions for contingent or non-contingent administration of drugs of abuse in order to determine effects of chronic drug exposure on subsequent drug-related behaviors. Therefore, in our studies testing hypotheses related to interactions between sexual experience and drugs of abuse, we also utilized daily consecutive administration sessions, but of sexual behavior. This was done in an attempt to parallel the drug addiction research and allow for direct comparisons in effects of these experiences on drug-related behavior and the underlying neural circuitry. In these studies, male rats were allowed to mate until the first ejaculation and first intromission after the post-ejaculatory interval, typically for 5 consecutive days. This end point was selected since it did not result in satiety-like effects on subsequent sexual behavior. In other words, males continued to engage in mating behavior each day, while demonstrating a significant experience-induced reduction in latencies to mount, intromission, and ejaculation across the five sessions. In some studies, additional experiments were included during which male rats displayed sexual activity in five sessions that took place over a period of 2.5 weeks, with similar results as the daily sessions (Pitchers et al., 2014). During all studies, males were housed in same sex pairs and thus never socially isolated, and received moderate enrichment in their cages, such as tubes.

First, studies in our lab showed that sexually-experienced male rats (five consecutive or five twice-weekly mating sessions) display a locomotor sensitization to Amph (Pitchers, Balfour, et al., 2010). This effect of sexual experience is independent of the context in which sex or drugs were experienced, i.e., it is not essential that mating activity occurs in the same environment in which drug effects are recorded. This is a key finding as the sensitized locomotor responses to Amph are not due to a contextual conditioned response, but rather suggest that sexual experience causes alterations within the neural circuitry that mediates drug-induced locomotor behavior. Furthermore, this effect is seen immediately after sexual experience and is long-lasting, as it persists up to a month after the last mating session (Pitchers, Balfour, et al., 2010). This effect of sexual experience may affect psychostimulant-induced locomotion in general, as sexual experience in male rats also leads to a locomotor sensitization to methamphetamine (Frohmader, Lehman, Laviolette, & Coolen, 2011). Lastly, ejaculation during sexual experience is required for the cross-sensitization to psychostimulant locomotion, which does not occur by exposure to female scents or display of mounts and intromissions without ejaculation (discussed in Beloate & Coolen, 2017). Of note, the effects of sexual experience on psychostimulant-induced locomotor behavior is similar to that observed following experience with repeated psychostimulant exposure (Robinson & Becker, 1986).

4.1.2 CPP Sensitization

Sexually-experienced male rats also develop a sensitized response to the rewarding effects of Amph. This effect of sexual experience was tested using the CPP paradigm and multiple doses of Amph. Sexually-experienced males spend an increased amount of time in the drug-paired chamber after conditioning for lower doses of Amph (0.5 and 1 mg/kg), compared to sexually-naïve males, who only formed a significant CPP for higher doses of Amph (2.5 and 5 mg/kg). Interestingly, sexually-experienced rats do not form a significant CPP for the higher doses of Amph, which may indicate a shift to aversion as associated with a highly sensitized response (Pitchers, Balfour, et al., 2010; Pitchers et al., 2013). It is currently unknown which elements of sexual activity specifically contribute to or are required for the cross-sensitization of Amph reward. It is noteworthy that the effect of sexual experience on sensitized Amph CPP requires a period of abstinence from sexual reward, as increased CPP is seen after 7 days of sexual behavior loss, but not immediately (1 day) after the last mating session. Furthermore, this effect is long-lasting and persists up to at least 4 weeks of abstinence (Pitchers et al., 2013). Hence, sensitization of Amph CPP or Amph reward-seeking behavior is dependent on the loss of sexual reward in experienced rats. It is of note that animals are not socially isolated during these experiments but are instead housed in standard conditions with another male under moderate environmental enrichment. Therefore, these findings suggest that sexual experience as well as a subsequent abstinence period cause long-lasting changes to the mesocorticolimbic system.

4.2 Morphine

Repeated exposure to opiates, both non-contingent and contingent, can lead to a behavioral tolerance (i.e., a dampened response) in rodents, in paradigms such as CPP (Russo et al., 2007; Shippenberg, Bals-Kubik, & Herz, 1987). This is seemingly the opposite of what is observed following chronic exposure to psychostimulants, although both drug classes can lead to similar endpoints: addiction and relapse in human users. Our lab has shown that sexual experience in male rats not only affects psychostimulant reward, but also influences opiate reward. Specifically, sexually-experienced males require a higher dose of morphine (5 and 10 mg/kg) to form a CPP than sexually-naïve males (0.5 mg/kg), which indicates that sexual experience leads to cross-tolerance to opiate reward (Pitchers et al., 2014). It remains unknown whether this effect requires a period of sexual abstinence, as for Amph CPP, or whether or not it generalizes to other opiates or opioids. Hence, further studies are needed to investigate the effects of different periods of loss of reward utilizing a similar approach as for the studies on cross-sensitization of Amph reward and investigate drug-seeking for other opioids/opiates, including heroin and oxycodone. Because experience with sexual reward affects both psychostimulant and opiate behavior, albeit in seemingly opposite directions, it may be beneficial to study the neural mechanisms by which these effects converge and diverge.

5. NEURAL MECHANISMS

It has long been known that drugs of abuse act on the neural circuitry that mediates natural reward behaviors, including sexual behavior. This circuitry consists of the mesocorticolimbic system, which includes the VTA, NAc, BLA, and mPFC (Cooper, Robison, & Mazei-Robison, 2017). The mesocorticolimbic system is critical for reward learning and prediction of reward errors (Keiflin & Janak, 2015). The common concept supported by a vast body of research is that repeated exposure to drugs of abuse causes alterations within this system, thereby leading to the development and maintenance of addiction (Cooper et al., 2017). However, the behavioral studies described above further extended this framework by leading to the overall hypothesis that experience with sociosexual behavior causes changes in drug-seeking behavior through long-term changes in the mesocorticolimbic system (Frohmader, Pitchers, et al., 2010). In the past several years, we have conducted numerous studies to support this hypothesis and have demonstrated that the mesocorticolimbic system is activated by sexual behavior and is involved in sexual reward-related learning. Moreover, we have delineated the neural plasticity mechanisms that contribute to the effects of sexual experience on drug-seeking behaviors. These studies are summarized below.

5.1 Mesolimbic Dopamine Pathway Activation During Sexual Behavior

Sexual behavior in male rats leads to activation of mu opioid receptors (MOR) in the VTA, which suggests that endogenous opioids are being released in the VTA during sexual behavior (Balfour, Yu, & Coolen, 2004). MOR are expressed both pre- and post-synaptically on GABA neurons (Balfour et al., 2004); hence, MOR activation likely results in inhibition of GABA neurons and disinhibition of dopamine neurons in the VTA. Indeed, mating causes activation of dopamine neurons in the VTA, as evidenced by expression of cFos in tyrosine hydroxylase (TH)+ cells (Balfour et al., 2004), and dopamine release in downstream areas, such as NAc (Damsma, Pfaus, Wenkstern, Phillips, & Fibiger, 1992; Fiorino & Phillips, 1999) and prefrontal cortex (Sanna et al., 2017).

Studies in our lab have shown that VTA MOR activation and dopamine activity in the mesolimbic pathway is not essential for the initiation of mating or performance of sexual behavior. Instead, it is important for the long-term maintenance of memory for sexual reward and cross-sensitization to Amph CPP via mechanisms of neural plasticity. This is consistent with the general role of the VTA dopamine system in reward learning and error prediction (Keiflin & Janak, 2015). Both systemic and intra-VTA administration of the opioid receptor antagonist, naloxone, during the initial acquisition of sexual behavior experience prevents the long-term facilitation of both initiation and completion of sexual performance, without having an effect on mating behavior during the acquisition of sexual experience (Pitchers et al., 2014). In order to directly test the role of VTA dopamine activity in the long-term effects of sexual experience, we expressed a Cre-driven G_i-coupled designer receptor exclusively activated by designer drugs (DREADD) in the VTA of *TH::Cre* adult male rats to block dopamine activity during sexual acquisition. Again, there were no effects of VTA dopamine activity blockade on the performance of sexual behavior. However, chemogenetic inhibition of VTA dopamine activity during mating behavior did prevent the cross-sensitization to Amph CPP following a period of sexual abstinence (Beloate, Omrani, et al., 2016). Furthermore, intra-NAc D1, but not D2, dopamine receptor antagonism during sexual experience additionally prevents cross-sensitization to Amph reward (Pitchers et al., 2013). Together, these findings indicate that, during sexual behavior, endogenous opiates are released that activate MORs on GABA interneurons in the VTA, leading to disinhibition of dopamine neurons, dopamine release in downstream areas, and activation of D1 dopamine receptors in the NAc. Blockade or antagonism of any of these events prevents the effects of sexual experience on cross-sensitization of Amph CPP, without disrupting the expression of mating behavior per se. Moreover, cross-sensitization of Amph CPP is dependent on a period of abstinence from sexual behavior; therefore, the additional neural plasticity mechanisms that develop during this period of reward loss are dependent on the initial activation of the VTA–NAc pathway during the acquisition of sexual experience.

5.2 cFos and deltaFosB Expression

cFos is a member of the Fos family of transcription factors that is commonly used as a marker for short-term neuronal activation. It is upregulated by many stimuli, including drugs of abuse, sexual behavior and mating-associated hormonal and olfactory cues in brain areas such as the mPFC and NAc (Bossert et al., 2011; Cruz et al., 2014; Pitchers, Frohmader, et al., 2010; Veening & Coolen, 2014). Chemogenetic inhibition of VTA dopamine signaling blocks mating-induced cFos, in the NAc and mPFC, but not the BLA, suggesting that dopamine activation and release in specific brain areas contributes to mating-induced cFos expression (Beloate, Omrani, et al., 2016). In addition, mating-induced cFos in the NAc is dependent on GluN activation, as it is prevented by NAc GluN antagonism (Beloate, Weems, et al., 2016). Infusion of the GluN antagonist, MK-801, in the NAc does not affect initiation or expression of sexual behavior (Beloate, Weems, et al., 2016). Finally, exposure to contextual cues associated with and predictive of sexual reward leads to cFos activation of VTA MORs and dopamine cells, and cFos and phosphorylated extracellular signal-regulated kinase (pERK) in neurons in VTA target areas, NAc, BLA, and mPFC (Balfour et al., 2004; Beloate, Omrani, et al., 2016; Pitchers et al., 2014). MOR antagonism during sexual experience prevents this mating-associated cue-induced activation of neurons in NAc, mPFC and BLA (Pitchers et al., 2014). Therefore, both dopamine and glutamate signaling in the mesocorticolimbic pathway regulate mating-induced cFos expression, or neural activation.

Although identification of cFos expression as a marker of neural activation has served an essential role in our understanding of the neural circuitry involved in social reward behavior, the exact functional contribution of cFos or the neurons expression cFos for sexual behavior and effects of experience remains unclear. Future studies are needed to test the functional significance of the neuronal ensembles that express cFos during sexual behavior, potentially using the Daun02-inactivation method (Koya et al., 2009). This approach relies on the use of cFos-lacZ transgenic rats and the local infusion of Daun02, which acts as a neurotoxin in cells expressing beta-galactosidase and, in turn, cFos (Whitaker et al., 2017). Specifically, inactivating drug-induced neuronal ensembles in the NAc blocks context-specific locomotor sensitization following repeated exposure to cocaine (Koya et al., 2009). We hypothesize that mating and drugs activate similar neuronal ensembles and that those ensembles play a functional role in the interaction between these two types of reward. Indeed, mating and methamphetamine co-activate the same neurons in the NAc, BLA and mPFC subregions, as measured by cFos and another immediate marker for neuronal activity, phosphorylated mitogen-activated protein kinase (MAP, also referred to as pERK) (Frohmader, Wiskerke, Wise, Lehman, & Coolen, 2010). However, the functional role of such co-activation remains to be determined.

DeltaFosB, another member of the Fos family, is upregulated in a long-term manner in response to various stimuli, including drugs of abuse, and is implicated in long-lasting neuroplasticity (Maze et al., 2010; Perrotti et al., 2008; Robison & Nestler, 2011). Sexual experience leads to robust deltaFosB expression in areas such as NAc, mPFC, BLA and VTA in both male and female rodents (Meisel & Mullins, 2006; Pitchers, Frohmader, et al., 2010). Furthermore, the functional blockade of deltaFosB in the NAc via a viral vector gene transfer of the dominant-negative binding partner for deltaFosB, deltaJunD, prevented both the effects of sexual experience on the long-term experience-induced facilitation of mating initiation and the cross-sensitization to Amph CPP (Pitchers, Frohmader, et al., 2010; Pitchers et al., 2013). Therefore, NAc deltaFosB expression is essential for the long-term memory for sexual performance and the long-term effects on psychostimulant reward. Mating-induced deltaFosB is, in turn, dependents on activation of VTA MOR and dopamine neurons, as chemogenetic inactivation of VTA dopamine activity prevents deltaFosB in the NAc and mPFC (Beloate, Omrani, et al., 2016). Furthermore, dopamine D1 receptor antagonism in the NAc during mating experience prevents sex-induced NAc deltaFosB (Pitchers, Frohmader, et al., 2010) and deltaFosB expression in turn regulates morphological changes of neurons in the NAc (see in Section 5.3; Beloate, Omrani, et al., 2016; Pitchers et al., 2014, 2013). Finally, sexual experience-induced deltaFosB expression in the NAc downregulates mating-induced cFos, suggesting that mating-induced cFos and deltaFosB are expressed in the same cells (Pitchers, Frohmader, et al., 2010). This is also consistent with a role for deltaFosB as a mediator of expression of other genes, including cFos and glutamate receptors (Nestler, 2013).

These results are in line with reports showing that overexpression of deltaFosB in the NAc leads to an increased cocaine CPP (Kelz et al., 1999), that repeated psychostimulant administration leads to an increase in deltaFosB in the NAc that is dependent on D1 dopamine receptor activation (Lee et al., 2006; Zhang et al., 2002), and that deltaFosB overexpression in D1 NAc neurons leads to a sensitized drug response in drug naïve animals (Kelz et al., 1999). NAc deltaFosB expression is also induced by optogenetic stimulation of the VTA (Lobo et al., 2013). Together, these findings support the hypothesis that deltaFosB contributes to the development of addiction (Nestler, 2013; Ruffle, 2014). Moreover, these findings suggest that deltaFosB is a mediator for reward learning in general, as it is induced by sociosexual reward and, thereby, is a common mediator for the interaction between social reward and drug responses.

5.3 Morphological and Synaptic Changes in the NAc

Sexual experience leads to an increase in dendritic arborization and in the number of spines in the NAc, shown using both Golgi impregnation and diOlistic visualization methods (Pitchers, Frohmader, et al., 2010; Pitchers et al., 2013). Activation of D1 receptors and deltaFosB in the NAc during sexual experience are essential for sexual experience-induced increases in dendritic arborization and number of spines (Pitchers et al., 2013). However, these morphological changes in dendritic spines are dependent on a period of sexual abstinence and are observed at 1 week, but not 1 day following last mating experience (Pitchers, Frohmader, et al., 2010). Since this NAc spinogenesis coincides with cross-sensitization of Amph CPP, we hypothesized that spinogenesis would play a functional role in the cross-sensitization to Amph CPP. However, we found that it is transient, with return back to baseline levels within 4 weeks after the last mating session, at a time point when Amph CPP cross-sensitization is still evident (Pitchers, Frohmader, et al., 2010; Pitchers et al., 2013). Therefore, we propose that NAc spinogenesis may be critical for the early development of cross-sensitization of drug reward following sexual experience and loss of sex reward, but not essential for the long-term maintenance of such effects.

Experience with psychostimulants similarly leads to increases in spinogenesis (Christian et al., 2017). Specifically, cocaine self-administration leads to increased spines only after 36 days of drug abstinence and returns to baseline at 60 days, although lever presses during reinstatement is still high at those time points (Cahill et al., 2016). Experimenter-administered cocaine leads to changes in spines that are associated with enhanced cocaine cue reinstatement (Christian et al., 2017). Therefore, NAc spinogenesis is associated with cocaine cue-induced reinstatement but does not seem to be essential for the maintenance of cross-sensitization or drug reinstatement, consistent with our earlier findings for spinogenesis after sexual experience.

5.4 Mesolimbic Glutamatergic Plasticity

As discussed above, sexual experience leads to changes in dendritic spines in the NAc. Previous studies have shown that changes in spines have been associated with alterations in synaptic strength and the localization of glutamate receptors (Gipson, Kupchik, & Kalivas, 2014). In line with these findings, sexual experience also leads to alterations in AMPA/NMDA ratios in prefrontal cortex-to-NAc post-synaptic neurons in the shell subregion (Pitchers et al., 2012). Specifically, AMPA/NMDA ratios are decreased in sexually-experienced male rats, and this effect is persistent until at least 28 days after the last mating session (Pitchers et al., 2012). Furthermore, using cross-linking biochemical assays, our lab has shown that mating leads to a transient GluN1 subunit upregulation in the NAc within 24 h after last sexual behavior exposure, and GluA-1 and -2 subunits are upregulated after 4 weeks of sexual reward abstinence (Pitchers et al., 2012). These results suggest that initial sexual experience leads to activation and an increase in GluN receptors, followed by synthesis and trafficking of GluA receptors (Pitchers et al., 2012). We tested the hypothesis that GluN activation during mating regulates sexual experience-induced neuroplasticity and the cross-sensitization to Amph reward. Intra-NAc GluN antagonism each of the 4 days of initial sexual experience prevents experience-induced Amph CPP and cFos expression in the NAc without affecting sexual behavior (Beloate, Weems, et al., 2016). Finally, sexual experience is associated with downregulation of type 5 metabotropic glutamate receptor (mGluR5) expression in the NAc (Pitchers et al., 2016). Similar to the changes in GluA receptor expression, this downregulation is dependent on abstinence from sexual reward is observed at 7 days after last mating session and lasts at least 28 days after last mating session. Blockade of mGluR5 receptors in the NAc with the antagonists, 2-methyl-6-(phenylethynyl)pyridine (MPEP) or 3-((2-methyl-4-thiazolyl)ethynyl)pyridine (MTEP), during mating experience did not affect the initiation nor expression of sexual behavior and did not prevent effects of sexual experience on facilitation of sexual behavior or cross-sensitization of Amph CPP. Instead, repeated daily antagonism of mGluR5 receptors in the NAc of sexually-naïve male rats caused an enhanced CPP for Amph and cross-sensitization of Amph locomotor activity. Hence, these findings suggest that cross-sensitization of Amph reward and locomotor responses after loss of sexual reward is partially caused by a reduction mGluR5 receptor activation in the NAc.

Experience with drugs of abuse also leads to changes in the mesolimbic glutamatergic pathway, and these changes contribute to subsequent self-administration, CPP and reinstatement (Caprioli, Justinova, Venniro, & Shaham, 2018; Scofield et al., 2016). Of note, the GluN1 subunit is increased following drug experience, although it occurs later in the abstinence period, rather than immediately after experience, as is seen after mating (Ghasemzadeh, Mueller, & Vasudevan, 2009; Pitchers et al., 2012). Furthermore, much like during sexual experience abstinence, GluA-1 and -2 receptor subunits are upregulated during the drug abstinence period (Boudreau & Wolf, 2005; Ghasemzadeh et al., 2009). However, during drug abstinence, GluA1 receptors remain elevated, while GluA2 expression decreases, leading to GluA2-lacking receptors (Conrad et al., 2008; Mameli et al., 2009). This was not observed in our studies following sexual experience. Sexual behavior only leads to a slight increase in GluA1 expression, and GluA2 membrane expression is upregulated after 1 month of sexual abstinence. This suggests that GluA receptors may regulate the cross-sensitization to psychostimulant CPP, but only after long periods of sexual abstinence (Pitchers et al., 2012). This hypothesis remains untested, and examination of later time points in the abstinence period is required to further explore this possibility. Finally, changes in AMPA/NMDA ratios following drug experience are dependent on both NAc subregion and cortical input (Kalivas, Volkow, & Seamans, 2005). However, this has not been determined for sexual experience.

5.5 Interactions Between Dopamine and Glutamate in the Mesolimbic Pathway

Our work has shown that both the dopamine and glutamate neurotransmitter systems in the mesolimbic pathway regulate the long-term effects of the natural reward, sexual behavior, such as long-term memory for initiation and performance of mating, mating-induced neuroplasticity, and the cross-sensitization to psychostimulant reward. We hypothesize that the two systems interact to play this role. Indeed, VTA neurons can co-release dopamine and GABA, dopamine and glutamate, or glutamate and GABA, although little is known about the functional role of this co-release in reward behavior (Lapish, Seamans, & Chandler, 2006; Morales & Margolis, 2017; Root et al., 2014; Seutin, 2005). Furthermore, glutamate synaptic plasticity has been proposed to regulate the roles of dopamine receptor signaling (Hikida et al., 2013), and glutamatergic synaptic plasticity at dopamine D1 receptor-expressing neurons facilitates cocaine cue-induced relapse (Pascoli et al., 2014). Studying the functional role of these interactions could be beneficial for a better understanding of the influence of natural rewards on drug rewards.

5.6 Morphological Changes in the VTA

Sexually-experienced male rats display a reduction in both area and perimeter of VTA dopamine neurons (Beloate, Omrani, et al., 2016; Pitchers et al., 2014). This effect is dependent on MOR activation in the VTA, is transient and essential for long-term maintenance of sexual reward memory (Beloate, Omrani, et al., 2016, Pitchers et al., 2014). This decrease in VTA dopamine neurons is similar to what is seen following chronic contingent or non-contingent opiate experience (Mazei-Robison & Nestler, 2012; Sklair-Tavron et al., 1996). Furthermore, opiate-induced morphological changes are associate with decreased dopamine trafficking and release and are associated with opiate reward tolerance following experience (Mazei-Robison et al., 2011; Russo et al., 2007). Together, these results suggest that sex experience-induced morphological changes in VTA dopamine neurons is at least part of the neural mechanism that regulates the cross-tolerance to opiate reward. In addition, sex experience-induced morphological changes in VTA dopamine neurons are also likely to contribute to the cross-sensitization to psychostimulant reward, even though psychostimulant exposure does not directly affect VTA dopamine soma size (Mazei-Robison et al., 2014).

6. SUMMARY AND CONCLUSIONS

In conclusion, experience with sexual behavior and the subsequent loss of sexual reward leads to changes in both psychostimulant- and opiate-induced behaviors in male rats. Work in our lab has uncovered the neural subpopulations that are activated during initial mating experience and the mechanisms by which sex experience induces neuroplasticity in the mesocorticolimbic pathway, as well the functional role those changes play. Furthermore, our lab has shown that MOR and dopamine cell activation in the VTA and downstream dopamine and glutamate receptor activation in the NAc is not essential for the initiation or expression mating behavior per se, but instead plays an important role for sexual reward memory, sex experience-induced neuroplasticity, and cross-sensitization to psychostimulants. Recent findings also indicate that dopamine cell morphological changes in the VTA play an important role in the regulation of the cross-tolerance to opiate reward. Together, these results may help elucidate the neural mechanisms underlying the interaction between social and drug rewards and the mechanisms by which loss of social rewards increases the vulnerability to drug addiction development.

REFERENCES

Adams, E., Klug, J., Quast, M., & Stairs, D. J. (2013). Effects of environmental enrichment on nicotine-induced sensitization and cross-sensitization to d-amphetamine in rats. *Drug and Alcohol Dependence*, *129*(3), 247–253.

Andrews, J., Tildesley, E., Hops, H., & Li, F. (2002). The influence of peers on young adult substance use. *Health Psychology*, *21*(4), 349–357.

Aragona, B. J., Detwiler, J. M., & Wang, Z. (2007). Amphetamine reward in the monogamous prairie vole. *Neuroscience Letters*, *418*(2), 190–194.

Bahr, S. J., Hoffmann, J. P., & Yang, X. (2005). Parental and peer influences on the risk of adolescent drug use. *The Journal of Primary Prevention*, *26*(6), 529–551.

Bai, Y., Belin, D., Zheng, X., Liu, Z., & Zhang, Y. (2017). Acute stress worsens the deficits in appetitive behaviors for social and sexual stimuli displayed by rats after long-term withdrawal from morphine. *Psychopharmacology*, *234*(11), 1693–1702.

Balfour, M. E., Yu, L., & Coolen, L. M. (2004). Sexual behavior and sex-associated environmental cues activate the mesolimbic system in male rats. *Neuropsychopharmacology*, *29*(4), 718–730.

Beckmeyer, J. J. (2015). Comparing the associations between three types of adolescents' romantic involvement and their engagement in substance use. *Journal of Adolescence*, *42*, 140–147.

Beloate, L. N., & Coolen, L. M. (2017). Influences of social reward experience on behavioral responses to drugs of abuse: Review of shared and divergent neural plasticity mechanisms for sexual reward and drugs of abuse. *Neuroscience and Biobehavioral Reviews*, *83*, 356–372.

Beloate, L. N., Omrani, A., Adan, R. A., Webb, I. C., & Coolen, L. M. (2016). Ventral tegmental area dopamine cell activation during male rat sexual behavior regulates neuroplasticity and d-amphetamine cross-sensitization following sex abstinence. *The Journal of Neuroscience*, *36*(38), 9949–9961.

Beloate, L. N., Weems, P. W., Casey, G. R., Webb, I. C., & Coolen, L. M. (2016). Nucleus accumbens NMDA receptor activation regulates amphetamine cross-sensitization and deltaFosB expression following sexual experience in male rats. *Neuropharmacology*, *101*, 154–164.

Bossert, J. M., Marchant, N. J., Calu, D. J., & Shaham, Y. (2013). The reinstatement model of drug relapse: Recent neurobiological findings, emerging research topics, and translational research. *Psychopharmacology*, *229*(3), 453–476.

Bossert, J. M., Stern, A. L., Theberge, F. R., Cifani, C., Koya, E., Hope, B. T., et al. (2011). Ventral medial prefrontal cortex neuronal ensembles mediate context-induced relapse to heroin. *Nature Neuroscience*, *14*(4), 420–422.

Boudreau, A. C., & Wolf, M. E. (2005). Behavioral sensitization to cocaine is associated with increased AMPA receptor surface expression in the nucleus accumbens. *The Journal of Neuroscience*, *25*(40), 9144–9151.

Cahill, M. E., Bagot, R. C., Gancarz, A. M., Walker, D. M., Sun, H., Wang, Z. J., et al. (2016). Bidirectional synaptic structural plasticity after chronic cocaine administration occurs through Rap1 small GTPase signaling. *Neuron*, *89*(3), 566–582.

Caprioli, D., Justinova, Z., Venniro, M., & Shaham, Y. (2018). Effect of novel allosteric modulators of metabotropic glutamate receptors on drug self-administration and relapse: A review of preclinical studies and their clinical implications. *Biological Psychiatry*, *84*(3), 180–192.

Christian, D. T., Wang, X., Chen, E. L., Sehgal, L. K., Ghassemlou, M. N., Miao, J. J., et al. (2017). Dynamic alterations of rat nucleus accumbens dendritic spines over 2 months of abstinence from extended-access cocaine self-administration. *Neuropsychopharmacology*, *42*(3), 748–756.

Conrad, K. L., Tseng, K. Y., Uejima, J. L., Reimers, J. M., Heng, L. J., Shaham, Y., et al. (2008). Formation of accumbens GluR2-lacking AMPA receptors mediates incubation of cocaine craving. *Nature*, *454*(7200), 118–121.

Cooper, S., Robison, A. J., & Mazei-Robison, M. S. (2017). Reward circuitry in addiction. *Neurotherapeutics*, *14*(3), 687–697.

Cruz, F. C., Babin, K. R., Leao, R. M., Goldart, E. M., Bossert, J. M., Shaham, Y., et al. (2014). Role of nucleus accumbens shell neuronal ensembles in context-induced reinstatement of cocaine-seeking. *The Journal of Neuroscience*, *34*(22), 7437–7446.

Cunningham, S. T., Finn, M., & Kelley, A. E. (1997). Sensitization of the locomotor response to psychostimulants after repeated opiate exposure: Role of the nucleus accumbens. *Neuropsychopharmacology*, *16*, 147–155.

Damsma, G., Pfaus, J. G., Wenkstern, D., Phillips, A. G., & Fibiger, H. C. (1992). Sexual behavior increases dopamine transmission in the nucleus accumbens and striatum of male rats: Comparison with novelty and locomotion. *Behavioral Neuroscience*, *106*(1), 181–191.

Fiorino, D. F., & Phillips, A. G. (1999). Facilitation of sexual behavior and enhanced dopamine efflux in the nucleus accumbens of male rats after d-amphetamine-induced behavioral sensitization. *The Journal of Neuroscience*, *19*(1), 456–463.

Frohmader, K. S., Lehman, M. N., Laviolette, S. R., & Coolen, L. M. (2011). Concurrent exposure to methamphetamine and sexual behavior enhances subsequent drug reward and causes compulsive sexual behavior in male rats. *The Journal of Neuroscience*, *31*(45), 16473–16482.

Frohmader, K. S., Pitchers, K. K., Balfour, M. E., & Coolen, L. M. (2010). Mixing pleasures: Review of the effects of drugs on sex behavior in humans and animal models. *Hormones and Behavior*, *58*(1), 149–162.

Frohmader, K. S., Wiskerke, J., Wise, R. A., Lehman, M. N., & Coolen, L. M. (2010). Methamphetamine acts on subpopulations of neurons regulating sexual behavior in male rats. *Neuroscience*, *166*(3), 771–784.

Ghasemzadeh, M. B., Mueller, C., & Vasudevan, P. (2009). Behavioral sensitization to cocaine is associated with increased glutamate receptor trafficking to the postsynaptic density after extended withdrawal period. *Neuroscience*, *159*(1), 414–426.

Gipson, C. D., Beckmann, J. S., El-Maraghi, S., Marusich, J. A., & Bardo, M. T. (2011). Effect of environmental enrichment on escalation of cocaine self-administration in rats. *Psychopharmacology*, *214*(2), 557 566.

Gipson, C. D., Kupchik, Y. M., & Kalivas, P. W. (2014). Rapid, transient synaptic plasticity in addiction. *Neuropharmacology*, *76*(Pt B), 276–286.

Heijkoop, R., Huijgens, P. T., & Snoeren, E. M. S. (2017). Assessment of sexual behavior in rats: The potentials and pitfalls. *Behavioural Brain Research*. pii: S0166-4328(17)31233-0.

Hernandez-Gonzalez, M., Prieto-Beracoechea, C. A., Arteaga-Silva, M., & Guevara, M. A. (2007). Different functionality of the medial and orbital prefrontal cortex during a sexually motivated task in rats. *Physiology & Behavior*, *90*(2–3), 450–458.

Hikida, T., Yawata, S., Yamaguchi, T., Danjo, T., Sasaoka, T., Wang, Y., et al. (2013). Pathway-specific modulation of nucleus accumbens in reward and aversive behavior via selective transmitter receptors. *Proceedings of the National Academy of Sciences of the United States of America*, *110*(1), 342–347.

Huston, J. P., Silva, M. A., Topic, B., & Muller, C. P. (2013). What's conditioned in conditioned place preference? *Trends in Pharmacological Sciences*, *34*(3), 162–166.

Kalivas, P. W., Volkow, N., & Seamans, J. (2005). Unmanageable motivation in addiction: A pathology in prefrontal-accumbens glutamate transmission. *Neuron*, *45*(5), 647–650.

Keiflin, R., & Janak, P. H. (2015). Dopamine prediction errors in reward learning and addiction: From theory to neural circuitry. *Neuron*, *88*(2), 247–263.

Kelz, M. B., Chen, J., Carlezon, W. A., Whisler, K., Gilden, L., Beckmann, A. M., et al. (1999). Expression of the transcription factor deltaFosB in the brain controls sensitivity to cocaine. *Nature*, *401*(6750), 272–276.

Koya, E., Golden, S. A., Harvey, B. K., Guez-Barber, D. H., Berkow, A., Simmons, D. E., et al. (2009). Targeted disruption of cocaine-activated nucleus accumbens neurons prevents context-specific sensitization. *Nature Neuroscience*, *12*(8), 1069–1073.

Lapish, C. C., Seamans, J. K., & Chandler, L. J. (2006). Glutamate-dopamine cotransmission and reward processing in addiction. *Alcoholism, Clinical and Experimental Research*, *30*(9), 1451–1465.

Lee, K. W., Kim, Y., Kim, A. M., Helmin, K., Nairn, A. C., & Greengard, P. (2006). Cocaine-induced dendritic spine formation in D1 and D2 dopamine receptor-containing medium spiny neurons in nucleus accumbens. *Proceedings of the National Academy of Sciences of the United States of America*, *103*(9), 3399–3404.

Lobo, M. K., Zaman, S., Damez-Werno, D. M., Koo, J. W., Bagot, R. C., DiNieri, J. A., et al. (2013). DeltaFosB induction in striatal medium spiny neuron subtypes in response to chronic pharmacological, emotional, and optogenetic stimuli. *The Journal of Neuroscience*, *33*(47), 18381–18395.

Mameli, M., Halbout, B., Creton, C., Engblom, D., Parkitna, J. R., Spanagel, R., et al. (2009). Cocaine-evoked synaptic plasticity: Persistence in the VTA triggers adaptations in the NAc. *Nature Neuroscience*, *12*(8), 1036–1041.

Maze, I., Covington, H. E., 3rd, Dietz, D. M., LaPlant, Q., Renthal, W., Russo, S. J., et al. (2010). Essential role of the histone methyltransferase G9a in cocaine-induced plasticity. *Science*, *327*(5962), 213–216.

Mazei-Robison, M. S., Appasani, R., Edwards, S., Wee, S., Taylor, S. R., Picciotto, M. R., et al. (2014). Self-administration of ethanol, cocaine, or nicotine does not decrease the soma size of ventral tegmental area dopamine neurons. *PLoS One*, *9*(4), e95962.

Mazei-Robison, M. S., Koo, J. W., Friedman, A. K., Lansink, C. S., Robison, A. J., Vinish, M., et al. (2011). Role for mTOR signaling and neuronal activity in morphine-induced adaptations in ventral tegmental area dopamine neurons. *Neuron*, *72*(6), 977–990.

Mazei-Robison, M. S., & Nestler, E. J. (2012). Opiate-induced molecular and cellular plasticity of ventral tegmental area and locus coeruleus catecholamine neurons. *Cold Spring Harbor Perspectives in Medicine*, *2*(7), a012070.

McIntosh, T. K., Barfield, R. J., & Thomas, D. (1984). Electrophysiological and ultrasonic correlates of reproductive behavior in the male rat. *Behavioral Neuroscience*, *98*(6), 1100–1103.

Meisel, R. L., & Mullins, A. J. (2006). Sexual experience in female rodents: Cellular mechanisms and functional consequences. *Brain Research*, *1126*(1), 56–65.

Morales, M., & Margolis, E. B. (2017). Ventral tegmental area: Cellular heterogeneity, connectivity and behaviour. *Nature Reviews. Neuroscience*, *18*(2), 73–85.

Nader, J., Claudia, C., El Rawas, R., Favot, L., Jaber, M., Thiriet, N., et al. (2014). Loss of environmental enrichment increases vulnerability to cocaine addiction. *Neuropsychopharmacology*, *39*(3), 780.

Napier, T. C., Herrold, A. A., & de Wit, H. (2013). Using conditioned place preference to identify relapse prevention medications. *Neuroscience and Biobehavioral Reviews*, *37*(9 Pt A), 2081–2086.

Nestler, E. J. (2013). Cellular basic of memory for addiction. *Dialogues in Clinical Neuroscience*, *15*(4), 431–443.

O'Brien, C. P., Childress, A. R., McLellan, A. T., & Ehrman, R. (1992). Classical conditioning in drug-dependent humans. *The Neurobiology of Drug and Alcohol Addiction*, *654*(1), 400–415.

Pascoli, V., Terrier, J., Espallergues, J., Valjent, E., O'Connor, E. C., & Luscher, C. (2014). Contrasting forms of cocaine-evoked plasticity control components of relapse. *Nature*, *509*(7501), 459–464.

Perrotti, L. I., Weaver, R. R., Robison, B., Renthal, W., Maze, I., Yazdani, S., et al. (2008). Distinct patterns of DeltaFosB induction in brain by drugs of abuse. *Synapse*, *62*(5), 358–369.

Phillips-Farfan, B. V., & Fernandez-Guasti, A. (2009). Endocrine, neural and pharmacological aspects of sexual satiety in male rats. *Neuroscience and Biobehavioral Reviews*, *33*(3), 442–455.

Pitchers, K. K., Balfour, M. E., Lehman, M. N., Richtand, N. M., Yu, L., & Coolen, L. M. (2010). Neuroplasticity in the mesolimbic system induced by natural reward and subsequent reward abstinence. *Biological Psychiatry*, *67*(9), 872–879.

Pitchers, K. K., Coppens, C. M., Beloate, L. N., Fuller, J., Van, S., Frohmader, K. S., et al. (2014). Endogenous opioid-induced neuroplasticity of dopaminergic neurons in the ventral tegmental area influences natural and opiate reward. *The Journal of Neuroscience*, *34*(26), 8825–8836.

Pitchers, K. K., Di Sebastiano, A. R., & Coolen, L. M. (2016). mGluR5 activation in the nucleus accumbens is not essential for sexual behavior or cross-sensitization of amphetamine responses by sexual experience. *Neuropharmacology*, *107*, 122–130.

Pitchers, K. K., Frohmader, K. S., Vialou, V., Mouzon, E., Nestler, E. J., Lehman, M. N., et al. (2010). DeltaFosB in the nucleus accumbens is critical for reinforcing effects of sexual reward. *Genes, Brain, and Behavior*, *9*(7), 831–840.

Pitchers, K. K., Schmid, S., Di Sebastiano, A. R., Wang, X., Laviolette, S. R., Lehman, M. N., et al. (2012). Natural reward experience alters AMPA and NMDA receptor distribution and function in the nucleus accumbens. *PLoS One*, 7(4), e34700.

Pitchers, K. K., Vialou, V., Nestler, E. J., Laviolette, S. R., Lehman, M. N., & Coolen, L. M. (2013). Natural and drug rewards act on common neural plasticity mechanisms with DeltaFosB as a key mediator. *The Journal of Neuroscience*, *33*(8), 3434–3442.

Robinson, T. E., & Becker, J. B. (1986). Enduring changes in brain and behavior produced by chronic amphetamine administration: A review and evaluation of animal models of amphetamine psychosis. *Brain Research Reviews*, *11*, 157–198.

Robison, A. J., & Nestler, E. J. (2011). Transcriptional and epigenetic mechanisms of addiction. *Nature Reviews. Neuroscience*, *12*(11), 623–637.

Root, D. H., Mejias-Aponte, C. A., Zhang, S., Wang, H. L., Hoffman, A. F., Lupica, C. R., et al. (2014). Single rodent mesohabenular axons release glutamate and GABA. *Nature Neuroscience*, *17*(11), 1543–1551.

Ruffle, J. K. (2014). Molecular neurobiology of addiction: What's all the deltaFosB about? *The American Journal of Drug and Alcohol Abuse*, *40*(6), 428–437.

Russo, S. J., Bolanos, C. A., Theobald, D. E., DeCarolis, N. A., Renthal, W., Kumar, A., et al. (2007). IRS2-Akt pathway in midbrain dopamine neurons regulates behavioral and cellular responses to opiates. *Nature Neuroscience*, *10*(1), 93–99.

Sanna, F., Bratzu, J., Piludu, M. A., Corda, M. G., Melis, M. R., Giorgi, O., et al. (2017). Dopamine, noradrenaline and differences in sexual behavior between roman high and low avoidance male rats: A microdialysis study in the medial prefrontal cortex. *Frontiers in Behavioral Neuroscience*, *11*, 108.

Scofield, M. D., Heinsbroek, J. A., Gipson, C. D., Kupchik, Y. M., Spencer, S., Smith, A. C., et al. (2016). The nucleus accumbens: Mechanisms of addiction across drug classes reflect the importance of glutamate homeostasis. *Pharmacological Reviews*, *68*(3), 816–871.

Seutin, V. (2005). Dopaminergic neurones: Much more than dopamine? *British Journal of Pharmacology*, *146*(2), 167–169.

Shippenberg, T. S., Bals-Kubik, R., & Herz, A. (1987). Motivational properties of opioids: Evidence that an activation of delta-receptors mediates reinforcement processes. *Brain Research*, *436*, 234–239.

Sinha, R., Shaham, Y., & Heilig, M. (2011). Translational and reverse translational research on the role of stress in drug craving and relapse. *Psychopharmacology*, *218*(1), 69–82.

Sklair-Tavron, L., Chi, W. X., Lane, S. B., Harris, H. W., Bunney, B. S., & Nestler, E. J. (1996). Chronic morphine induces visible changes in the morphology of mesolimbic dopamine neurons. *Proceedings of the National Academy of Sciences of the United States of America*, *93*(20), 11202–11207.

Solinas, M., Chauvet, C., Thiriet, N., Rawas, R. E., & Jaber, M. (2008). Reversal of cocaine addiction by environmental enrichment. *Proceedings of the National Academy of Sciences of the United States of America*, *105*(44), 17145–17150.

Tenk, C. M., Wilson, H., Zhang, Q., Pitchers, K. K., & Coolen, L. M. (2009). Sexual reward in male rats: Effects of sexual experience on conditioned place preferences associated with ejaculation and intromissions. *Hormones and Behavior*, *55*(1), 93–97.

Thiel, K. J., Sanabria, F., Pentkowski, N. S., & Neisewander, J. L. (2009). Anti-craving effects of environmental enrichment. *The International Journal of Neuropsychopharmacology*, *12*(9), 1151–1156.

Tzschentke, T. M. (2007). Measuring reward with the conditioned place preference (CPP) paradigm: Update of the last decade. *Addiction Biology*, *12*(3–4), 227–462.

Veening, J. G., & Coolen, L. M. (2014). Neural mechanisms of sexual behavior in the male rat: Emphasis on ejaculation-related circuits. *Pharmacology, Biochemistry, and Behavior*, *121*, 170–183.

Veening, J. G., Coolen, L. M., & Gerrits, P. O. (2014). Neural mechanisms of female sexual behavior in the rat; comparison with male ejaculatory control. *Pharmacology, Biochemistry, and Behavior*, *121*, 16–30.

Whitaker, L. R., Warren, B. L., Venniro, M., Harte, T. C., McPherson, K. B., Beidel, J., et al. (2017). Bidirectional modulation of intrinsic excitability in rat prelimbic cortex neuronal ensembles and non-ensembles after operant learning. *The Journal of Neuroscience*, *37*(36), 8845–8856.

Zeinali, A., Sharifi, H., Enayati, M., Asgari, P., & Pasha, G. (2011). The mediational pathway among parenting styles, attachment styles and self-regulation with addiction susceptibility of adolescents. *Journal of Research in Medical Sciences*, *16*(9), 1105–1121.

Zhang, D., Zhang, L., Lou, D. W., Nakabeppu, Y., Zhang, J., & Xu, M. (2002). The dopamine D1 receptor is a critical mediator for cocaine-induced gene expression. *Journal of Neurochemistry*, *82*(6), 1453–1464.

CHAPTER TEN

Social Defeat Stress, Sex, and Addiction-Like Behaviors

Akiko Shimamoto[1]
Department of Biochemistry, Cancer Biology, Neuroscience, and Pharmacology, Meharry Medical College, Nashville, TN, United States
[1]Corresponding author: e-mail address: ashimamoto@mmc.edu

Contents

Abstract

Social confrontation is a form of social interaction in animals where two conspecific individuals confront each other in dispute over territory, during the formation of hierarchies, and during breeding seasons. Typically, a social confrontation involves a prevailing individual and a yielding individual. The prevailing individual often exhibits aggressive postures and launches attacks, whereas the yielding individual often adopts postures of defeat. The yielding or defeated animals experience a phenomenon known as social defeat stress, in which they show exaggerated stress as well as autonomic and endocrine responses that cause impairment of both the brain and body. In laboratory

International Review of Neurobiology, Volume 140
ISSN 0074-7742
https://doi.org/10.1016/bs.irn.2018.07.009

settings, one can reliably generate social defeat stress by allowing a naïve (or already defeated) animal to intrude into a home cage in which its resident has already established a territory or is nursing. This resident-intruder paradigm has been widely used in both males and females to study mechanisms in the brain that underlie the stress responses.

Stress has profound effects on drug reward for cocaine, methamphetamine, alcohol, and opioids. Particularly, previous experiences with social defeat can exaggerate subsequent addiction-like behaviors. The extent of these addiction-like behaviors depends on the intensity, duration, frequency, and intermittency of the confrontation episodes. This chapter describes four types of social defeat stress: acute, repeated, intermittent, and chronic. Specifically, it focuses on social defeat stress models used in laboratories to study individual, sex, and animal strain differences in addiction-like behaviors.

1. INTRODUCTION

In society, a group of individuals within a common territory share interaction and culture, and must come in contact with one another. This social interaction is well preserved among both vertebrate and invertebrate animals, and allows them to form aggregations, cooperate in sexual and parental behavior, engage in disputes over territory, or simply to communicate across space (Koenig & Dickinson, 2009). Unpredictable or uncontrollable situations, such as isolation from society or during the formation of hierarchies or breeding seasons, can induce normal or exaggerated stress responses within individuals. Stress responses caused by unsettling social interaction, characterized by either the absence of an anticipatory response or prolonged neuroendocrine reaction, constitute social stress (Fish, DeBold, & Miczek, 2005; Koolhaas et al., 2011; Sgoifo, Koolhaas, Alleva, Musso, & Parmigiani, 2001). For this reason, stress responses or other associated impairments caused by social stress do not always show linearity like those caused by other types of stressors, such as physiological, mechanical, or pharmacological stressors (Koolhaas et al., 2011; Sapolsky, 2004). Thus, social stress has a unique, complex, and global impact on both brain and body.

Social stress is strongly linked to central nervous system disorders (McEwen & Stellar, 1993). Particularly, social stress can increase the prevalence, interfere with treatments, or even exaggerate the symptoms of a number of psychiatric disorders, including major depressive disorder (Turner & Lloyd, 1999; Turner & Marino, 1994), post-traumatic stress disorder (Brewin, Andrews, & Valentine, 2000), and drug addiction (Sinha, 2001).

In addition, social stress can worsen a person's physical health (Turner, Wheaton, & Lloyd, 1995). For example, it is well known that stress increases the risk of cardiovascular disease (Bartolomucci, 2007; Dimsdale, 2008; Rozanski, Blumenthal, & Kaplan, 1999; Steptoe & Kivimaki, 2013), compromises immune system in order to facilitate cancer cell progression, and promotes tumor growth (Kiecolt-Glaser et al., 2003; Reiche, Nunes, & Morimoto, 2004; Thaker et al., 2006). On the other hand, social ties and social support can maintain the psychological well-being, prevent mental illnesses from progressing, and buffer the impact of such illnesses (Kawachi & Berkman, 2001).

Some social stressors that occur in society can be mimicked in laboratory settings (Blanchard, McKittrick, & Blanchard, 2001; Toth & Neumann, 2013). Such social stressors include social deprivation (i.e., social isolation and maternal separation) (Cirulli et al., 2009; Hall, 1998), social instability (i.e., a cycle of social isolation and crowding) (Baranyi, Bakos, & Haller, 2005; Green & McCormick, 2013), witnessing social confrontation (Church, 1959; Ramsey & Van Ree, 1993), and territorial/maternal aggression that results in social defeat (Haney, Maccari, Le Moal, Simon, & Piazza, 1995; Kudryavtseva, 1994; Kudryavtseva & Bakshtanovskaya, 1989; Lucion & De Almeida, 1996; Miczek, Yap, & Covington, 2008; Shimamoto, Debold, Holly, & Miczek, 2011). Among these social stressors, social defeat produces the most severe stress responses, resulting in the most dramatic rises in plasma corticosterone (CORT) levels (Koolhaas et al., 2011). Thus, this review will focus on social defeat stress used in laboratory settings. Particularly, it will discuss how the intensity, duration, intermittency, and frequency of social defeat impact subsequent addiction-like behaviors. The review will also discuss methods used in laboratories to predict or categorize stress vulnerability behaviorally, and its outcomes on addiction-like responses. Additionally, this review will also present evidence that indicate potential sex differences in social defeat stress-induced addiction-like behaviors. As such, this review will not cover mechanical and physical stressors, such as restraint stress, foot-shock stress, swim stress, or cold stress, unless otherwise necessary.

2. SOCIAL DEFEAT STRESS IN LABORATORY ANIMALS

2.1 Social Confrontations as a Social Stressor

Social confrontations occur when two males engage in a pervasive pattern of continuous interactions, with a prevailing individual and a yielding individual (Miczek, Covington, Nikulina, & Hammer, 2004). While both prevailing and yielding animals show an increase in plasma CORT levels (Covington & Miczek, 2001; Koolhaas et al., 2011), their behaviors can be precisely differentiated. Specifically, the prevailing (i.e., dominant) male shows threat displays, as indicated by his commanding posture that pushes the yielding (i.e., subordinate) male aside with his flanks, pursuits, attack bites, and upright postures (i.e., boxing) (Miczek et al., 2004). On the other hand, the subordinate male exhibits defensive or submissive supine posture and flight responses such as those seen in rats, or upright posture with retracted ears and limp forepaws seen in mice (Miczek et al., 2004). While both dominant and subordinate males show a robust increase in blood pressure during confrontations (Fokkema & Koolhaas, 1985; Raab, Seizinger, & Herz, 1985), it is only the subordinate animals that decrease nociception and increase analgesia that are otherwise blocked by opioid antagonists (Miczek, Thompson, & Shuster, 1982; Teskey & Kavaliers, 1987; Teskey, Kavaliers, & Hirst, 1984). These conspecific encounters frequently occur while forming hierarchy or during breeding seasons, and rarely occur otherwise.

2.2 Social Defeat Stress in Laboratory Settings

2.2.1 Males: Territorial Aggression

Social defeat stress is defined as stress responses in subordinate animals caused by social defeat during social confrontations (Blanchard, Fukunaga-Stinson, Takahashi, Flannelly, & Blanchard, 1984; Blanchard et al., 2001; Miczek et al., 2008). Social defeat stress can be reliably produced in male laboratory animals by allowing a confrontation between a naïve male and a larger male who has been co-housed with a receptive female for a period of time and has experience victory in prior confrontations (Goyens & Noirot, 1975; Miczek et al., 2008; Pryce & Fuchs, 2017). The introduction of a naïve male to the pair's homecage (the receptive female is removed from the homecage during confrontations) causes the naïve male (i.e., intruder) to escape from the dominant male and prompts the dominant male to attack the intruder. This laboratory-induced social confrontation is known as the resident-intruder paradigm and has been extensively used to produce social defeat stress in rats (Miczek et al., 2008), hamsters (Huhman, 2006), both inbred and outbred mice (Goyens & Noirot, 1975; Kudryavtseva, 1994; Kudryavtseva & Bakshtanovskaya, 1989), and tree shrews (Fuchs et al., 2001). In addition, a modified resident-intruder paradigm has been developed for inbred C57BL/6 mouse strain where these mice receive attacks by experienced heterospecific CD-1 mice, in order to induce more robust depressive-like effects (Berton et al., 2006; Golden, Covington, Berton, & Russo, 2011).

2.2.2 Females: Maternal Aggression

Female animals of some species, including California mice (Trainor et al., 2011), Syrian hamsters (Huhman et al., 2003; Solomon, Karom, & Huhman, 2007), prairie voles (Smith, Lieberwirth, & Wang, 2013), and mandarin voles (Wang et al., 2018), show reliable territorial aggression toward conspecific naïve female intruders in general laboratory settings. Female rats and mice that are more commonly used in biomedical research can also develop territorial aggression when they are group-housed in a relatively large cage (DeBold & Miczek, 1984; Haug & Mandel, 1978), but exhibit less aggression when they are singly-housed (Edwards & Rowe, 1975). It has also been proposed that females from both outbred and inbred mouse strains may not show reliable territorial aggression (Palanza, 2001). Lactating dams, on the other hand, exhibit the most intense and reliable aggression toward naïve female intruders in order to protect their pups from the intruders, regardless of the housing condition (Erskine, Barfield, & Goldman, 1978; Gandelman, 1972). This maternal aggression is widely exploited in rats (Haney & Miczek, 1993; Holly, Shimamoto, Debold, & Miczek, 2012; Shimamoto et al., 2011; Shimamoto, Holly, Boyson, DeBold, & Miczek, 2015), outbred mouse (Jacobson-Pick, Audet, McQuaid, Kalvapalle, & Anisman, 2013), and Mongolian gerbils (Lonstein & Gammie, 2002) to induce social stress in naïve female intruders. It should be noted, however, that maternal aggression is ethologically distinct from territorial aggression. While maternal aggression is a form of defensive aggression where intruders receive less offensive attacks from the dams, territorial aggression is a form of offensive aggression where the intruders receive bouts of intensive attacks from the aggressors (Miczek et al., 2008). In addition, studies have shown that females are more vulnerable to social instability than to aggression since aggression does not reduce body weight gain, adrenal weights, or plasma CORT levels in females, regardless of the number of attack bites they receive (Dunn & Critchlow, 1973; Haller, Fuchs, Halasz, & Makara, 1999; Haller, Kiem, & Makara, 1996). Thus, from an ethological perspective, it is important to perform a variety of assessments to justify or address potential confounding effects underlying the sex differences in social defeat stress-induced outcomes when using territorial and maternal aggression as a social stressor for males and females, respectively.

2.3 Stress and Endocrine Responses to Social Defeat by Different Time Schedules

Both the hypothalamic–pituitary–adrenal (Palanisamy et al., 2010) axis and the sympathetic adrenomedullary (SA) system are responsible for stress responses. It is widely accepted that the SA system is more involved in "fight-or-flight" responses by releasing norepinephrine (NE), while the HPA axis is involved in all dynamics of stress by releasing glucocorticoids (GCs) (Sapolsky, Romero, & Munck, 2000). Thus, both the HPA and SA systems are the primary indicators for evaluating stress responses to social defeat. Like any other social stressors, social defeat causes non-linear stress responses that are dependent on the defeat duration, the frequency of social defeat episodes, and the intermittency between each social defeat. Thus, this review identifies four types of social defeat stress: acute, repeated, intermittent, and chronic. While there is no written criteria for the definition, this review will identify a single social defeat as "acute," a single social defeat for three to seven consecutive days as "repeated," repeated social defeat episodes with an interval of more than 24 h between each episode as "intermittent," and daily social defeat for more than 10 days as "chronic" (Fig. 1).

2.3.1 Acute Social Defeat Stress

Acute social defeat stress refers to a situation where an animal is exposed to a brief (1–15 min) territorial or maternal aggression from conspecifics. In males, acute social defeat as a result of territorial aggression increases plasma CORT levels (Bhatnagar, Vining, Iyer, & Kinni, 2006; Keeney, Hogg, & Marsden, 2001; Kollack-Walker, Don, Watson, & Akil, 1999; Walker, Masters, Dielenberg, & Day, 2009) and adrenocorticotropic hormone (ACTH) levels (Backstrom & Winberg, 2013; Bhatnagar et al., 2006; Keeney et al., 2006; Wotjak et al., 1996), and autonomic responses (Chung, Martinez, & Herbert, 2000). It also facilitates the expression of corticotropin releasing

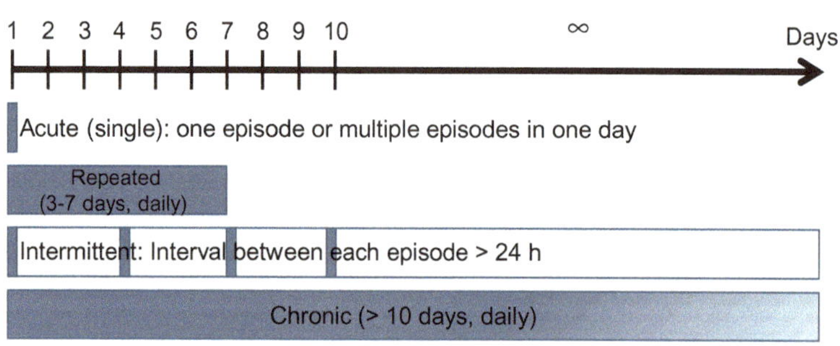

Fig. 1 Description of social defeat stress by frequency.

factor (CRF), a 41-amino acid peptide secreted from the paraventricular nucleus (PVN) of the hypothalamus in response to stress (Fekete et al., 2009; Keeney et al., 2006; Marini et al., 2006). Among the social stressors, acute social defeat yields the highest CORT responses (Koolhaas et al., 2011). Behaviorally, males do not show apparent stress-induced despair, such as impaired locomotor activity or excessive weight gain or loss (Bhatnagar et al., 2006). However, a single defeat for 30min suppressed CORT responses and increased food intake during the light phase (Berton, Durand, Aguerre, Mormede, & Chaouloff, 1999; Bhatnagar et al., 2006). These animals also showed sensitized ACTH and autonomic responses to an additional acute stressor (Berton et al., 1999). Although the number of studies examined the effect of acute social defeat in females is small (Haller et al., 1999; Solomon, 2017), it has been shown that exposure to maternal aggression (Neumann, Toschi, Ohl, Torner, & Kromer, 2001) or to territorial aggression (Smith et al., 2013) can increase CORT and/or ACTH. Acute social defeat also increased cellular responses in cells producing arginine-vasopressin (AVP), a neuropeptide responsible for stress responses (Steinman et al., 2015).

2.3.2 Repeated Social Defeat Stress

Repeated social defeat refers a situation where animals are exposed to episodes of territorial or maternal aggression in consecutive days ranging from a few days to a week. While most studies perform one aggressive encounter a day, some studies perform bouts of aggressive encounters within a day (Trainor et al., 2011, 2013). Repeated exposure to territorial or maternal aggression induces habituation in stress responses and does not induce ACTH or CORT responses in defeated males (Bhatnagar et al., 2006; Trainor et al., 2013), but it increases their autonomic responses (Keeney et al., 2006; Meehan, Tornatzky, & Miczek, 1995; Tornatzky & Miczek, 1993, 1994) and food intake during the light phase (Bhatnagar et al., 2006). Furthermore, repeated social defeat is shown to decrease locomotor activity and weight gain (Bhatnagar et al., 2006). In females, repeated territorial aggression increased CORT levels (Trainor et al., 2013).

2.3.3 Intermittent Social Defeat Stress

Intermittent or episodic social defeat refers to a situation where animals are exposed to repeated social defeat episodes with an interval of more than 24h between each episode. Intermittency in social defeat produces long-lasting and non-habituation effects on autonomic responses, body temperature, and heart rate (Sgoifo, Pozzato, et al., 2001; Sgoifo et al., 2002).

Furthermore, intermittent social defeat increases plasma CORT levels and decreases glucocorticoid receptor (GR) mRNA expression in the hippocampus (Covington & Miczek, 2001; Patel, Katz, Karssen, & Lyons, 2008). Intermittent social defeat has been shown to increase body mass and food intake in Syrian hamsters (Foster, Solomon, Huhman, & Bartness, 2006). Intermittent social defeat has been well-adopted in exaggerating behavioral and neuronal responses to psychostimulants (see Section 3.3). Intermittent social defeat increased CORT levels to a similar extent in both males and females (Jacobson-Pick et al., 2013).

2.3.4 Chronic Social Defeat Stress

In contrast to animals exposed to acute, repeated, and intermittent social defeat, animals that are chronically exposed to territorial or maternal aggression (daily social defeat exposure for 10–35 days) show increased or unaltered stress responses, and their responses appear to be dependent on species and laboratory settings. For example, inbred C57BL/6 mice confronted by heterospecific outbred CD-1 mice for 10 days showed increased plasma CORT levels (Krishnan et al., 2007) but no change in CRF mRNA expression (Han, Yang, et al., 2017; Keeney et al., 2006). Outbred CD-1 mice exposed to chronic social defeat stress for 21 days also show an increased CORT levels in both males and females (Jacobson-Pick et al., 2013). California mice also showed increased CORT responses in both males and females and their degree of response was indistinguishable particularly during the short photoperiod (Ribble & Salvioni, 1990; Trainor et al., 2011). On the other hand, our previous findings did not reveal any changes in CORT levels in outbred Long–Evans male and female rats exposed to daily defeat by conspecifics for 21 days (Rappeneau, Blaker, Petro, Yamamoto, & Shimamoto, 2016). Other chronic stressors such as chronic mild stress or chronic variable stress also did not cause changes in CORT levels (Dalla et al., 2005; Hodes et al., 2015). Nonetheless, chronic social defeat stress has been widely used as a model of major depression in both inbred and outbred rodents since these animals show some cardinal features of human depression, such as anhedonia-like responses, decreased social interaction, and attenuated weight gain (Krishnan et al., 2007; Krishnan & Nestler, 2008; Rappeneau et al., 2016; Shimamoto et al., 2011, 2015).

One of the unique features of chronic social defeat is that it disrupts the estrous cycle in females. Similar to chronic mild stress protocol by Dalla and colleagues (Dalla et al., 2005, 2008; Dalla, Pitychoutis, Kokras, & Papadopoulou-Daifoti, 2010, 2011), our chronic social defeat stress protocol has successfully quantitated the disruption of estrous cycle by counting the number of reproductive or non-reproductive phases using a cytological examination of vaginal smears (Rappeneau et al., 2016; Shimamoto et al., 2011, 2015). Using this method, we have shown that a 21-day chronic social defeat stress exposure reduces the reproductive phases in Long–Evans female rats (Rappeneau et al., 2016; Shimamoto et al., 2011, 2015). This disruption in the estrous cycle is in line with irregularity in menstrual cycle in women (Barron, Flick, Cook, Homan, & Campbell, 2008). Reduced or erratic secretion of reproductive hormones is known to mediate depressive symptoms in women, and hormone replacement therapy has been successful in alleviating these symptoms to some depressive patients (Berlanga & Flores-Ramos, 2006; Bloch et al., 2000; Cohen, Soares, Vitonis, Otto, & Harlow, 2006; Freeman, Sammel, Lin, & Nelson, 2006; Martenyi, Dossenbach, Mraz, & Metcalfe, 2001; Thase, Entsuah, Cantillon, & Kornstein, 2005).

2.4 Individual Differences in Stress Reactivity

2.4.1 Social Defeat to Characterize Individual Differences in Stress Reactivity

Humans as well as non-human primates and rodents react to stress differently, even when they are exposed to the same intensity of stress (McEwen & Stellar, 1993; Sapolsky, 2015). As previously described, social confrontations between conspecifics induce stress responses in both individuals in feral, farm, and laboratory animals, as indicated by their endocrine and behavioral responses (Koolhaas et al., 1999). The individual differences in stress reactivity and/or stress responses can be determined phenotypically by observing the animals' behaviors during or post-confrontation in acute, repeated, or chronic exposure to social defeat. Two statistical models have been used to characterize behavioral phenotypes with defeat behaviors as a variable: cut-off behavioral criteria (CBC) and clustering analysis. The CBC model extracts two extreme datasets within the same variable (Cohen & Zohar, 2004). Using this model, one study showed that a single episode of territorial aggression in Sprague-Dawley rats can cause the naïve intruders to exhibit either "active" or "passive" behavior (Walker et al., 2009). An "active" behavior, as characterized by the increased number of fighting and guarding behaviors toward an aggressor, results in lower CORT levels in the intruders, whereas a "passive" behavior, as characterized by less fighting and guarding and more immobility, results in higher CORT levels in the intruders (Walker et al., 2009). The passive or "low" fighters also show significant increase in immunoreactivity to transcription factor c-Fos in several of their brain regions, including prefrontal cortex (PFC) and extended amygdala, compared to active or "high" fighters. These observations suggest that individual differences in stress reactivity or coping style can be determined by a single episode of social encounter that affects both endocrine and neuronal changes (Koolhaas et al., 1999; Walker et al., 2009). In addition to the CBC model, clustering analysis has been used to characterize individual differences in defeat-associated stress reactivity. In one case, the latency for an intruder to show a spine (submissive) posture during a social confrontation is used as a variable (Wood, Walker, Valentino, & Bhatnagar, 2010). Intruders whose latency to show a spine posture for longer than 300 s by the third defeat episode are known as long latency (LL) intruders, whereas those whose latency is less than 300 s are known as short latency (SL) intruders. In a study using rats, Wood et al. showed that the normalization of ACTH and CORT post-defeat was significantly faster in the LL rats compared to the SL rats (Wood et al., 2010). In this study, the SL rats also showed decreased mRNA levels for CRF and CRF receptor subtype 1 (CRFR1) in the PVN, suggesting that the SL phenotype is more vulnerable to repeated social defeat compared to the LL phenotype (Wood et al., 2010). In mice, strain differences, rather than individual differences, have been well-documented with respect to both acute and chronic stress response (Crawley et al., 1997). When exposed to acute or chronic inescapable foot shock, BALB/cByJ mouse strain has shown the highest ACTH and CORT responses compared to other strains, such as CD-1 and C57Bl/6J mouse strains (Shanks, Griffiths, Zalcman, Zacharko, & Anisman, 1990).

Chronic social defeat is known to produce two distinct behavioral phenotypes within the same strain in both inbred and outbred animals. In an outbred OF1 male mouse, a 21-day exposure to chronic social defeat produces either an "active" or a "passive" behavioral phenotype, as determined by clustering analysis with immobility duration being most discriminative variable (Gomez-Lazaro et al., 2011). The "passive" mice that showed an extended immobility time had higher CORT and lower hippocampal brain-derived neurotrophic factor (BDNF) protein levels compared to "active" mice that had a shorter immobility duration (Gomez-Lazaro et al., 2011). We have previously used clustering analysis with intake of a sweet-tasting solution (0.02% saccharin) as a variable to classify two subpopulations within the stressed Long–Evans female rats during exposure to 21 days of chronic social defeat stress (Shimamoto et al., 2015). Specifically, we showed that the stress-vulnerable group had a lower tendency to self-administer cocaine than the stress-resilient or non-stressed control group (Shimamoto et al., 2015). Furthermore, the stress-resilient group showed a blunted dopamine response in the nucleus accumbens (NAc) to acute cocaine (Shimamoto et al., 2015). Thus, both male and female outbred strains show endocrine and behavioral profiles that are distinguishable within a group when exposed to chronic social defeat stress.

When chronically exposed, inbred mouse strains are relatively more resilient to social confrontations by conspecifics compared to outbred strains (Crawley et al., 1997). For example, in C57Bl/6J mice, a prolonged exposure to intermittent stress for 60 days does not produce depressive- or anxiety-like symptoms, nor does it alter monoamine turnovers in brain areas associated with stress (Tannenbaum & Anisman, 2003). On the other hand, in the NAc in BALB/cByJ mice, the same type of stress exposure results in pronounced depression- and anxiety-like symptoms, increased NE and serotonin (5-HT) metabolite levels in the PVN and hippocampus, and increased dopamine (DA) metabolite level (Tannenbaum & Anisman, 2003). The differences in stress reactivity observed for BALB/cByJ and C57Bl6/J mice were also apparent in a social interaction test when both strains of mice were exposed to 10 days of chronic social defeat, such that BALB/cByJ mice spent less time interacting with an aggressor than the C57Bl/6J mice (Razzoli et al., 2011). These strain differences in behavioral profile appear to be correlated with CRF, such that its mRNA expression in C57Bl/6J is increased compared to that in BALB/cByJ mice during exposure to unpredictable stress (Anisman, Lacosta, Kent, McIntyre, & Merali, 1998; Anisman, Prakash, Merali, & Poulter, 2007; Belzung, El Hage, Moindrot, & Griebel, 2001; Shanks, Griffiths, & Anisman, 1994; Shanks, Zalcman, Zacharko, & Anisman, 1991). Similarly, C57Bl/6J mice showed a shorter immobility time than CBA/Lac mice (Kudryavtseva & Bakshtanovskaya, 1989). The stress resiliency in C57Bl/6J mice is possibly due to their very low aggressiveness that does not produce stress responses during social confrontations (Crawley et al., 1997; Selmanoff, Maxson, & Ginsburg, 1976). However, when each social confrontation was followed by an indirect contact in which the two conspecifics were allowed olfactory and visual communication through a partition with small holes or a meshed screen without physical contacts (Kudryavtseva, 1994), C57Bl/6J mice exhibit depressive-like symptoms as indicated by their extended immobility time in forced swim test as well as their reduced social interaction with an unfamiliar mouse (Kudryavtseva, 1994). This partition method has been used in outbred Long–Evans male and female rats as well to induce depressive-like symptoms (Miczek, Nikulina, Shimamoto, & Covington, 2011; Rappeneau et al., 2016; Shimamoto et al., 2011, 2015).

While conspecific confrontations do not produce intense stress effects in both individuals for the C57Bl/6J mouse strain, 10 days of social defeat experience by a heterospecific CD-1 mouse results in two phenotypes in the interaction time when facing an unfamiliar naïve mouse. Compared to the interaction time exhibited by the non-stressed group, a subpopulation of the defeated mice spends less time in the interacting zone with the unfamiliar naïve mouse. Roughly 50% of this subpopulation, termed the "susceptible" group, showed anhedonia- and anxiety-like responses, increased serum CORT levels, increased body temperature during a social interaction test, increased immune responses and mRNA expression of CRF in hypothalamus, and a disrupted circadian rhythm, compared to non-defeated controls (Han, Yang, et al., 2017; Hodes et al., 2014; Joana et al., 2016; Krishnan et al., 2007). The "unsusceptible" group that did not show any deficits in social interaction exhibited anxiety-like responses and increase in serum CORT levels, but did not show depression-like responses or disrupted circadian rhythm (Krishnan et al., 2007, 2008).

2.4.2 Prediction of Individual Differences in Stress Response and Reactivity

In both animals and humans, the individual differences in stress response and reactivity can be predicted by exposing a subject to a subtle or moderate stressor (Bardo, Neisewander, & Kelly, 2013; Dellu, Piazza, Mayo, Le Moal, & Simon, 1996). For example, when placed in a new cage for the first time, Wistar rats have shown differential duration in locomotor activity, below or above the median locomotor activity within a cohort. The subgroup above the median was termed as the high-responder (HR) group and below the median was termed as the low-responder (LR) group (Hooks, Jones, Smith, Neill, & Justice, 1991a, 1991b; Piazza, Deminiere, Le Moal, & Simon, 1989). These two subgroups showed striking differences in anxiety-like responses as well as in CORT response to an additional acute stressor, with the HR group being more active and vulnerable to those responses and more prone to obtain psychostimulant rewards (Bardo et al., 2013; Brennan, Roberts, Anisman, & Merali, 2001; Hooks et al., 1991a, 1991b; Kabbaj, Devine, Savage, & Akil, 2000; Kabbaj, Evans, Watson, & Akil, 2004). The heightened responses in HR animals corresponded to low GR mRNA expression in dentate gyrus, low CRF levels in central amygdala, and high mRNA levels of CRH in the PVN (Kabbaj et al., 2000). A microarray studies showed that HR and LR animals were uniquely different in their mRNA expression levels, such as genes associated with extracellular cell signaling, neurotransmitter receptors-related genes, and genes implicated in synaptic function (Kabbaj et al., 2004). Moreover, repeated exposure to social defeat resulted in depressive-like symptoms in HR, but not LR, animals, as assessed by sucrose preference test (Duclot, Hollis, Darcy, & Kabbaj, 2011; Hollis, Duclot, Gunjan, & Kabbaj, 2011).

3. ANIMAL MODELS OF DRUG ADDICTION

Nearly 27 million people suffer from drug addiction in the United States (Hedden, Kennet, Lipari, Medley, & Tice, 2014). It is a progressive disease that causes a person's drug taking behavior to shift from a casual recreational use to a more impulsive and compulsive use, followed by a withdrawal that facilitates drug craving and relapsing behaviors (for comprehensive review see Becker & Koob, 2016; Carroll, Lynch, Roth, Morgan, & Cosgrove, 2004; Koob & Le Moal, 2001; Koob & Volkow, 2010). Drug addiction is thus often referred to as a chronic relapsing disorder. Several neuronal networks, including dopaminergic, serotonergic, glutamatergic, and GABAergic neurons across discrete midbrain and forebrain regions, are involved in facilitating drug reward and dependence. In laboratory settings, several experimental models have been used to study various stages of addiction-like behaviors and possible underlying mechanisms.

3.1 Conditioned Place Preference

The conditioned place preference (CPP) paradigm is widely used to test an animal's preference for an addictive drug over a non-addictive drug (for comprehensive review see Tzschentke, 2007). First, an animal is conditioned to associate the texture of the wall and floor of a chamber with drug experiences. Eventually, the animal will prefer to spend its time in this drug-associated chamber in the absence of an actual drug. CPP can effectively test animals conditioned with psychostimulants (cocaine, methamphetamine, and amphetamine) as well as alcohol and opioids. The increased preference can be also seen in animals conditioned with DA agonists; such preference can be reversed using DA antagonists (Tzschentke, 2007). Thus, CPP assesses a motivational component of drug reward (Tzschentke, 2007).

3.2 Behavioral Sensitization

Behavioral sensitization refers to sensitized locomotor responses to a low-to-moderate dose of stimulant in animals, preceded by their experiences with the repeated administration of a higher dose of the same or a different kind of stimulant. The behavioral sensitization is particularly prominent when psychomotor stimulants that directly affect the animals' motor responses, such as cocaine or methamphetamine, are administered. Behavioral sensitization is also observed for non-psychomotor stimulants, such as nicotine, opioids, or alcohol (for comprehensive review see Robinson & Berridge, 1993). Effects of behavioral sensitization are also seen in healthy non-addicted human volunteers, who exhibit increased eye blinks, amounts of speech, activity/energy levels, and energetic mood (Boileau et al., 2006; Strakowski, Sax, Setters, & Keck, 1996; but also see Rothman et al., 1994; Wachtel & de Wit, 1999). Both human and animal models of behavioral sensitization show neuronal sensitization, namely, increased DA and glutamate release in the NAc and PFC projecting from the ventral tegmental area (VTA) (Boileau et al., 2006; Steketee & Kalivas, 2011; Wolf, 1998). Behavioral sensitization is thought to reinforce incentive salience to the drugs and thus may play a role in relapse (Robinson & Berridge, 1993; Vanderschuren & Kalivas, 2000).

3.3 Operant Self-Administration

By far, operant self-administration protocol in animals provides the most accurate and precise information regarding the behavioral, molecular, and biological effects of drugs abuse (Sanchis-Segura & Spanagel, 2006). As the name implies, the self-administration protocol tests an animal's voluntary intake of a drug by performing an operant response (i.e., lever press), often in conjunction with discrete cues such as a light or sound. In this test, the animal presses a lever extended from an operant panel or emits a nosepoke that breaks an infrared beam, prompting a delivery of drug either via an intravenous injection or through a receptacle for oral consumption. Drug delivery often occurs with simultaneous presentation of a discrete cue, and this cue-response cycle reinforces the animal's drug taking behavior. A variety of reinforcement schedule can recapitulate any addiction stages typically observed in humans, including an acquisition of drug experiences, maintenance of drug intake, motivational intake, a binge-like drug taking behavior, withdrawal, and relapse-like behaviors. Operant self-administration protocol has been established for the use of psychostimulants (Farrell, Schoch, & Mahler, 2018; Roberts, Morgan, & Liu, 2007; Stefanski, Ladenheim, Lee, Cadet, & Goldberg, 1999), alcohol (McBride & Li, 1998; Weiss, Lorang, Bloom, & Koob, 1993), nicotine (Corrigall, Franklin, Coen, & Clarke, 1992; Pontieri, Tanda, Orzi, & Di Chiara, 1996), opioids (Mello & Negus, 1996), and cannabinoids (Tanda & Goldberg, 2003).

4. SOCIAL DEFEAT STRESS-INDUCED ADDICTION-LIKE BEHAVIORS

4.1 Acute (Single) Exposure to Social Defeat

An acute or single exposure to social defeat immediately increases CORT levels, as well as *fos* gene expression and immunoreactivity to c-Fos protein in both males and females (Haney et al., 1995; Martinez, Phillips, & Herbert, 1998; Miczek, Nikulina, Kream, Carter, & Espejo, 1999; Nikulina, Marchand, Kream, & Miczek, 1998). During a confrontation, the defeated animals showed profound increase in the accumbal DA (Haney et al., 1995; Tidey & Miczek, 1997). When psychostimulants were administered immediately after the social defeat, however, the defeated animal's locomotor activity was transiently suppressed; this suppression corresponded with a reduction in c-Fos protein expression, an indirect marker for neuronal activity (Miczek et al., 1999; Nikulina et al., 1998). Acute social defeat also transiently decreased mRNA expression of BDNF, a protein that assists neuronal development and plasticity (Pizarro et al., 2004). However, these transient suppressions were normalized and exaggerated when the psychostimulants were administered few days after the one episode of social defeat (de Jong, Wasilewski, van der Vegt, Buwalda, & Koolhaas, 2005; Miczek et al., 1999; Nikulina et al., 1998; Wang et al., 2013). Moreover, one study showed that cocaine-induced increase in c-Fos protein expression was observed 7 days after the acute social defeat (Miczek et al., 1999). However, the effects of acute social defeat did extinguish 14 days after the stress, as indicated by normalized locomotor responses to cocaine (Wang et al., 2013). The social defeat-induced increase in c-Fos immunoreactivity were particularly prominent in the nuclei of dopaminergic, serotonergic, noradrenergic, and enkephalinergic neurons (Martinez et al., 1998; Miczek et al., 1999; Nikulina et al., 1998; Nikulina, Hammer, Miczek, & Kream, 1999). Thus, acute social defeat is sufficient to modulate neuronal activities that mediate drug reward. In fact, acutely defeated animals acquired cocaine faster (Haney et al., 1995; Tidey & Miczek, 1997), spent more time in cocaine-paired chamber in CPP paradigm, and showed exaggerated reinstatement-like responses (Reguilon et al., 2017). On the other hand, a single exposure to social defeat stress did not alter alcohol preference or intake (Croft, Brooks, Cole, & Little, 2005).

4.2 Repeated Exposure to Social Defeat

Repeated exposure to social defeat for 3–7 consecutive days exaggerates cocaine preference (McLaughlin, Li, Valdez, Chavkin, & Chavkin, 2006; Stelly, Pomrenze, Cook, & Morikawa, 2016), possibly due to an enhanced learning in contextual cue that is mediated by long-term potentiation in the VTA neurons expressing glutamate receptors, such as *N*-methyl-D-aspartate receptors (NMDARs) (Stelly et al., 2016). Repeated exposure to social defeat also exaggerated locomotor activity induced by morphine (Capriles, Rodaros, Sorge, & Stewart, 2003; Stohr et al., 1999). Furthermore, repeated social defeat induced analgesia, followed by increases in mRNA expression of opioid peptides, such as dynorphin and enkephalin (Berube, Laforest, Bhatnagar, & Drolet, 2013; McLaughlin et al., 2006). These behavioral changes are attenuated by a κ-opioid receptor (KOR) antagonism or prevented by deletion of prodynorphin, suggesting that repeated exposure to social defeat can induce reward-like behaviors and analgesia that may be mediated by opioid peptides (McLaughlin et al., 2006). Furthermore, one study has shown that repeated exposure to social defeat increased immunoreactivity for μ-opioid receptor (MOR), and that the activation of MOR in the VTA enhanced general locomotor activity (Nikulina, Miczek, & Hammer, 2005). Unlike cocaine and morphine preference, repeated exposure to social defeat stress reduced operant self-administration and voluntary consumption for alcohol (Funk, Harding, Juzytsch, & Le, 2005; van Erp & Miczek, 2001; also see Albrechet-Souza, Viola, Grassi-Oliveira, Miczek, & de Almeida, 2017; Lopez, Anderson, & Becker, 2016). However, when combined with an additional stressor, rats showed an exaggerated operant response to alcohol and a more pronounced relapsing behavior compared to non-stressed controls (Riga et al., 2014). Repeated exposure to social defeat also reduced nicotine self-administration (Zou, Funk, Shram, & Le, 2014).

4.3 Intermittent Exposure to Social Defeat

Activation of the HPA axis is a prerequisite for escalation in drug reward, reinforcement, and reinstatement (Goeders, 2002; Koob, 2008). As described in Section 2.3.2, repeated stress normalizes the animals' CORT and ACTH responses (Bhatnagar et al., 2006). Intermittency in each social defeat episode, however, prevents the habituation of stress responses and thus maintains HPA responses with high levels of circulating CORT (Covington & Miczek, 2001). Under this condition, animals show behavioral sensitization to cocaine (Covington & Miczek, 2001, 2005; Quadros & Miczek, 2009; Yap et al., 2015), *d*-amphetamine (Johnston, Herschel, Lasek, Hammer, & Nikulina, 2015; Nikulina, Covington, Ganschow, Hammer, & Miczek, 2004; Nikulina, Lacagnina, Fanous, Wang, & Hammer, 2012; Yap, Covington, Gale, Datta, & Miczek, 2005; Yap & Miczek, 2007), morphine (Covington & Miczek, 2001), and 3,4-methylenedioxymethamphetamine (Garcia-Pardo et al., 2015). Intermittent social defeat exposure also increased cocaine CPP (Ferrer-Perez et al., 2018). For operant self-administration, animals that are exposed to intermittent social defeat acquire operant response faster, self-administer more cocaine, and show an escalation in binge-like cocaine self-administration (Boyson, Miguel, Quadros, Debold, & Miczek, 2011; Covington & Miczek, 2001, 2005; Holly et al., 2016; Quadros & Miczek, 2009; Yap et al., 2015; Yap & Miczek, 2007). These animals also showed an exaggerated cocaine-seeking responses after a period of abstinence, compared to non-stressed controls (Holly et al., 2016).

One particular molecule associated with exaggerated cocaine reinforcement in intermittently defeated animals is CRF that has been involved in the neurobiology of drug addiction (Zorrilla, Logrip, & Koob, 2014). It has been shown that exposure to single social defeat facilitated CRF transmission at posterior VTA, while exposure to intermittent social defeat expanded its transmission to both posterior and anterior VTA (Holly et al., 2016). Alternatively, a prevention of CRF transmission in the VTA during each episode of intermittent social defeat attenuated later behavioral sensitization and binge-like self-administration of cocaine (Boyson et al., 2011). In this scenario, both CRFR1 and CRF receptor subtype 2 (CRFR2) mediated the intermittent stress-induced exaggeration in the binge (Boyson et al., 2014; Holly et al., 2016), but only CRFR1 mediated behavioral sensitization and cocaine CPP (Boyson et al., 2014; Ferrer-Perez et al., 2018). In addition, intermittent social defeat increased protein levels of α-amino-3-hydroxy-5-methyl-4-isoxazolepropionic acid receptor subtype 1 (GluA1) in the VTA (Covington, Tropea, Rajadhyaksha, Kosofsky, & Miczek, 2008). Moreover, a blockade of glutamate neurotransmission via NMDAR and metabotropic glutamate receptor subtype 5 during each episode of social defeat stress attenuated stress-induced behavioral sensitization to cocaine and *d*-amphetamine, as well as cocaine self-administration (Covington et al., 2008; Yap et al., 2005). In contrast to cocaine, exaggerated response to alcohol and opioids due to intermittent social defeat appears to be less common. For example, prior exposure to intermittent social defeat did not exaggerate binge-like escalation in heroin self-administration (Cruz, Quadros, Hogenelst, Planeta, & Miczek, 2011). Moreover, only a small increase in alcohol intake was observed when the stress was given during a withdrawal period (Funk, Vohra, & Le, 2004). Thus, intermittent exposure to social defeat may only have the most robust effects on cocaine reward but not on other drugs.

4.4 Chronic Exposure to Social Defeat

As described in Section 2.3.4, exposure to chronic social defeat is used as a model for major depression. Daily defeat ranging from a period of 10 days to 5 weeks has induced anhedonia-like responses, decreased social interaction, increased social avoidance, reduced weight gain, and disrupted estrous cycle. Some earlier studies (particularly those applying chronic social defeat for 10 days) do not test the animals' depressive-like symptoms, whereas more recent articles conducted a battery of behavioral tests to confirm the depressive-like state. For the purpose of this chapter, any articles that expose animals to daily defeat for more than 10 days are categorized as studies on chronic social defeat stress.

In C57Bl/6J male mice, chronic social defeat increased c-Fos and BDNF protein levels in the dopaminergic neurons projecting from the VTA to the NAc (Berton et al., 2006; Wook Koo et al., 2016). The increase in BDNF protein level in the VTA coincided with a decreased activity in protein kinase B and increased phosphorylation of extracellular signal-regulated kinases subtype 1/2 (ERK 1/2) (Krishnan et al., 2008).This increased ERK 1/2 phosphorylation was followed by an increased phosphorylation of its downstream targets, ribosomal S6 kinase and mitogen- and stress-activated protein kinase 1 (Iniguez et al., 2010). Alternatively, knocking down BDNF expression attenuated VTA-DA neuronal firing and ameliorated the anhedonia-like responses (Krishnan et al., 2007). One study showed that this chronic stress-induced attenuation in VTA-DA neuronal activity was mediated by NE neurons projecting from locus coeruleus (Isingrini et al., 2016). In fact, activation of NA neurons reversed a susceptible phenotype to an "unsusceptible"-like phenotype (Isingrini et al., 2016). In line with the BDNF signaling cascade, exposure to chronic social defeat increased the overall levels of proteins in the c-Fos family (Perrotti et al., 2004). Particularly, ΔFosB has been implicated in enhancing cocaine reward (Kelz et al., 1999; Nestler, Kelz, & Chen, 1999; Vialou, Robinson, et al., 2010), and chronic social defeat has shown to increase ΔFosB protein specifically in the susceptible phenotype (Perrotti et al., 2004;

Vialou, Robison, et al., 2010). This increase in ΔFosB protein levels was mediated by another transcription factor, the serum response factor that is activity-dependent (Vialou, Maze, et al., 2010). Furthermore, studies have shown that chronic social defeat can alter gene activity through histone protein modulations that epigenetically control behaviors (Tsankova, Renthal, Kumar, & Nestler, 2007). For example, chronic social defeat decreased mRNA expression of class II histone deacetylase, HDAC5, in the NAc in susceptible mice (Renthal et al., 2007). HDACs family is responsible for repressing gene expression by removing acetyl group (i.e., deacetylation) from histones to produce a less accessible chromatin structure. The decreased mRNA expression of HDAC5 was restored by administration of an antidepressant, followed by an attenuation of depressive-like symptoms in susceptible mice (Renthal et al., 2007). Chronic social defeat also increased acetylation of histone H3 by decreasing histone deacetylase 2 (HDAC2), histone variant H3.3, and the class III histone deacetylase sirtuin 1 in the NAc (Covington et al., 2009; Kim et al., 2016; Lepack et al., 2016). These HDACs studies indicate that chronic stress produces long-lasting epigenetic changes in the NAc that can be prevented by HDAC inhibitors (Covington et al., 2009; Renthal et al., 2007; Schroeder, Lin, Crusio, & Akbarian, 2007; Tsankova et al., 2006). Also, chronic social defeat decreased levels of a DNA dioxygenase protein, ten-eleven translocation methylcytosine protein I (Tet1), in animals with a susceptible phenotype (Feng et al., 2017). In addition to BDNF, chronic social defeat increased the protein level of another neurotrophic factor, VGF nerve growth factor in the NAc in susceptible mice (Jiang et al., 2017). Furthermore, exposure to social defeat for 10 days had decreased GluA2 protein levels in the NAc and increased inward rectification of NAc neurons. Overexpression of GluA2 normalized their social interaction behavior (Vialou, Robison, et al., 2010). Also, "susceptible" mice showed an increased GluA1 protein levels in the NAc (Vialou, Robison, et al., 2010).

Despite all the strong scientific evidence supporting the activation of BDNF signaling, ΔFosB accumulation, and epigenetic changes confirmed by histone deacetylation in VTA-NAc reward pathway, the consequences of exposure to chronic social defeat stress in addiction-like behaviors vary. For example, 10 days of exposure to chronic social defeat induced cross-sensitization to amphetamine along with exaggerated DA response in the NAc in outbred mouse strain (Han, DeBold, & Miczek, 2017; Yap & Miczek, 2007). However, experience with an amphetamine challenge reduced subsequent cocaine reinforcements in the same strain of animals (Yap & Miczek, 2007), but omission of the amphetamine challenge did show an increase in cocaine reinforcements (Han et al., 2015; Han, DeBold, et al., 2017). Amphetamine can increase ΔFosB protein levels in the NAc when BDNF is overexpressed (Wang et al., 2013). Because ΔFosB in the NAc has been implicated in cocaine reward (Kelz et al., 1999; Nestler et al., 1999; Vialou, Robison, et al., 2010), one can speculate that amphetamine challenge would exaggerate later cocaine reinforcements via increased ΔFosB. However, one study showed that an overexpression of wildtype ΔFosB in the NAc in chronically stressed animals did not alter the extent of cocaine reward; on the other hand, the overexpression accumbal ΔFosB lacking the 78 aa N-terminal point mutation did (Ohnishi et al., 2015). One possible explanation for this paradox is the fact that most of neurons in the NAc are GABAergic medium spiny neurons (MSNs). There are two distinct types of MSNs: MSNs that express mostly either D1 (D1-MSNs) or D2 (D2-MSNs) receptors (Lobo & Nestler, 2011). ΔFosB can localize on both D1- and D2-MSNs, but more ΔFosB are localized on D2-MSNs when animals are exposed to chronic social defeat stress (Lobo et al., 2013). It has been suggested that D1-MSNs are more involved in cocaine reward, whereas D2-MSNs are more involved in natural reward or stress (Francis & Lobo, 2017). Thus, the paradox between the role of ΔFosB in prior challenge with or without amphetamine on cocaine reward may be explained by the differential localization of ΔFosB in accumbal MSNs as a result of chronic stress. In fact, our previous studies in Long–Evans male and female rats showed 3–5 weeks of chronic social defeat attenuated both behavioral sensitization to cocaine and motivational and "binge"-like cocaine self-administration (Miczek et al., 2011; Shimamoto et al., 2015). These observations corresponded to reduced spontaneous and cocaine-induced DA responses and reduced expression of c-Fos/BDNF in the NAc (Miczek et al., 2011).

As described in Section 3.2, both MOR and KOR can alter stress-induced behaviors (Bruchas, Land, & Chavkin, 2010; Knoll & Carlezon, 2010; Van't Veer & Carlezon, 2013). For example, social avoidance induced by chronic social defeat was ameliorated in MOR, but not KOR, knockout mice (Donahue et al., 2015; Komatsu et al., 2011). In one study that administered immobilization stress to Sprague-Dawley rats for 10 days revealed more MOR localization in the terminals of PVN neurons in females but not in males (Milner et al., 2013). In addition, chronic social defeat reduced mRNA expression of prodynorphin in the NAc (Donahue et al., 2015), and the exhibition of anhedonia-like responses was delayed in KOR knockout mice (Donahue et al., 2015). A more recent study showed the behavioral deficits associated with depression-like symptoms in chronically stressed animals had reduced KOR mRNA expression in the amygdala and increased KOR mRNA in the frontal cortex (Browne, Falcon, Robinson, Berton, & Lucki, 2018).

Exposure to chronic social defeat for 10 days resulted in a higher alcohol consumption in moderately defeated CFW mice than in mildly defeated mice when the animals were chronically exposed to alcohol using a two-bottle choice test (Norman et al., 2015). This behavioral difference in alcohol consumption corresponded to their plasma CORT levels, with higher CORT levels in moderately defeated mice than in mildly defeated mice (Norman et al., 2015). Similarly, the increase in plasma CORT levels in Swiss mice as a result of a 10-day exposure to chronic social defeat led to anhedonia-like responses that corresponded to an increase in alcohol preference and alcohol-induced exaggeration in cross-sensitization (Macedo et al., 2018). The exaggeration in alcohol drinking in chronically defeated animals also showed a positive correlation with mRNA expression of AVP, a neuropeptide that mediates stress response (Keeney et al., 2006; Nelson, Sequeira, & Schank, 2018). Intermittent access to alcohol, which induces excess alcohol drinking, further amplified alcohol consumption in chronically stressed mice (Hwa, Holly, DeBold, & Miczek, 2016). These observations suggest a crucial role of stress hormones in exaggerated alcohol consumption (Koob, 2008; Koob & Heinrichs, 1999). In fact, exaggerated alcohol consumption in the chronically stressed animals was successfully attenuated by microinjection of the CRFR1 antagonist CP376,395 into the VTA (Hwa et al., 2016). Interestingly, extracellular DA in the NAc was significantly increased in stressed animals when CP376,395 was injected in the VTA (Hwa et al., 2016). Intuitively, this significant increase in accumbal DA as a result of CRFR1 antagonism leading to the decrease in alcohol drinking seems contradictory, in that DA release in the NAc is the primary initiating factor of drug reward and withdrawal-induced reinstatement, and that CRF antagonism successfully reduces both drug reward and reinstatement. Furthermore, extensive studies using electrophysiological studies (Ungless et al., 2003; Wanat, Hopf, Stuber, Phillips, & Bonci, 2008), as well as in vivo microdialysis (Wang et al., 2005) have confirmed that CRF can act on VTA-DA neurons through glutamatergic inputs to release DA in the NAc (but also see Wanat, Bonci, & Phillips, 2013). It has been suggested that glutamatergic inputs projecting from bed nucleus of the stria terminalis (BNST) play a crucial role in mediating stress-induced excessive alcohol consumption (Silberman, Matthews, & Winder, 2013; Silberman & Winder, 2013). In this mechanism, CRF interneurons within BNST control its excitability to VTA-DA neurons, particularly when exposed to alcohol or stress (Silberman et al., 2013; Silberman & Winder, 2013). However, given the fact that DA release was observed in response to a CRF antagonist microinjected into the VTA, it is less likely that the antagonist had a distinct effect outside of the VTA. In this regard, the VTA-DA neurons receive inputs from both the glutamatergic neurons and GABAergic interneurons (Nie et al., 2004; Silberman et al., 2013). While VTA-DA neurons express both CRF R1 and R2 in the soma, it is not clear whether CRFRs are expressed on other neurons in the VTA, particularly GABAergic interneurons (Morales & Margolis, 2017). It is well known that in the amygdala, alcohol enhances GABAergic activity and that CRF antagonism can attenuate its activity (Nie et al., 2004, 2009; Silberman & Winder, 2015). Thus, it is possible that intra-VTA microinjection of the CRF antagonist can reduce GABAergic interneuron activity and subsequently disinhibit DA activity in VTA-DA neurons.

5. INDIVIDUAL DIFFERENCES IN SOCIAL DEFEAT-INDUCED ADDICTION-LIKE BEHAVIORS

As described in Section 2.4.1, individual differences in animals that are exposed to chronic social defeat are well characterized in the C57Bl/6 mouse strain. The susceptible phenotype consistently showed increase in BDNF signaling in the NAc, and when BDNF was microinjected into the NAc in unsusceptible mice, they showed less social interaction, similar to susceptible mice (Krishnan et al., 2007). By contrast, unsusceptible mice showed an increase in ΔFosB in the NAc compared to non-defeated controls (Vialou, Maze, et al., 2010). The increase in ΔFosB in the unsusceptible phenotype corresponded to an increase in expression of a transcription factor, serum response factor (SRF). Knocking out the SRF in the NAc in this phenotype reduced the social interaction and led to susceptible-like phenotype (Vialou, Maze, et al., 2010). Furthermore, the unsusceptible mice showed an upregulation in synaptic strength at the large mushroom-shaped dendritic spines of D1-MSNs and a concomitant downregulation at those of D2-MSNs (Khibnik et al., 2016). They also showed an upregulation in Baz1b, an accessory subunit of the imitation of switch family of chromatin remodeling complexes (Sun et al., 2016). In rats, the LL phenotype showed an increase in dynorphin mRNA in dorsal striatum, while the SL phenotype showed decreased enkephalin and dynorphin mRNAs in basolateral amygdala and in the NAc, respectively (Berube et al., 2013). Yet, few studies so far have tested how individual differences in responses to social defeat could influence subsequent drug reward. One earlier study using a novelty test that differentiates HR from LR rats revealed striking behavioral difference in *d*-amphetamine self-administration (Piazza et al., 1991). This difference corresponded with the animals' plasma CORT levels, such that HR rats with high CORT response self-administered more *d*-amphetamine than LR rats with low CORT response (Piazza et al., 1991). When HR and LR rats are subsequently exposed to 4 days of repeated social defeat, however, their differences in *d*-amphetamine responses were abolished, possibly due to the recovery of mRNA expression of D2 receptors in the NAc shell and dorsal striatum (Dietz, Dietz, Moore, Ouimet, & Kabbaj, 2008). Yet, defeated HR rats did acquire cocaine relatively faster than defeated LR rats when tested with the self-administration protocol (Kabbaj et al., 2001). As described in Section 2.3.2, exposure to repeated social defeat ranging from three to seven consecutive days normalized both ACTH and CORT levels (Bhatnagar et al., 2006). Thus, it is possible that exposure to repeated social defeat may have abolished CORT reactivity that would otherwise differentiate HR from LR rats and produce robust stress effects on subsequent responses to drugs. In addition to the novelty test that differentiated HR/LR phenotypes, a more recent study used the latency for a socially defeated male rat to enter a threat zone that had been cordoned off by a meshed screen wall and in which the auditory and olfactory signals of an aggressor were present (Boyson et al., 2016). A principal component analysis revealed that defeated animals that showed shorter latency to enter the threat zone took more cocaine during the "binge" test in cocaine self-administration, while the defeated animals that showed longer latency to enter the zone self-administered similar amounts of cocaine to non-stressed controls during the "binge," suggesting that risky behavior as assessed with the latency to enter a threat zone can predict subsequent cocaine taking behavior (Boyson et al., 2016). In alcohol, a recent study showed that susceptible C57Bl/6 mice consumed more alcohol than the unsusceptible mice (Nelson et al., 2018).

Ongoing social defeat experiences do produce phenotypic differences in natural reward in animals prior to their reward-like behaviors in response to abused drugs. Using an intracranial self-stimulation (ICSS) protocol that measures brain reward or anhedonic responses, one study in male Sprague-Dawley rats showed that exposure to chronic social defeat for

21 days produced two phenotypes (Der-Avakian & Markou, 2012). Specifically, when tested during the 21-day period, 50% of the animals showed a high ICSS threshold, suggesting that animals required more electrical stimulation to gain positive reinforcement, which is indicative of an anhedonia-like state. However, this phenotypic difference was abolished when ICSS was performed after the last social defeat. The disappearance of the phenotypic difference corresponded to VTA BDNF signaling and its downstream signaling cascade (Der-Avakian & Markou, 2012). In our previous study using hierarchical cluster analysis with the intake of sweetened taste solution (i.e., 0.02% saccharin), we found in female Long–Evans rats that a 21-day exposure to social defeat caused by daily maternal aggression produced two phenotypes (Shimamoto et al., 2015). The "stress-vulnerable" phenotype that caused a reduction in consumption of saccharin solution during the 21-day period showed blunted DA levels in response to acute cocaine, followed by reduced cocaine consumption during the "binge" phase (Shimamoto et al., 2015). By contrast, the "stress-resilient" phenotype that did not reduce their intake for saccharin consumed more cocaine than the stress-vulnerable and non-stressed controls (Shimamoto et al., 2015).

6. SEX-RELATED DIFFERENCES IN SOCIAL DEFEAT-INDUCED ADDICTION-LIKE BEHAVIORS

As described in Section 2.2, the ethological difference between territorial and maternal aggression in rodents poses some argument that the direct comparison between brain and behavioral changes of males and females in socially defeated animals may be inadequate, except for some species or strains. One such exception is the use of California mice. Studies using California mice have shown that females but not males showed social withdrawal, that is, a decrease in interacting time with a novel mouse, after exposure to repeated social defeat (Trainor et al., 2011, 2013). A recent study showed that DA-D1 receptors were particularly necessary for mediating defeat-induced social withdrawal in females but not in males (Campi, Greenberg, Kapoor, Ziegler, & Trainor, 2014). Furthermore, these socially withdrawn females showed an increase in BDNF protein levels in the anterior part of BNST that could be ameliorated by blocking the BDNF receptor, TrKB (Greenberg, Howerton, & Trainor, 2014). When Long–Evans rats were exposed to intermittent social defeat due to either territorial or maternal aggression, however, sex-related differences were seen in cross-sensitization,

DA response in the NAc, and duration of the "binge" phase (Holly et al., 2012). Furthermore, stressed females showed increased locomotor activity compared to both non-stressed females and respective stressed males, particularly during the first 5 min after cocaine injection. This locomotor activity was independent of their estrous phase. Also, these females showed prolonged DA accumulation in the NAc followed by an extended "binge" period, suggesting that intermittent social defeat produced profound sex differences in cocaine consumption and DA response. In this particular series of study, the authors did not see any difference in latency to the first bite, total number of bites received, or duration of defeat encounter between males and females (Holly et al., 2012). In line with these studies, my laboratory has not observed any differences in anhedonia-like responses between males and females when exposed to chronic social defeat (Rappeneau et al., 2016).

7. CONCLUDING REMARKS

Social defeat stress as a result of social confrontations leads to a spectrum of stress responses that relies on the intensity, duration, frequency, and intermittency of the confrontation episodes. Prevailing males often use territorial aggression to produce social stress in yielding males. The females of some species also use territorial aggression against female intruders; however, maternal aggression is more common among female laboratory animals. Both genetic makeup and biological sex may diversify the stress responses within the same population. While social confrontations typically occur between conspecific individuals in feral, farm, and laboratory environments, they can also occur between heterospecific individuals under specific conditions.

In general, experiences with repeated or intermittent social defeat facilitate the neuronal activity and initiate the signaling cascade in postsynaptic neurons in the NAc, thereby sensitizing neurochemical and behavioral responses to psychostimulants that facilitate reward processing. CRF plays a significant role in social defeat-induced addiction-like behaviors in this scenario. There is mounting evidence suggesting that chronic social defeat suppresses the tendency to consume cocaine in both males and females; yet, the underlying mechanisms for this suppression need further examination. Unlike effects on psychostimulant use, the effects of repeated or intermittent social defeat on alcohol and opioid consumption are not uniform. Also, few studies have examined the alcohol or opioid taking behaviors following chronic social defeat.

This chapter does not include other social stressor models that alter addiction-like behaviors. Such stressors include social isolation (maternal separation) (Nylander & Roman, 2013) and "witness" or "emotional" stress (Crestani, 2017; Sial, Warren, Alcantara, Parise, & Bolanos-Guzman, 2016). Like those of social confrontations, the interpretation, confounding effects, and limitations of these social stressor models should be clearly defined. For instance, some authors use "emotional" stress as an anthropomorphic expression without testing their emotionality. As stated by Cabanac (2002), there is no consensus on the definition of "emotion." Therefore, to objectify the measurements of "emotion," researchers have used both neurological and somatic signs to assess an individual's emotional state. Interestingly, some studies have demonstrated that both laboratory rats and mice show distinct facial expressions under stress, although there is currently no scientific proof that these expressions are linked to "emotion" (Defensor, Corley, Blanchard, & Blanchard, 2012; Finlayson, Lampe, Hintze, Wurbel, & Melotti, 2016; Leach, 2012; Sotocinal et al., 2011). In fact, these "emotional" states are often inferred and interpreted from several behavioral tests, including ultrasonic vocalizations (Burgdorf, Panksepp, & Moskal, 2011). While it is well established that emotional circuits exist in animals (LeDoux, 2000), it is still crucial for researchers to perform a battery of behavioral tests or manipulations of the emotional circuits when interpreting social stress as an "emotional" stress.

ACKNOWLEDGMENTS

The author's research is supported by the Meharry Endowment Award. The author would like to thank Dr. Sheila T. Yong (Meharry Office for Scientific Editing and Publications) for her scientific and language editing support.

REFERENCES

Albrechet-Souza, L., Viola, T. W., Grassi-Oliveira, R., Miczek, K. A., & de Almeida, R. M. M. (2017). Corticotropin releasing factor in the bed nucleus of the stria terminalis in socially defeated and non-stressed mice with a history of chronic alcohol intake. *Frontiers in Pharmacology*, *8*, 762. https://doi.org/10.3389/fphar.2017.00762.

Anisman, H., Lacosta, S., Kent, P., McIntyre, D. C., & Merali, Z. (1998). Stressor-induced corticotropin-releasing hormone, bombesin, ACTH and corticosterone variations in strains of mice differentially responsive to stressors. *Stress*, *2*(3), 209–220.

Anisman, H., Prakash, P., Merali, Z., & Poulter, M. O. (2007). Corticotropin releasing hormone receptor alterations elicited by acute and chronic unpredictable stressor challenges in stressor-susceptible and resilient strains of mice. *Behavioural Brain Research*, *181*(2), 180–190. https://doi.org/10.1016/j.bbr.2007.04.002.

Backstrom, T., & Winberg, S. (2013). Central corticotropin releasing factor and social stress. *Frontiers in Neuroscience*, 7, 117. https://doi.org/10.3389/fnins.2013.00117.

Baranyi, J., Bakos, N., & Haller, J. (2005). Social instability in female rats: The relationship between stress-related and anxiety-like consequences. *Physiology & Behavior*, *84*(4), 511–518. https://doi.org/10.1016/j.physbeh.2005.01.005.

Bardo, M. T., Neisewander, J. L., & Kelly, T. H. (2013). Individual differences and social influences on the neurobehavioral pharmacology of abused drugs. *Pharmacological Reviews*, *65*(1), 255–290. https://doi.org/10.1124/pr.111.005124.

Barron, M. L., Flick, L. H., Cook, C. A., Homan, S. M., & Campbell, C. (2008). Associations between psychiatric disorders and menstrual cycle characteristics. *Archives of Psychiatric Nursing*, *22*(5), 254–265. https://doi.org/10.1016/j.apnu.2007.11.001.

Bartolomucci, A. (2007). Social stress, immune functions and disease in rodents. *Frontiers in Neuroendocrinology*, *28*(1), 28–49. https://doi.org/10.1016/j.yfrne.2007.02.001.

Becker, J. B., & Koob, G. F. (2016). Sex differences in animal models: Focus on addiction. *Pharmacological Reviews*, *68*(2), 242–263. https://doi.org/10.1124/pr.115.011163.

Belzung, C., El Hage, W., Moindrot, N., & Griebel, G. (2001). Behavioral and neurochemical changes following predatory stress in mice. *Neuropharmacology*, *41*(3), 400–408.

Berlanga, C., & Flores-Ramos, M. (2006). Different gender response to serotonergic and noradrenergic antidepressants. A comparative study of the efficacy of citalopram and reboxetine. *Journal of Affective Disorders*, *95*(1–3), 119–123. https://doi.org/10.1016/j.jad.2006.04.029.

Berton, O., Durand, M., Aguerre, S., Mormede, P., & Chaouloff, F. (1999). Behavioral, neuroendocrine and serotonergic consequences of single social defeat and repeated fluoxetine pretreatment in the Lewis rat strain. *Neuroscience*, *92*(1), 327–341.

Berton, O., McClung, C. A., Dileone, R. J., Krishnan, V., Renthal, W., Russo, S. J., et al. (2006). Essential role of BDNF in the mesolimbic dopamine pathway in social defeat stress. *Science*, *311*(5762), 864–868. https://doi.org/10.1126/science.1120972.

Berube, P., Laforest, S., Bhatnagar, S., & Drolet, G. (2013). Enkephalin and dynorphin mRNA expression are associated with resilience or vulnerability to chronic social defeat stress. *Physiology & Behavior*, *122*, 237–245. https://doi.org/10.1016/j.physbeh.2013.04.009.

Bhatnagar, S., Vining, C., Iyer, V., & Kinni, V. (2006). Changes in hypothalamic–pituitary-adrenal function, body temperature, body weight and food intake with repeated social stress exposure in rats. *Journal of Neuroendocrinology*, *18*(1), 13–24. https://doi.org/10.1111/j.1365-2826.2005.01375.x.

Blanchard, D. C., Fukunaga-Stinson, C., Takahashi, L. K., Flannelly, K. J., & Blanchard, R. J. (1984). Dominance and aggression in social groups of male and female rats. *Behavioural Processes*, *9*(1), 31–48. https://doi.org/10.1016/0376-6357(84)90006-8.

Blanchard, R. J., McKittrick, C. R., & Blanchard, D. C. (2001). Animal models of social stress: Effects on behavior and brain neurochemical systems. *Physiology & Behavior*, *73*(3), 261–271.

Bloch, M., Schmidt, P. J., Danaceau, M., Murphy, J., Nieman, L., & Rubinow, D. R. (2000). Effects of gonadal steroids in women with a history of postpartum depression. *The American Journal of Psychiatry*, *157*(6), 924–930. https://doi.org/10.1176/appi.ajp.157.6.924.

Boileau, I., Dagher, A., Leyton, M., Gunn, R. N., Baker, G. B., Diksic, M., et al. (2006). Modeling sensitization to stimulants in humans: An [11C]raclopride/positron emission tomography study in healthy men. *Archives of General Psychiatry*, *63*(12), 1386–1395. https://doi.org/10.1001/archpsyc.63.12.1386.

Boyson, C. O., Holly, E. N., Burke, A. R., Montagud-Romero, S., DeBold, J. F., & Miczek, K. A. (2016). Maladaptive choices by defeated rats: Link between rapid approach to social threat and escalated cocaine self-administration. *Psychopharmacology, 233*(17), 3173–3186. https://doi.org/10.1007/s00213-016-4363-1.

Boyson, C. O., Holly, E. N., Shimamoto, A., Albrechet-Souza, L., Weiner, L. A., DeBold, J. F., et al. (2014). Social stress and CRF-dopamine interactions in the VTA: Role in long-term escalation of cocaine self-administration. *The Journal of Neuroscience, 34*(19), 6659–6667. https://doi.org/10.1523/JNEUROSCI.3942-13.2014.

Boyson, C. O., Miguel, T. T., Quadros, I. M., Debold, J. F., & Miczek, K. A. (2011). Prevention of social stress-escalated cocaine self-administration by CRF-R1 antagonist in the rat VTA. *Psychopharmacology, 218*(1), 257–269. https://doi.org/10.1007/s00213-011-2266-8.

Brennan, K., Roberts, D. C., Anisman, H., & Merali, Z. (2001). Individual differences in sucrose consumption in the rat: Motivational and neurochemical correlates of hedonia. *Psychopharmacology, 157*(3), 269–276. https://doi.org/10.1007/s002130100805.

Brewin, C. R., Andrews, B., & Valentine, J. D. (2000). Meta-analysis of risk factors for posttraumatic stress disorder in trauma-exposed adults. *Journal of Consulting and Clinical Psychology, 68*(5), 748–766.

Browne, C. A., Falcon, E., Robinson, S. A., Berton, O., & Lucki, I. (2018). Reversal of stress-induced social interaction deficits by buprenorphine. *The International Journal of Neuropsychopharmacology, 21*(2), 164–174. https://doi.org/10.1093/ijnp/pyx079.

Bruchas, M. R., Land, B. B., & Chavkin, C. (2010). The dynorphin/kappa opioid system as a modulator of stress-induced and pro-addictive behaviors. *Brain Research, 1314*, 44–55. https://doi.org/10.1016/j.brainres.2009.08.062.

Burgdorf, J., Panksepp, J., & Moskal, J. R. (2011). Frequency-modulated 50 kHz ultrasonic vocalizations: A tool for uncovering the molecular substrates of positive affect. *Neuroscience and Biobehavioral Reviews, 35*(9), 1831–1836. https://doi.org/10.1016/j.neubiorev.2010.11.011.

Cabanac, M. (2002). What is emotion? *Behavioural Processes, 60*(2), 69–83.

Campi, K. L., Greenberg, G. D., Kapoor, A., Ziegler, T. E., & Trainor, B. C. (2014). Sex differences in effects of dopamine D1 receptors on social withdrawal. *Neuropharmacology*, 77, 208–216. https://doi.org/10.1016/j.neuropharm.2013.09.026.

Capriles, N., Rodaros, D., Sorge, R. E., & Stewart, J. (2003). A role for the prefrontal cortex in stress- and cocaine-induced reinstatement of cocaine seeking in rats. *Psychopharmacology, 168*(1–2), 66–74. https://doi.org/10.1007/s00213-002-1283-z.

Carroll, M. E., Lynch, W. J., Roth, M. E., Morgan, A. D., & Cosgrove, K. P. (2004). Sex and estrogen influence drug abuse. *Trends in Pharmacological Sciences, 25*(5), 273–279. https://doi.org/10.1016/j.tips.2004.03.011.

Chung, K. K., Martinez, M., & Herbert, J. (2000). c-Fos expression, behavioural, endocrine and autonomic responses to acute social stress in male rats after chronic restraint: Modulation by serotonin. *Neuroscience, 95*(2), 453–463.

Church, R. M. (1959). Emotional reactions of rats to the pain of others. *Journal of Comparative and Physiological Psychology, 52*(2), 132–134.

Cirulli, F., Francia, N., Berry, A., Aloe, L., Alleva, E., & Suomi, S. J. (2009). Early life stress as a risk factor for mental health: Role of neurotrophins from rodents to non-human primates. *Neuroscience and Biobehavioral Reviews, 33*(4), 573–585. https://doi.org/10.1016/j.neubiorev.2008.09.001.

Cohen, L. S., Soares, C. N., Vitonis, A. F., Otto, M. W., & Harlow, B. L. (2006). Risk for new onset of depression during the menopausal transition: The Harvard study of moods and cycles. *Archives of General Psychiatry, 63*(4), 385–390. https://doi.org/10.1001/archpsyc.63.4.385.

Cohen, H., & Zohar, J. (2004). An animal model of posttraumatic stress disorder: The use of cut-off behavioral criteria. *Annals of the New York Academy of Sciences, 1032*, 167–178. https://doi.org/10.1196/annals.1314.014.

Corrigall, W. A., Franklin, K. B., Coen, K. M., & Clarke, P. B. (1992). The mesolimbic dopaminergic system is implicated in the reinforcing effects of nicotine. *Psychopharmacology, 107*(2–3), 285–289.

Covington, H. E., 3rd, Maze, I., LaPlant, Q. C., Vialou, V. F., Ohnishi, Y. N., Berton, O., et al. (2009). Antidepressant actions of histone deacetylase inhibitors. *The Journal of Neuroscience, 29*(37), 11451–11460. https://doi.org/10.1523/jneurosci.1758-09.2009.

Covington, H. E., 3rd, & Miczek, K. A. (2001). Repeated social-defeat stress, cocaine or morphine. Effects on behavioral sensitization and intravenous cocaine self-administration "binges". *Psychopharmacology, 158*(4), 388–398. https://doi.org/10.1007/s002130100858.

Covington, H. E., 3rd, & Miczek, K. A. (2005). Intense cocaine self-administration after episodic social defeat stress, but not after aggressive behavior: Dissociation from corticosterone activation. *Psychopharmacology, 183*(3), 331–340. https://doi.org/10.1007/s00213-005-0190-5.

Covington, H. E., 3rd, Tropea, T. F., Rajadhyaksha, A. M., Kosofsky, B. E., & Miczek, K. A. (2008). NMDA receptors in the rat VTA: A critical site for social stress to intensify cocaine taking. *Psychopharmacology, 197*(2), 203–216. https://doi.org/10.1007/s00213-007-1024-4.

Crawley, J. N., Belknap, J. K., Collins, A., Crabbe, J. C., Frankel, W., Henderson, N., et al. (1997). Behavioral phenotypes of inbred mouse strains: Implications and recommendations for molecular studies. *Psychopharmacology, 132*(2), 107–124.

Crestani, C. C. (2017). Adolescent vulnerability to cardiovascular consequences of chronic emotional stress: Review and perspectives for future research. *Neuroscience and Biobehavioral Reviews, 74*(Pt B), 466–475. https://doi.org/10.1016/j.neubiorev.2016.03.027.

Croft, A. P., Brooks, S. P., Cole, J., & Little, H. J. (2005). Social defeat increases alcohol preference of C57BL/10 strain mice; effect prevented by a CCKB antagonist. *Psychopharmacology, 183*(2), 163–170. https://doi.org/10.1007/s00213-005-0165-6.

Cruz, F. C., Quadros, I. M., Hogenelst, K., Planeta, C. S., & Miczek, K. A. (2011). Social defeat stress in rats: Escalation of cocaine and "speedball" binge self-administration, but not heroin. *Psychopharmacology, 215*(1), 165–175. https://doi.org/10.1007/s00213-010-2139-6.

Dalla, C., Antoniou, K., Drossopoulou, G., Xagoraris, M., Kokras, N., Sfikakis, A., et al. (2005). Chronic mild stress impact: Are females more vulnerable? *Neuroscience, 135*(3), 703–714. https://doi.org/10.1016/j.neuroscience.2005.06.068.

Dalla, C., Antoniou, K., Kokras, N., Drossopoulou, G., Papathanasiou, G., Bekris, S., et al. (2008). Sex differences in the effects of two stress paradigms on dopaminergic neurotransmission. *Physiology & Behavior, 93*(3), 595–605. https://doi.org/10.1016/j.physbeh.2007.10.020.

Dalla, C., Pitychoutis, P. M., Kokras, N., & Papadopoulou-Daifoti, Z. (2010). Sex differences in animal models of depression and antidepressant response. *Basic & Clinical Pharmacology & Toxicology, 106*(3), 226–233. https://doi.org/10.1111/j.1742-7843.2009.00516.x.

Dalla, C., Pitychoutis, P. M., Kokras, N., & Papadopoulou-Daifoti, Z. (2011). Sex differences in response to stress and expression of depressive-like behaviours in the rat. *Current Topics in Behavioral Neurosciences, 8*, 97–118. https://doi.org/10.1007/7854_2010_94.

de Jong, J. G., Wasilewski, M., van der Vegt, B. J., Buwalda, B., & Koolhaas, J. M. (2005). A single social defeat induces short-lasting behavioral sensitization to amphetamine. *Physiology & Behavior, 83*(5), 805–811. https://doi.org/10.1016/j.physbeh.2004.10.002.

DeBold, J. F., & Miczek, K. A. (1984). Aggression persists after ovariectomy in female rats. *Hormones and Behavior, 18*(2), 177–190.

Defensor, E. B., Corley, M. J., Blanchard, R. J., & Blanchard, D. C. (2012). Facial expressions of mice in aggressive and fearful contexts. *Physiology & Behavior, 107*(5), 680–685. https://doi.org/10.1016/j.physbeh.2012.03.024.

Dellu, F., Piazza, P. V., Mayo, W., Le Moal, M., & Simon, H. (1996). Novelty-seeking in rats—Biobehavioral characteristics and possible relationship with the sensation-seeking trait in man. *Neuropsychobiology, 34*(3), 136–145. https://doi.org/10.1159/000119305.

Der-Avakian, A., & Markou, A. (2012). The neurobiology of anhedonia and other reward-related deficits. *Trends in Neurosciences, 35*(1), 68–77. https://doi.org/10.1016/j.tins.2011.11.005.

Dietz, D. M., Dietz, K. C., Moore, S., Ouimet, C. C., & Kabbaj, M. (2008). Repeated social defeat stress-induced sensitization to the locomotor activating effects of d-amphetamine: Role of individual differences. *Psychopharmacology, 198*(1), 51–62. https://doi.org/10.1007/s00213-008-1078-y.

Dimsdale, J. E. (2008). Psychological stress and cardiovascular disease. *Journal of the American College of Cardiology, 51*(13), 1237–1246. https://doi.org/10.1016/j.jacc.2007.12.024.

Donahue, R. J., Landino, S. M., Golden, S. A., Carroll, F. I., Russo, S. J., & Carlezon, W. A., Jr. (2015). Effects of acute and chronic social defeat stress are differentially mediated by the dynorphin/kappa-opioid receptor system. *Behavioural Pharmacology, 26*(7 Spec), 654–663. https://doi.org/10.1097/FBP.0000000000000155.

Duclot, F., Hollis, F., Darcy, M. J., & Kabbaj, M. (2011). Individual differences in novelty-seeking behavior in rats as a model for psychosocial stress-related mood disorders. *Physiology & Behavior, 104*(2), 296–305. https://doi.org/10.1016/j.physbeh.2010.12.014.

Dunn, J., & Critchlow, V. (1973). Pituitary–adrenal function following ablation of medial basal hypothalamus. *Proceedings of the Society for Experimental Biology and Medicine, 142*(3), 749–754.

Edwards, D. A., Rowe, F. A., Eleftheriou, B. E., & Sprott, R. L. (1975). Neuronal and endocrine control of aggressive behavior. In: *Hormonal correlates of behavior: Vol. 1* (pp. 275–303). New York: Plenum.

Erskine, M. S., Barfield, R. J., & Goldman, B. D. (1978). Intraspecific fighting during late pregnancy and lactation in rats and effects of litter removal. *Behavioral Biology, 23*(2), 206–218.

Farrell, M. R., Schoch, H., & Mahler, S. V. (2018). Modeling cocaine relapse in rodents: Behavioral considerations and circuit mechanisms. *Progress in Neuropsychopharmacology and Biological Psychiatry*, pii: S0278-5846(17)30561-4. https://doi.org/10.1016/j.pnpbp.2018.01.002 (Epub ahead of print).

Fekete, E. M., Zhao, Y., Li, C., Sabino, V., Vale, W. W., & Zorrilla, E. P. (2009). Social defeat stress activates medial amygdala cells that express type 2 corticotropin-releasing factor receptor mRNA. *Neuroscience, 162*(1), 5–13. https://doi.org/10.1016/j.neuroscience.2009.03.078.

Feng, J., Pena, C. J., Purushothaman, I., Engmann, O., Walker, D., Brown, A. N., et al. (2017). Tet1 in nucleus accumbens opposes depression- and anxiety-like behaviors. *Neuropsychopharmacology, 42*(8), 1657–1669. https://doi.org/10.1038/npp.2017.6.

Ferrer-Perez, C., Reguilon, M. D., Manzanedo, C., Aguilar, M. A., Minarro, J., & Rodriguez-Arias, M. (2018). Antagonism of corticotropin-releasing factor CRF1 receptors blocks the enhanced response to cocaine after social stress. *European Journal of Pharmacology, 823*, 87–95. https://doi.org/10.1016/j.ejphar.2018.01.052.

Finlayson, K., Lampe, J. F., Hintze, S., Wurbel, H., & Melotti, L. (2016). Facial indicators of positive emotions in rats. *PLoS One, 11*(11), e0166446. https://doi.org/10.1371/journal.pone.0166446.

Fish, E. W., DeBold, J. F., & Miczek, K. A. (2005). Escalated aggression as a reward: Corticosterone and GABA(A) receptor positive modulators in mice. *Psychopharmacology, 182*(1), 116–127. https://doi.org/10.1007/s00213-005-0064-x.

Fokkema, D. S., & Koolhaas, J. M. (1985). Acute and conditioned blood pressure changes in relation to social and psychosocial stimuli in rats. *Physiology & Behavior, 34*(1), 33–38.

Foster, M. T., Solomon, M. B., Huhman, K. L., & Bartness, T. J. (2006). Social defeat increases food intake, body mass, and adiposity in Syrian hamsters. *American Journal of Physiology. Regulatory, Integrative and Comparative Physiology, 290*(5), R1284–R1293. https://doi.org/10.1152/ajpregu.00437.2005.

Francis, T. C., & Lobo, M. K. (2017). Emerging role for nucleus accumbens medium spiny neuron subtypes in depression. *Biological Psychiatry, 81*(8), 645–653. https://doi.org/10.1016/j.biopsych.2016.09.007.

Freeman, E. W., Sammel, M. D., Lin, H., & Nelson, D. B. (2006). Associations of hormones and menopausal status with depressed mood in women with no history of depression. *Archives of General Psychiatry, 63*(4), 375–382. https://doi.org/10.1001/archpsyc.63.4.375.

Fuchs, E., Flugge, G., Ohl, F., Lucassen, P., Vollmann-Honsdorf, G. K., & Michaelis, T. (2001). Psychosocial stress, glucocorticoids, and structural alterations in the tree shrew hippocampus. *Physiology & Behavior, 73*(3), 285–291.

Funk, D., Harding, S., Juzytsch, W., & Le, A. D. (2005). Effects of unconditioned and conditioned social defeat on alcohol self-administration and reinstatement of alcohol seeking in rats. *Psychopharmacology, 183*(3), 341–349. https://doi.org/10.1007/s00213-005-0194-1.

Funk, D., Vohra, S., & Le, A. D. (2004). Influence of stressors on the rewarding effects of alcohol in Wistar rats: Studies with alcohol deprivation and place conditioning. *Psychopharmacology, 176*(1), 82–87. https://doi.org/10.1007/s00213-004-1859-x.

Gandelman, R. (1972). Mice: Postpartum aggression elicited by the presence of an intruder. *Hormones and Behavior, 3*(1), 23–28.

Garcia-Pardo, M. P., Blanco-Gandia, M. C., Valiente-Lluch, M., Rodriguez-Arias, M., Minarro, J., & Aguilar, M. A. (2015). Long-term effects of repeated social stress on the conditioned place preference induced by MDMA in mice. *Progress in Neuro-Psychopharmacology & Biological Psychiatry, 63*, 98–109. https://doi.org/10.1016/j.pnpbp.2015.06.006.

Goeders, N. E. (2002). Stress and cocaine addiction. *The Journal of Pharmacology and Experimental Therapeutics, 301*(3), 785–789.

Golden, S. A., Covington, H. E., 3rd, Berton, O., & Russo, S. J. (2011). A standardized protocol for repeated social defeat stress in mice. *Nature Protocols, 6*(8), 1183–1191. https://doi.org/10.1038/nprot.2011.361.

Gomez-Lazaro, E., Arregi, A., Beitia, G., Vegas, O., Azpiroz, A., & Garmendia, L. (2011). Individual differences in chronically defeated male mice: Behavioral, endocrine, immune, and neurotrophic changes as markers of vulnerability to the effects of stress. *Stress, 14*(5), 537–548. https://doi.org/10.3109/10253890.2011.562939.

Goyens, J., & Noirot, E. (1975). Effects of cohabitation with females on aggressive behavior between male mice. *Developmental Psychobiology, 8*(1), 79–84. https://doi.org/10.1002/dev.420080111.

Green, M. R., & McCormick, C. M. (2013). Effects of social instability stress in adolescence on long-term, not short-term, spatial memory performance. *Behavioural Brain Research, 256*, 165–171. https://doi.org/10.1016/j.bbr.2013.08.011.

Greenberg, G. D., Howerton, C. L., & Trainor, B. C. (2014). Fighting in the home cage: Agonistic encounters and effects on neurobiological markers within the social decision-making network of house mice (Mus musculus). *Neuroscience Letters, 566*, 151–155. https://doi.org/10.1016/j.neulet.2014.02.051.

Hall, F. S. (1998). Social deprivation of neonatal, adolescent, and adult rats has distinct neurochemical and behavioral consequences. *Critical Reviews in Neurobiology*, *12*(1-2), 129–162.

Haller, J., Fuchs, E., Halasz, J., & Makara, G. B. (1999). Defeat is a major stressor in males while social instability is stressful mainly in females: Towards the development of a social stress model in female rats. *Brain Research Bulletin*, *50*(1), 33–39.

Haller, J., Kiem, D. T., & Makara, G. B. (1996). The physiology of social conflict in rats: What is particularly stressful? *Behavioral Neuroscience*, *110*(2), 353–359.

Han, X., Albrechet-Souza, L., Doyle, M. R., Shimamoto, A., DeBold, J. F., & Miczek, K. A. (2015). Social stress and escalated drug self-administration in mice II. Cocaine and dopamine in the nucleus accumbens. *Psychopharmacology*, *232*(6), 1003–1010. https://doi.org/10.1007/s00213-014-3734-8.

Han, X., DeBold, J. F., & Miczek, K. A. (2017). Prevention and reversal of social stress-escalated cocaine self-administration in mice by intra-VTA CRFR1 antagonism. *Psychopharmacology*, *234*(18), 2813–2821. https://doi.org/10.1007/s00213-017-4676-8.

Han, Q. Q., Yang, L., Huang, H. J., Wang, Y. L., Yu, R., Wang, J., et al. (2017). Differential GR expression and translocation in the hippocampus mediates susceptibility vs. resilience to chronic social defeat stress. *Frontiers in Neuroscience*, *11*, 287. https://doi.org/10.3389/fnins.2017.00287.

Haney, M., Maccari, S., Le Moal, M., Simon, H., & Piazza, P. V. (1995). Social stress increases the acquisition of cocaine self-administration in male and female rats. *Brain Research*, *698*(1–2), 46–52.

Haney, M., & Miczek, K. A. (1993). Ultrasounds during agonistic interactions between female rats (Rattus norvegicus). *Journal of Comparative Psychology*, *107*(4), 373–379.

Haug, M., & Mandel, P. (1978). Strain differences in aggressive behaviour of female mice against lactating and non-lactating individuals. *Neuroscience Letters*, 7(2–3), 235–238.

Hedden, S. L., Kennet, J., Lipari, R., Medley, G., & Tice, P. (2014). *Behavioral health trends in the United States: Results from the 2014 National Survey on Drug Use and Health*. Substance Abuse and Mental Health Services Administration.

Hodes, G. E., Pfau, M. L., Leboeuf, M., Golden, S. A., Christoffel, D. J., Bregman, D., et al. (2014). Individual differences in the peripheral immune system promote resilience versus susceptibility to social stress. *Proceedings of the National Academy of Sciences of the United States of America*, *111*(45), 16136–16141. https://doi.org/10.1073/pnas.1415191111.

Hodes, G. E., Pfau, M. L., Purushothaman, I., Ahn, H. F., Golden, S. A., Christoffel, D. J., et al. (2015). Sex differences in nucleus accumbens transcriptome profiles associated with susceptibility versus resilience to subchronic variable stress. *The Journal of Neuroscience*, *35*(50), 16362–16376. https://doi.org/10.1523/JNEUROSCI.1392-15.2015.

Hollis, F., Duclot, F., Gunjan, A., & Kabbaj, M. (2011). Individual differences in the effect of social defeat on anhedonia and histone acetylation in the rat hippocampus. *Hormones and Behavior*, *59*(3), 331–337. https://doi.org/10.1016/j.yhbeh.2010.09.005.

Holly, E. N., Boyson, C. O., Montagud-Romero, S., Stein, D. J., Gobrogge, K. L., DeBold, J. F., et al. (2016). Episodic social stress-escalated cocaine self-administration: Role of phasic and tonic corticotropin releasing factor in the anterior and posterior ventral tegmental area. *The Journal of Neuroscience*, *36*(14), 4093–4105. https://doi.org/10.1523/JNEUROSCI.2232-15.2016.

Holly, E. N., Shimamoto, A., Debold, J. F., & Miczek, K. A. (2012). Sex differences in behavioral and neural cross-sensitization and escalated cocaine taking as a result of episodic social defeat stress in rats. *Psychopharmacology*, *224*(1), 179–188. https://doi.org/10.1007/s00213-012-2846-2.

Hooks, M. S., Jones, G. H., Smith, A. D., Neill, D. B., & Justice, J. B., Jr. (1991a). Individual differences in locomotor activity and sensitization. *Pharmacology, Biochemistry, and Behavior*, *38*(2), 467–470.

Hooks, M. S., Jones, G. H., Smith, A. D., Neill, D. B., & Justice, J. B., Jr. (1991b). Response to novelty predicts the locomotor and nucleus accumbens dopamine response to cocaine. *Synapse*, *9*(2), 121–128. https://doi.org/10.1002/syn.890090206.

Huhman, K. L. (2006). Social conflict models: Can they inform us about human psychopathology? *Hormones and Behavior*, *50*(4), 640–646. https://doi.org/10.1016/j.yhbeh.2006.06.022.

Huhman, K. L., Solomon, M. B., Janicki, M., Harmon, A. C., Lin, S. M., Israel, J. E., et al. (2003). Conditioned defeat in male and female Syrian hamsters. *Hormones and Behavior*, *44*(3), 293–299.

Hwa, L. S., Holly, E. N., DeBold, J. F., & Miczek, K. A. (2016). Social stress-escalated intermittent alcohol drinking: Modulation by CRF-R1 in the ventral tegmental area and accumbal dopamine in mice. *Psychopharmacology*, *233*(4), 681–690. https://doi.org/10.1007/s00213-015-4144-2.

Iniguez, S. D., Vialou, V., Warren, B. L., Cao, J. L., Alcantara, L. F., Davis, L. C., et al. (2010). Extracellular signal-regulated kinase-2 within the ventral tegmental area regulates responses to stress. *The Journal of Neuroscience*, *30*(22), 7652–7663. https://doi.org/10.1523/JNEUROSCI.0951-10.2010.

Isingrini, E., Perret, L., Rainer, Q., Amilhon, B., Guma, E., Tanti, A., et al. (2016). Resilience to chronic stress is mediated by noradrenergic regulation of dopamine neurons. *Nature Neuroscience*, *19*(4), 560–563. https://doi.org/10.1038/nn.4245.

Jacobson-Pick, S., Audet, M. C., McQuaid, R. J., Kalvapalle, R., & Anisman, H. (2013). Social agonistic distress in male and female mice: Changes of behavior and brain monoamine functioning in relation to acute and chronic challenges. *PLoS One*, *8*(4), e60133. https://doi.org/10.1371/journal.pone.0060133.

Jiang, C., Lin, W. J., Sadahiro, M., Labonte, B., Menard, C., Pfau, M. L., et al. (2017). VGF function in depression and antidepressant efficacy. *Molecular Psychiatry*. https://doi.org/10.1038/mp.2017.233 (Epub ahead of print).

Joana, P. T., Amaia, A., Arantza, A., Garikoitz, B., Eneritz, G. L., & Larraitz, G. (2016). Central immune alterations in passive strategy following chronic defeat stress. *Behavioural Brain Research*, 298(Pt. B), 291–300. https://doi.org/10.1016/j.bbr.2015.11.015.

Johnston, C. E., Herschel, D. J., Lasek, A. W., Hammer, R. P., Jr., & Nikulina, E. M. (2015). Knockdown of ventral tegmental area mu-opioid receptors in rats prevents effects of social defeat stress: Implications for amphetamine cross-sensitization, social avoidance, weight regulation and expression of brain-derived neurotrophic factor. *Neuropharmacology*, *89*, 325–334. https://doi.org/10.1016/j.neuropharm.2014.10.010.

Kabbaj, M., Devine, D. P., Savage, V. R., & Akil, H. (2000). Neurobiological correlates of individual differences in novelty-seeking behavior in the rat: Differential expression of stress-related molecules. *The Journal of Neuroscience*, *20*(18), 6983–6988.

Kabbaj, M., Evans, S., Watson, S. J., & Akil, H. (2004). The search for the neurobiological basis of vulnerability to drug abuse: Using microarrays to investigate the role of stress and individual differences. *Neuropharmacology*, *47*(Suppl. 1), 111–122. https://doi.org/10.1016/j.neuropharm.2004.07.021.

Kabbaj, M., Norton, C. S., Kollack-Walker, S., Watson, S. J., Robinson, T. E., & Akil, H. (2001). Social defeat alters the acquisition of cocaine self-administration in rats: Role of individual differences in cocaine-taking behavior. *Psychopharmacology*, *158*(4), 382–387. https://doi.org/10.1007/s002130100918.

Kawachi, I., & Berkman, L. F. (2001). Social ties and mental health. *Journal of Urban Health*, *78*(3), 458–467. https://doi.org/10.1093/jurban/78.3.458.

Keeney, A. J., Hogg, S., & Marsden, C. A. (2001). Alterations in core body temperature, locomotor activity, and corticosterone following acute and repeated social defeat of male NMRI mice. *Physiology & Behavior*, *74*(1–2), 177–184.

Keeney, A., Jessop, D. S., Harbuz, M. S., Marsden, C. A., Hogg, S., & Blackburn-Munro, R. E. (2006). Differential effects of acute and chronic social defeat stress on hypothalamic-pituitary-adrenal axis function and hippocampal serotonin release in mice. *Journal of Neuroendocrinology*, *18*(5), 330–338. https://doi.org/10.1111/j.1365-2826.2006.01422.x.

Kelz, M. B., Chen, J., Carlezon, W. A., Jr., Whisler, K., Gilden, L., Beckmann, A. M., et al. (1999). Expression of the transcription factor deltaFosB in the brain controls sensitivity to cocaine. *Nature*, *401*(6750), 272–276. https://doi.org/10.1038/45790.

Khibnik, L. A., Beaumont, M., Doyle, M., Heshmati, M., Slesinger, P. A., Nestler, E. J., et al. (2016). Stress and cocaine trigger divergent and cell type-specific regulation of synaptic transmission at single spines in nucleus accumbens. *Biological Psychiatry*, *79*(11), 898–905. https://doi.org/10.1016/j.biopsych.2015.05.022.

Kiecolt-Glaser, J. K., Preacher, K. J., MacCallum, R. C., Atkinson, C., Malarkey, W. B., & Glaser, R. (2003). Chronic stress and age-related increases in the proinflammatory cytokine IL-6. *Proceedings of the National Academy of Sciences of the United States of America*, *100*(15), 9090–9095. https://doi.org/10.1073/pnas.1531903100.

Kim, H. D., Hesterman, J., Call, T., Magazu, S., Keeley, E., Armenta, K., et al. (2016). SIRT1 mediates depression-like behaviors in the nucleus accumbens. *The Journal of Neuroscience*, *36*(32), 8441–8452. https://doi.org/10.1523/JNEUROSCI.0212-16.2016.

Knoll, A. T., & Carlezon, W. A., Jr. (2010). Dynorphin, stress, and depression. *Brain Research*, *1314*, 56–73. https://doi.org/10.1016/j.brainres.2009.09.074.

Koenig, W., & Dickinson, J. (2009). *Animal social behaviour*. United States: Encyclopædia Britannica, Inc. https://www.britannica.com/topic/animal-social-behaviour.

Kollack-Walker, S., Don, C., Watson, S. J., & Akil, H. (1999). Differential expression of c-fos mRNA within neurocircuits of male hamsters exposed to acute or chronic defeat. *Journal of Neuroendocrinology*, *11*(7), 547–559.

Komatsu, H., Ohara, A., Sasaki, K., Abe, H., Hattori, H., Hall, F. S., et al. (2011). Decreased response to social defeat stress in mu-opioid-receptor knockout mice. *Pharmacology, Biochemistry, and Behavior*, *99*(4), 676–682. https://doi.org/10.1016/j.pbb.2011.06.008.

Koob, G. F. (2008). A role for brain stress systems in addiction. *Neuron*, *59*(1), 11–34. https://doi.org/10.1016/j.neuron.2008.06.012.

Koob, G. F., & Heinrichs, S. C. (1999). A role for corticotropin releasing factor and urocortin in behavioral responses to stressors. *Brain Research*, *848*(1–2), 141–152.

Koob, G. F., & Le Moal, M. (2001). Drug addiction, dysregulation of reward, and allostasis. *Neuropsychopharmacology*, *24*(2), 97–129. https://doi.org/10.1016/S0893-133X(00)00195-0.

Koob, G. F., & Volkow, N. D. (2010). Neurocircuitry of addiction. *Neuropsychopharmacology*, *35*(1), 217–238. https://doi.org/10.1038/npp.2009.110.

Koolhaas, J. M., Bartolomucci, A., Buwalda, B., de Boer, S. F., Flugge, G., Korte, S. M., et al. (2011). Stress revisited: A critical evaluation of the stress concept. *Neuroscience and Biobehavioral Reviews*, *35*(5), 1291–1301. https://doi.org/10.1016/j.neubiorev.2011.02.003.

Koolhaas, J. M., Korte, S. M., De Boer, S. F., Van Der Vegt, B. J., Van Reenen, C. G., Hopster, H., et al. (1999). Coping styles in animals: Current status in behavior and stress-physiology. *Neuroscience and Biobehavioral Reviews*, *23*(7), 925–935.

Krishnan, V., Han, M. H., Graham, D. L., Berton, O., Renthal, W., Russo, S. J., et al. (2007). Molecular adaptations underlying susceptibility and resistance to social defeat in brain reward regions. *Cell*, *131*(2), 391–404. https://doi.org/10.1016/j.cell.2007.09.018.

Krishnan, V., Han, M. H., Mazei-Robison, M., Iniguez, S. D., Ables, J. L., Vialou, V., et al. (2008). AKT signaling within the ventral tegmental area regulates cellular and behavioral responses to stressful stimuli. *Biological Psychiatry*, *64*(8), 691–700. https://doi.org/10.1016/j.biopsych.2008.06.003.

Krishnan, V., & Nestler, E. J. (2008). The molecular neurobiology of depression. *Nature, 455*(7215), 894–902. https://doi.org/10.1038/nature07455.

Kudryavtseva, N. N. (1994). Experience of defeat decreases the behavioural reactivity to conspecifics in the partition test. *Behavioural Processes, 32*(3), 297–304. https://doi.org/10.1016/0376-6357(94)90049-3.

Kudryavtseva, N. N., & Bakshtanovskaya, I. V. (1989). Experience of defeat increases the susceptibility to catatonic-like state in mice. *Behavioural Processes, 20*(1–3), 139–149. https://doi.org/10.1016/0376-6357(89)90019-3.

Leach, M. (2012). The assessment of pain using facial expressions in laboratory rodents. *Proceedings of measuring behavior 2012. 8th international conference on methods and techniques in behavioral research* (pp. 238–240). Utrecht, The Netherlands, August 28–31.

LeDoux, J. E. (2000). Emotion circuits in the brain. *Annual Review of Neuroscience, 23*, 155–184. https://doi.org/10.1146/annurev.neuro.23.1.155.

Lepack, A. E., Bagot, R. C., Pena, C. J., Loh, Y. E., Farrelly, L. A., Lu, Y., et al. (2016). Aberrant H3.3 dynamics in NAc promote vulnerability to depressive-like behavior. *Proceedings of the National Academy of Sciences of the United States of America, 113*(44), 12562–12567. https://doi.org/10.1073/pnas.1608270113.

Lobo, M. K., & Nestler, E. J. (2011). The striatal balancing act in drug addiction: Distinct roles of direct and indirect pathway medium spiny neurons. *Frontiers in Neuroanatomy, 5*, 41. https://doi.org/10.3389/fnana.2011.00041.

Lobo, M. K., Zaman, S., Damez-Werno, D. M., Koo, J. W., Bagot, R. C., DiNieri, J. A., et al. (2013). DeltaFosB induction in striatal medium spiny neuron subtypes in response to chronic pharmacological, emotional, and optogenetic stimuli. *The Journal of Neuroscience, 33*(47), 18381–18395. https://doi.org/10.1523/JNEUROSCI.1875-13.2013.

Lonstein, J. S., & Gammie, S. C. (2002). Sensory, hormonal, and neural control of maternal aggression in laboratory rodents. *Neuroscience and Biobehavioral Reviews, 26*(8), 869–888.

Lopez, M. F., Anderson, R. I., & Becker, H. C. (2016). Effect of different stressors on voluntary ethanol intake in ethanol-dependent and nondependent C57BL/6J mice. *Alcohol, 51*, 17–23. https://doi.org/10.1016/j.alcohol.2015.11.010.

Lucion, A. B., & De Almeida, R. M. M. (1996). On the dual nature of maternal aggression in rats. *Aggressive Behavior, 22*, 365–373.

Macedo, G. C., Morita, G. M., Domingues, L. P., Favoretto, C. A., Suchecki, D., & Quadros, I. M. H. (2018). Consequences of continuous social defeat stress on anxiety- and depressive-like behaviors and ethanol reward in mice. *Hormones and Behavior, 97*, 154–161. https://doi.org/10.1016/j.yhbeh.2017.10.007.

Marini, F., Pozzato, C., Andreetta, V., Jansson, B., Arban, R., Domenici, E., et al. (2006). Single exposure to social defeat increases corticotropin releasing factor and glucocorticoid receptor mRNA expression in rat hippocampus. *Brain Research, 1067*(1), 25–35. https://doi.org/10.1016/j.brainres.2005.10.002.

Martenyi, F., Dossenbach, M., Mraz, K., & Metcalfe, S. (2001). Gender differences in the efficacy of fluoxetine and maprotiline in depressed patients: A double-blind trial of antidepressants with serotonergic or norepinephrinergic reuptake inhibition profile. *European Neuropsychopharmacology, 11*(3), 227–232.

Martinez, M., Phillips, P. J., & Herbert, J. (1998). Adaptation in patterns of c-fos expression in the brain associated with exposure to either single or repeated social stress in male rats. *The European Journal of Neuroscience, 10*(1), 20–33.

McBride, W. J., & Li, T. K. (1998). Animal models of alcoholism: Neurobiology of high alcohol-drinking behavior in rodents. *Critical Reviews in Neurobiology, 12*(4), 339–369.

McEwen, B. S., & Stellar, E. (1993). Stress and the individual. Mechanisms leading to disease. *Archives of Internal Medicine, 153*(18), 2093–2101.

McLaughlin, J. P., Li, S., Valdez, J., Chavkin, T. A., & Chavkin, C. (2006). Social defeat stress-induced behavioral responses are mediated by the endogenous kappa opioid system. *Neuropsychopharmacology, 31*(6), 1241–1248. https://doi.org/10.1038/sj.npp.1300872.

Meehan, W. P., Tornatzky, W., & Miczek, K. A. (1995). Blood pressure via telemetry during social confrontations in rats: Effects of clonidine. *Physiology & Behavior*, *58*(1), 81–88.

Mello, N. K., & Negus, S. S. (1996). Preclinical evaluation of pharmacotherapies for treatment of cocaine and opioid abuse using drug self-administration procedures. *Neuropsychopharmacology*, *14*(6), 375–424. https://doi.org/10.1016/0893-133X(95)00274-H.

Miczek, K. A., Covington, H. E., 3rd, Nikulina, E. M., Jr., & Hammer, R. P. (2004). Aggression and defeat: Persistent effects on cocaine self-administration and gene expression in peptidergic and aminergic mesocorticolimbic circuits. *Neuroscience and Biobehavioral Reviews*, *27*(8), 787–802. https://doi.org/10.1016/j.neubiorev.2003.11.005.

Miczek, K. A., Nikulina, E., Kream, R. M., Carter, G., & Espejo, E. F. (1999). Behavioral sensitization to cocaine after a brief social defeat stress: c-Fos expression in the PAG. *Psychopharmacology*, *141*(3), 225–234.

Miczek, K. A., Nikulina, E. M., Shimamoto, A., & Covington, H. E., 3rd. (2011). Escalated or suppressed cocaine reward, tegmental BDNF, and accumbal dopamine caused by episodic versus continuous social stress in rats. *The Journal of Neuroscience*, *31*(27), 9848–9857. https://doi.org/10.1523/JNEUROSCI.0637-11.2011.

Miczek, K. A., Thompson, M. L., & Shuster, L. (1982). Opioid-like analgesia in defeated mice. *Science*, *215*(4539), 1520–1522.

Miczek, K. A., Yap, J. J., & Covington, H. E., 3rd. (2008). Social stress, therapeutics and drug abuse: Preclinical models of escalated and depressed intake. *Pharmacology & Therapeutics*, *120*(2), 102–128. https://doi.org/10.1016/j.pharmthera.2008.07.006.

Milner, T. A., Burstein, S. R., Marrone, G. F., Khalid, S., Gonzalez, A. D., Williams, T. J., et al. (2013). Stress differentially alters mu opioid receptor density and trafficking in parvalbumin-containing interneurons in the female and male rat hippocampus. *Synapse*, *67*(11), 757–772. https://doi.org/10.1002/syn.21683.

Morales, M., & Margolis, E. B. (2017). Ventral tegmental area: Cellular heterogeneity, connectivity and behaviour. *Nature Reviews. Neuroscience*, *18*(2), 73–85. https://doi.org/10.1038/nrn.2016.165.

Nelson, B. S., Sequeira, M. K., & Schank, J. R. (2018). Bidirectional relationship between alcohol intake and sensitivity to social defeat: Association with Tacr1 and Avp expression. *Addiction Biology*, *23*(1), 142–153. https://doi.org/10.1111/adb.12494.

Nestler, E. J., Kelz, M. B., & Chen, J. (1999). DeltaFosB: A molecular mediator of long-term neural and behavioral plasticity. *Brain Research*, *835*(1), 10–17.

Neumann, I. D., Toschi, N., Ohl, F., Torner, L., & Kromer, S. A. (2001). Maternal defence as an emotional stressor in female rats: Correlation of neuroendocrine and behavioural parameters and involvement of brain oxytocin. *The European Journal of Neuroscience*, *13*(5), 1016–1024.

Nie, Z., Schweitzer, P., Roberts, A. J., Madamba, S. G., Moore, S. D., & Siggins, G. R. (2004). Ethanol augments GABAergic transmission in the central amygdala via CRF1 receptors. *Science*, *303*(5663), 1512–1514. https://doi.org/10.1126/science.1092550.

Nie, Z., Zorrilla, E. P., Madamba, S. G., Rice, K. C., Roberto, M., & Siggins, G. R. (2009). Presynaptic CRF1 receptors mediate the ethanol enhancement of GABAergic transmission in the mouse central amygdala. *TheScientificWorldJournal*, *9*, 68–85. https://doi.org/10.1100/tsw.2009.1.

Nikulina, E. M., Covington, H. E., 3rd, Ganschow, L., Hammer, R. P., Jr., & Miczek, K. A. (2004). Long-term behavioral and neuronal cross-sensitization to amphetamine induced by repeated brief social defeat stress: Fos in the ventral tegmental area and amygdala. *Neuroscience*, *123*(4), 857–865.

Nikulina, E. M., Hammer, R. P., Jr., Miczek, K. A., & Kream, R. M. (1999). Social defeat stress increases expression of mu-opioid receptor mRNA in rat ventral tegmental area. *Neuroreport*, *10*(14), 3015–3019.

Nikulina, E. M., Lacagnina, M. J., Fanous, S., Wang, J., & Hammer, R. P., Jr. (2012). Intermittent social defeat stress enhances mesocorticolimbic DeltaFosB/BDNF co-expression and persistently activates corticotegmental neurons: Implication for vulnerability to psychostimulants. *Neuroscience*, *212*, 38–48. https://doi.org/10.1016/j.neuroscience.2012.04.012.

Nikulina, E. M., Marchand, J. E., Kream, R. M., & Miczek, K. A. (1998). Behavioral sensitization to cocaine after a brief social stress is accompanied by changes in fos expression in the murine brainstem. *Brain Research*, *810*(1-2), 200–210.

Nikulina, E. M., Miczek, K. A., & Hammer, R. P., Jr. (2005). Prolonged effects of repeated social defeat stress on mRNA expression and function of mu-opioid receptors in the ventral tegmental area of rats. *Neuropsychopharmacology*, *30*(6), 1096–1103. https://doi.org/10.1038/sj.npp.1300658.

Norman, K. J., Seiden, J. A., Klickstein, J. A., Han, X., Hwa, L. S., DeBold, J. F., et al. (2015). Social stress and escalated drug self-administration in mice I. Alcohol and corticosterone. *Psychopharmacology*, *232*(6), 991–1001. https://doi.org/10.1007/s00213-014-3733-9.

Nylander, I., & Roman, E. (2013). Is the rodent maternal separation model a valid and effective model for studies on the early-life impact on ethanol consumption? *Psychopharmacology*, *229*(4), 555–569. https://doi.org/10.1007/s00213-013-3217-3.

Ohnishi, Y. N., Ohnishi, Y. H., Vialou, V., Mouzon, E., LaPlant, Q., Nishi, A., et al. (2015). Functional role of the N-terminal domain of DeltaFosB in response to stress and drugs of abuse. *Neuroscience*, *284*, 165–170. https://doi.org/10.1016/j.neuroscience.2014.10.002.

Palanisamy, V., Sharma, S., Deshpande, A., Zhou, H., Gimzewski, J., & Wong, D. T. (2010). Nanostructural and transcriptomic analyses of human saliva derived exosomes. *PLoS One*, *5*, e8577.

Palanza, P. (2001). Animal models of anxiety and depression: How are females different? *Neuroscience and Biobehavioral Reviews*, *25*(3), 219–233.

Patel, P. D., Katz, M., Karssen, A. M., & Lyons, D. M. (2008). Stress-induced changes in corticosteroid receptor expression in primate hippocampus and prefrontal cortex. *Psychoneuroendocrinology*, *33*(3), 360–367. https://doi.org/10.1016/j.psyneuen.2007.12.003.

Perrotti, L. I., Hadeishi, Y., Ulery, P. G., Barrot, M., Monteggia, L., Duman, R. S., et al. (2004). Induction of deltaFosB in reward-related brain structures after chronic stress. *The Journal of Neuroscience*, *24*(47), 10594–10602. https://doi.org/10.1523/JNEUROSCI.2542-04.2004.

Piazza, P. V., Deminiere, J. M., Le Moal, M., & Simon, H. (1989). Factors that predict individual vulnerability to amphetamine self-administration. *Science*, *245*(4925), 1511–1513.

Piazza, P. V., Maccari, S., Deminiere, J. M., Le Moal, M., Mormede, P., & Simon, H. (1991). Corticosterone levels determine individual vulnerability to amphetamine self-administration. *Proceedings of the National Academy of Sciences of the United States of America*, *88*(6), 2088–2092.

Pizarro, J. M., Lumley, L. A., Medina, W., Robison, C. L., Chang, W. E., Alagappan, A., et al. (2004). Acute social defeat reduces neurotrophin expression in brain cortical and subcortical areas in mice. *Brain Research*, *1025*(1-2), 10–20. https://doi.org/10.1016/j.brainres.2004.06.085.

Pontieri, F. E., Tanda, G., Orzi, F., & Di Chiara, G. (1996). Effects of nicotine on the nucleus accumbens and similarity to those of addictive drugs. *Nature*, *382*(6588), 255–257. https://doi.org/10.1038/382255a0.

Pryce, C. R., & Fuchs, E. (2017). Chronic psychosocial stressors in adulthood: Studies in mice, rats and tree shrews. *Neurobiology of Stress*, *6*, 94–103. https://doi.org/10.1016/j.ynstr.2016.10.001.

Quadros, I. M., & Miczek, K. A. (2009). Two modes of intense cocaine bingeing: Increased persistence after social defeat stress and increased rate of intake due to extended access conditions in rats. *Psychopharmacology*, *206*(1), 109–120. https://doi.org/10.1007/s00213-009-1584-6.

Raab, A., Seizinger, B. R., & Herz, A. (1985). Continuous social defeat induces an increase of endogenous opioids in discrete brain areas of the mongolian gerbil. *Peptides*, *6*(3), 387–391.

Ramsey, N. F., & Van Ree, J. M. (1993). Emotional but not physical stress enhances intravenous cocaine self-administration in drug-naive rats. *Brain Research*, *608*(2), 216–222.

Rappeneau, V., Blaker, A., Petro, J. R., Yamamoto, B. K., & Shimamoto, A. (2016). Disruption of the glutamate-glutamine cycle involving astrocytes in an animal model of depression for males and females. *Frontiers in Behavioral Neuroscience*, *10*, 231. https://doi.org/10.3389/fnbeh.2016.00231.

Razzoli, M., Carboni, L., Andreoli, M., Michielin, F., Ballottari, A., & Arban, R. (2011). Strain-specific outcomes of repeated social defeat and chronic fluoxetine treatment in the mouse. *Pharmacology, Biochemistry, and Behavior*, *97*(3), 566–576. https://doi.org/10.1016/j.pbb.2010.09.010.

Reguilon, M. D., Montagud-Romero, S., Ferrer-Perez, C., Roger-Sanchez, C., Aguilar, M. A., Minarro, J., et al. (2017). Dopamine D2 receptors mediate the increase in reinstatement of the conditioned rewarding effects of cocaine induced by acute social defeat. *European Journal of Pharmacology*, *799*, 48–57. https://doi.org/10.1016/j.ejphar.2017.01.039.

Reiche, E. M., Nunes, S. O., & Morimoto, H. K. (2004). Stress, depression, the immune system, and cancer. *The Lancet. Oncology*, *5*(10), 617–625. https://doi.org/10.1016/S1470-2045(04)01597-9.

Renthal, W., Maze, I., Krishnan, V., Covington, H. E., 3rd, Xiao, G., Kumar, A., et al. (2007). Histone deacetylase 5 epigenetically controls behavioral adaptations to chronic emotional stimuli. *Neuron*, *56*(3), 517–529. https://doi.org/10.1016/j.neuron.2007.09.032.

Ribble, D. O., & Salvioni, M. (1990). Social organization and nest cooccupancy in Peromyscus californicus, a monogamous rodent. *Behavioral Ecology and Sociobiology*, *26*, 9–15.

Riga, D., Schmitz, L. J., van der Harst, J. E., van Mourik, Y., Hoogendijk, W. J., Smit, A. B., et al. (2014). A sustained depressive state promotes a guanfacine reversible susceptibility to alcohol seeking in rats. *Neuropsychopharmacology*, *39*(5), 1115–1124. https://doi.org/10.1038/npp.2013.311.

Roberts, D. C., Morgan, D., & Liu, Y. (2007). How to make a rat addicted to cocaine. *Progress in Neuro-Psychopharmacology & Biological Psychiatry*, *31*(8), 1614–1624. https://doi.org/10.1016/j.pnpbp.2007.08.028.

Robinson, T. E., & Berridge, K. C. (1993). The neural basis of drug craving: An incentive-sensitization theory of addiction. *Brain Research. Brain Research Reviews*, *18*(3), 247–291.

Rothman, R. B., Gorelick, D. A., Baumann, M. H., Guo, X. Y., Herning, R. I., Pickworth, W. B., et al. (1994). Lack of evidence for context-dependent cocaine-induced sensitization in humans: Preliminary studies. *Pharmacology, Biochemistry, and Behavior*, *49*(3), 583–588.

Rozanski, A., Blumenthal, J. A., & Kaplan, J. (1999). Impact of psychological factors on the pathogenesis of cardiovascular disease and implications for therapy. *Circulation*, *99*(16), 2192–2217.

Sanchis-Segura, C., & Spanagel, R. (2006). Behavioural assessment of drug reinforcement and addictive features in rodents: An overview. *Addiction Biology*, *11*(1), 2–38. https://doi.org/10.1111/j.1369-1600.2006.00012.x.

Sapolsky, R. M. (2004). Social status and health in humans and other animals. *Annual Review of Anthropology*, *33*, 393–418.

Sapolsky, R. M. (2015). Stress and the brain: Individual variability and the inverted-U. *Nature Neuroscience*, *18*(10), 1344–1346. https://doi.org/10.1038/nn.4109.

Sapolsky, R. M., Romero, L. M., & Munck, A. U. (2000). How do glucocorticoids influence stress responses? Integrating permissive, suppressive, stimulatory, and preparative actions. *Endocrine Reviews*, *21*(1), 55–89. https://doi.org/10.1210/edrv.21.1.0389.

Schroeder, F. A., Lin, C. L., Crusio, W. E., & Akbarian, S. (2007). Antidepressant-like effects of the histone deacetylase inhibitor, sodium butyrate, in the mouse. *Biological Psychiatry*, *62*(1), 55–64. https://doi.org/10.1016/j.biopsych.2006.06.036.

Selmanoff, M. K., Maxson, S. C., & Ginsburg, B. E. (1976). Chromosomal determinants of intermale aggressive behavior in inbred mice. *Behavior Genetics*, *6*(1), 53–69.

Sgoifo, A., Koolhaas, J., Alleva, E., Musso, E., & Parmigiani, S. (2001). Social stress. Acute and long-term effects on physiology and behavior. *Physiology & Behavior*, *73*(3), 253–254.

Sgoifo, A., Pozzato, C., Costoli, T., Manghi, M., Stilli, D., Ferrari, P. F., et al. (2001). Cardiac autonomic responses to intermittent social conflict in rats. *Physiology & Behavior*, *73*(3), 343–349.

Sgoifo, A., Pozzato, C., Meerlo, P., Costoli, T., Manghi, M., Stilli, D., et al. (2002). Intermittent exposure to social defeat and open-field test in rats: Acute and long-term effects on ECG, body temperature and physical activity. *Stress*, *5*(1), 23–35. https://doi.org/10.1080/102538902900012387.

Shanks, N., Griffiths, J., & Anisman, H. (1994). Central catecholamine alterations induced by stressor exposure: Analyses in recombinant inbred strains of mice. *Behavioural Brain Research*, *63*(1), 25–33.

Shanks, N., Griffiths, J., Zalcman, S., Zacharko, R. M., & Anisman, H. (1990). Mouse strain differences in plasma corticosterone following uncontrollable footshock. *Pharmacology, Biochemistry, and Behavior*, *36*(3), 515–519.

Shanks, N., Zalcman, S., Zacharko, R. M., & Anisman, H. (1991). Alterations of central norepinephrine, dopamine and serotonin in several strains of mice following acute stressor exposure. *Pharmacology, Biochemistry, and Behavior*, *38*(1), 69–75.

Shimamoto, A., Debold, J. F., Holly, E. N., & Miczek, K. A. (2011). Blunted accumbal dopamine response to cocaine following chronic social stress in female rats: Exploring a link between depression and drug abuse. *Psychopharmacology*, *218*(1), 271–279. https://doi.org/10.1007/s00213-011-2364-7.

Shimamoto, A., Holly, E. N., Boyson, C. O., DeBold, J. F., & Miczek, K. A. (2015). Individual differences in anhedonic and accumbal dopamine responses to chronic social stress and their link to cocaine self-administration in female rats. *Psychopharmacology*, *232*(4), 825–834. https://doi.org/10.1007/s00213-014-3725-9.

Sial, O. K., Warren, B. L., Alcantara, L. F., Parise, E. M., & Bolanos-Guzman, C. A. (2016). Vicarious social defeat stress: Bridging the gap between physical and emotional stress. *Journal of Neuroscience Methods*, *258*, 94–103. https://doi.org/10.1016/j.jneumeth.2015.10.012.

Silberman, Y., Matthews, R. T., & Winder, D. G. (2013). A corticotropin releasing factor pathway for ethanol regulation of the ventral tegmental area in the bed nucleus of the stria terminalis. *The Journal of Neuroscience*, *33*(3), 950–960. https://doi.org/10.1523/JNEUROSCI.2949-12.2013.

Silberman, Y., & Winder, D. G. (2013). Emerging role for corticotropin releasing factor signaling in the bed nucleus of the stria terminalis at the intersection of stress and reward. *Frontiers in Psychiatry*, *4*, 42. https://doi.org/10.3389/fpsyt.2013.00042.

Silberman, Y., & Winder, D. G. (2015). Ethanol and corticotropin releasing factor receptor modulation of central amygdala neurocircuitry: An update and future directions. *Alcohol, 49*(3), 179–184. https://doi.org/10.1016/j.alcohol.2015.01.006.

Sinha, R. (2001). How does stress increase risk of drug abuse and relapse? *Psychopharmacology, 158*(4), 343–359. https://doi.org/10.1007/s002130100917.

Smith, A. S., Lieberwirth, C., & Wang, Z. (2013). Behavioral and physiological responses of female prairie voles (Microtus ochrogaster) to various stressful conditions. *Stress, 16*(5), 531–539. https://doi.org/10.3109/10253890.2013.794449.

Solomon, M. B. (2017). Evaluating social defeat as a model for psychopathology in adult female rodents. *Journal of Neuroscience Research, 95*(1–2), 763–776. https://doi.org/10.1002/jnr.23971.

Solomon, M. B., Karom, M. C., & Huhman, K. L. (2007). Sex and estrous cycle differences in the display of conditioned defeat in Syrian hamsters. *Hormones and Behavior, 52*(2), 211–219. https://doi.org/10.1016/j.yhbeh.2007.04.007.

Sotocinal, S. G., Sorge, R. E., Zaloum, A., Tuttle, A. H., Martin, L. J., Wieskopf, J. S., et al. (2011). The Rat Grimace Scale: A partially automated method for quantifying pain in the laboratory rat via facial expressions. *Molecular Pain*, 7, 55. https://doi.org/10.1186/1744-8069-7-55.

Stefanski, R., Ladenheim, B., Lee, S. H., Cadet, J. L., & Goldberg, S. R. (1999). Neuroadaptations in the dopaminergic system after active self-administration but not after passive administration of methamphetamine. *European Journal of Pharmacology, 371*(2–3), 123–135.

Steinman, M. Q., Laredo, S. A., Lopez, E. M., Manning, C. E., Hao, R. C., Doig, I. E., et al. (2015). Hypothalamic vasopressin systems are more sensitive to the long term effects of social defeat in males versus females. *Psychoneuroendocrinology, 51*, 122–134. https://doi.org/10.1016/j.psyneuen.2014.09.009.

Steketee, J. D., & Kalivas, P. W. (2011). Drug wanting: Behavioral sensitization and relapse to drug-seeking behavior. *Pharmacological Reviews, 63*(2), 348–365. https://doi.org/10.1124/pr.109.001933.

Stelly, C. E., Pomrenze, M. B., Cook, J. B., & Morikawa, H. (2016). Repeated social defeat stress enhances glutamatergic synaptic plasticity in the VTA and cocaine place conditioning. *eLife, 5*, e15448. https://doi.org/10.7554/eLife.15448.

Steptoe, A., & Kivimaki, M. (2013). Stress and cardiovascular disease: An update on current knowledge. *Annual Review of Public Health, 34*, 337–354. https://doi.org/10.1146/annurev-publhealth-031912-114452.

Stohr, T., Almeida, O. F., Landgraf, R., Shippenberg, T. S., Holsboer, F., & Spanagel, R. (1999). Stress- and corticosteroid-induced modulation of the locomotor response to morphine in rats. *Behavioural Brain Research, 103*(1), 85–93.

Strakowski, S. M., Sax, K. W., Setters, M. J., & Keck, P. E., Jr. (1996). Enhanced response to repeated d-amphetamine challenge: Evidence for behavioral sensitization in humans. *Biological Psychiatry, 40*(9), 872–880. https://doi.org/10.1016/0006-3223(95)00497-1.

Sun, H., Martin, J. A., Werner, C. T., Wang, Z. J., Damez-Werno, D. M., Scobie, K. N., et al. (2016). BAZ1B in nucleus accumbens regulates reward-related behaviors in response to distinct emotional stimuli. *The Journal of Neuroscience, 36*(14), 3954–3961. https://doi.org/10.1523/JNEUROSCI.3254-15.2016.

Tanda, G., & Goldberg, S. R. (2003). Cannabinoids: Reward, dependence, and underlying neurochemical mechanisms—A review of recent preclinical data. *Psychopharmacology, 169*(2), 115–134. https://doi.org/10.1007/s00213-003-1485-z.

Tannenbaum, B., & Anisman, H. (2003). Impact of chronic intermittent challenges in stressor-susceptible and resilient strains of mice. *Biological Psychiatry, 53*(4), 292–303.

Teskey, G. C., & Kavaliers, M. (1987). Aggression, defeat and opioid activation in mice: Influences of social factors, size and territory. *Behavioural Brain Research, 23*(1), 77–84.

Teskey, G. C., Kavaliers, M., & Hirst, M. (1984). Social conflict activates opioid analgesic and ingestive behaviors in male mice. *Life Sciences, 35*(3), 303–315.

Thaker, P. H., Han, L. Y., Kamat, A. A., Arevalo, J. M., Takahashi, R., Lu, C., et al. (2006). Chronic stress promotes tumor growth and angiogenesis in a mouse model of ovarian carcinoma. *Nature Medicine, 12*(8), 939–944. https://doi.org/10.1038/nm1447.

Thase, M. E., Entsuah, R., Cantillon, M., & Kornstein, S. G. (2005). Relative antidepressant efficacy of venlafaxine and SSRIs: Sex-age interactions. *Journal of Women's Health (2002), 14*(7), 609–616. https://doi.org/10.1089/jwh.2005.14.609.

Tidey, J. W., & Miczek, K. A. (1997). Acquisition of cocaine self-administration after social stress: Role of accumbens dopamine. *Psychopharmacology, 130*(3), 203–212.

Tornatzky, W., & Miczek, K. A. (1993). Long-term impairment of autonomic circadian rhythms after brief intermittent social stress. *Physiology & Behavior, 53*(5), 983–993.

Tornatzky, W., & Miczek, K. A. (1994). Behavioral and autonomic responses to intermittent social stress: Differential protection by clonidine and metoprolol. *Psychopharmacology, 116*(3), 346–356.

Toth, I., & Neumann, I. D. (2013). Animal models of social avoidance and social fear. *Cell and Tissue Research, 354*(1), 107–118. https://doi.org/10.1007/s00441-013-1636-4.

Trainor, B. C., Pride, M. C., Villalon Landeros, R., Knoblauch, N. W., Takahashi, E. Y., Silva, A. L., et al. (2011). Sex differences in social interaction behavior following social defeat stress in the monogamous California mouse (Peromyscus californicus). *PLoS One, 6*(2), e17405. https://doi.org/10.1371/journal.pone.0017405.

Trainor, B. C., Takahashi, E. Y., Campi, K. L., Florez, S. A., Greenberg, G. D., Laman-Maharg, A., et al. (2013). Sex differences in stress-induced social withdrawal: Independence from adult gonadal hormones and inhibition of female phenotype by corncob bedding. *Hormones and Behavior, 63*(3), 543–550. https://doi.org/10.1016/j.yhbeh.2013.01.011.

Tsankova, N. M., Berton, O., Renthal, W., Kumar, A., Neve, R. L., & Nestler, E. J. (2006). Sustained hippocampal chromatin regulation in a mouse model of depression and antidepressant action. *Nature Neuroscience, 9*(4), 519–525. https://doi.org/10.1038/nn1659.

Tsankova, N., Renthal, W., Kumar, A., & Nestler, E. J. (2007). Epigenetic regulation in psychiatric disorders. *Nature Reviews. Neuroscience, 8*(5), 355–367. https://doi.org/10.1038/nrn2132.

Turner, R. J., & Lloyd, D. A. (1999). The stress process and the social distribution of depression. *Journal of Health and Social Behavior, 40*(4), 374–404.

Turner, R. J., & Marino, F. (1994). Social support and social structure: A descriptive epidemiology. *Journal of Health and Social Behavior, 35*(3), 193–212.

Turner, R. J., Wheaton, B., & Lloyd, D. A. (1995). The epidemiology of social stress. *American Sociological Review, 60*(1), 104–125.

Tzschentke, T. M. (2007). Measuring reward with the conditioned place preference (CPP) paradigm: Update of the last decade. *Addiction Biology, 12*(3-4), 227–462. https://doi.org/10.1111/j.1369-1600.2007.00070.x.

Ungless, M. A., Singh, V., Crowder, T. L., Yaka, R., Ron, D., & Bonci, A. (2003). Corticotropin-releasing factor requires CRF binding protein to potentiate NMDA receptors via CRF receptor 2 in dopamine neurons. *Neuron, 39*(3), 401–407.

van Erp, A. M., & Miczek, K. A. (2001). Persistent suppression of ethanol self-administration by brief social stress in rats and increased startle response as index of withdrawal. *Physiology & Behavior, 73*(3), 301–311.

Vanderschuren, L. J., & Kalivas, P. W. (2000). Alterations in dopaminergic and glutamatergic transmission in the induction and expression of behavioral sensitization: A critical review of preclinical studies. *Psychopharmacology, 151*(2–3), 99–120.

Van't Veer, A., & Carlezon, W. A., Jr. (2013). Role of kappa-opioid receptors in stress and anxiety-related behavior. *Psychopharmacology, 229*(3), 435–452. https://doi.org/10.1007/s00213-013-3195-5.

Vialou, V., Maze, I., Renthal, W., LaPlant, Q. C., Watts, E. L., Mouzon, E., et al. (2010). Serum response factor promotes resilience to chronic social stress through the induction of DeltaFosB. *The Journal of Neuroscience*, *30*(43), 14585–14592. https://doi.org/10.1523/JNEUROSCI.2496-10.2010.

Vialou, V., Robison, A. J., Laplant, Q. C., Covington, H. E., 3rd, Dietz, D. M., Ohnishi, Y. N., et al. (2010). DeltaFosB in brain reward circuits mediates resilience to stress and antidepressant responses. *Nature Neuroscience*, *13*(6), 745–752. https://doi.org/10.1038/nn.2551.

Wachtel, S. R., & de Wit, H. (1999). Subjective and behavioral effects of repeated d-amphetamine in humans. *Behavioural Pharmacology*, *10*(3), 271–281.

Walker, F. R., Masters, L. M., Dielenberg, R. A., & Day, T. A. (2009). Coping with defeat: Acute glucocorticoid and forebrain responses to social defeat vary with defeat episode behaviour. *Neuroscience*, *162*(2), 244–253. https://doi.org/10.1016/j.neuroscience.2009.04.041.

Wanat, M. J., Bonci, A., & Phillips, P. E. (2013). CRF acts in the midbrain to attenuate accumbens dopamine release to rewards but not their predictors. *Nature Neuroscience*, *16*(4), 383–385. https://doi.org/10.1038/nn.3335.

Wanat, M. J., Hopf, F. W., Stuber, G. D., Phillips, P. E., & Bonci, A. (2008). Corticotropin-releasing factor increases mouse ventral tegmental area dopamine neuron firing through a protein kinase C-dependent enhancement of Ih. *The Journal of Physiology*, *586*(8), 2157–2170. https://doi.org/10.1113/jphysiol.2007.150078.

Wang, J., Fanous, S., Terwilliger, E. F., Bass, C. E., Hammer, R. P., Jr., & Nikulina, E. M. (2013). BDNF overexpression in the ventral tegmental area prolongs social defeat stress-induced cross-sensitization to amphetamine and increases DeltaFosB expression in mesocorticolimbic regions of rats. *Neuropsychopharmacology*, *38*(11), 2286–2296. https://doi.org/10.1038/npp.2013.130.

Wang, L., Hou, W., He, Z., Yuan, W., Yang, J., Yang, Y., et al. (2018). Effects of chronic social defeat on social behaviors in adult female mandarin voles (Microtus mandarinus): Involvement of the oxytocin system in the nucleus accumbens. *Progress in Neuro-Psychopharmacology & Biological Psychiatry*, *82*, 278–288. https://doi.org/10.1016/j.pnpbp.2017.11.002.

Wang, B., Shaham, Y., Zitzman, D., Azari, S., Wise, R. A., & You, Z. B. (2005). Cocaine experience establishes control of midbrain glutamate and dopamine by corticotropin-releasing factor: A role in stress-induced relapse to drug seeking. *The Journal of Neuroscience*, *25*(22), 5389–5396. https://doi.org/10.1523/JNEUROSCI.0955-05.2005.

Weiss, F., Lorang, M. T., Bloom, F. E., & Koob, G. F. (1993). Oral alcohol self-administration stimulates dopamine release in the rat nucleus accumbens: Genetic and motivational determinants. *The Journal of Pharmacology and Experimental Therapeutics*, *267*(1), 250–258.

Wolf, M. E. (1998). The role of excitatory amino acids in behavioral sensitization to psychomotor stimulants. *Progress in Neurobiology*, *54*(6), 679–720.

Wood, S. K., Walker, H. E., Valentino, R. J., & Bhatnagar, S. (2010). Individual differences in reactivity to social stress predict susceptibility and resilience to a depressive phenotype: Role of corticotropin-releasing factor. *Endocrinology*, *151*(4), 1795–1805. https://doi.org/10.1210/en.2009-1026.

Wook Koo, J., Labonte, B., Engmann, O., Calipari, E. S., Juarez, B., Lorsch, Z., et al. (2016). Essential role of mesolimbic brain-derived neurotrophic factor in chronic social stress-induced depressive behaviors. *Biological Psychiatry*, *80*(6), 469–478. https://doi.org/10.1016/j.biopsych.2015.12.009.

Wotjak, C. T., Kubota, M., Liebsch, G., Montkowski, A., Holsboer, F., Neumann, I., et al. (1996). Release of vasopressin within the rat paraventricular nucleus in response to emotional stress: A novel mechanism of regulating adrenocorticotropic hormone secretion? *The Journal of Neuroscience*, *16*(23), 7725–7732.

Yap, J. J., Chartoff, E. H., Holly, E. N., Potter, D. N., Carlezon, W. A., Jr., & Miczek, K. A. (2015). Social defeat stress-induced sensitization and escalated cocaine self-administration: The role of ERK signaling in the rat ventral tegmental area. *Psychopharmacology*, *232*(9), 1555–1569. https://doi.org/10.1007/s00213-014-3796-7.
Yap, J. J., Covington, H. E., 3rd, Gale, M. C., Datta, R., & Miczek, K. A. (2005). Behavioral sensitization due to social defeat stress in mice: Antagonism at mGluR5 and NMDA receptors. *Psychopharmacology*, *179*(1), 230–239. https://doi.org/10.1007/s00213-004-2023-3.
Yap, J. J., & Miczek, K. A. (2007). Social defeat stress, sensitization, and intravenous cocaine self-administration in mice. *Psychopharmacology*, *192*(2), 261–273. https://doi.org/10.1007/s00213-007-0712-4.
Zorrilla, E. P., Logrip, M. L., & Koob, G. F. (2014). Corticotropin releasing factor: A key role in the neurobiology of addiction. *Frontiers in Neuroendocrinology*, *35*(2), 234–244. https://doi.org/10.1016/j.yfrne.2014.01.001.
Zou, S., Funk, D., Shram, M. J., & Le, A. D. (2014). Effects of stressors on the reinforcing efficacy of nicotine in adolescent and adult rats. *Psychopharmacology*, *231*(8), 1601–1614. https://doi.org/10.1007/s00213-013-3314-3.

FSC
www.fsc.org
MIX
Papier | Fördert gute Waldnutzung
FSC® C083411

Kolibri 360 GmbH
Ferdinand-Jühlke-Straße 7
99095 Erfurt, Deutschland
produktsicherheit@kolibri360.de